SOLAR ACTIVE REGIONS

SOLAR ACTIVE REGIONS

A Monograph from Skylab Solar Workshop III

Edited by Frank Q. Orrall

COLORADO ASSOCIATED UNIVERSITY PRESS

Boulder, Colorado 80309
International Standard Book Number 0-87081-085-5
Library of Congress Card Catalog Number 79-56357
Printed in the United States of America

SOLAR ACTIVE REGIONS

PREFACE *Frank Q. Orrall* ix

INTRODUCTION *Frank Q. Orrall*
1.1 Solar Active Regions 1
1.2 The Third Skylab Solar Workshop 2
1.3 Achievements of the Workshop—An Overview 5
References 15

THE OVERALL STRUCTURE AND EVOLUTION OF ACTIVE REGIONS *Neil R. Sheeley, Jr.*
2.1 Introduction 17
2.2 Background 17
2.3 Surface Fields 19
2.4 Sunspots and Faculae 27
2.5 Chromospheric Tracers 30
2.6 The Transition Region and Lower Corona 32
2.7 Epilogue 38
References 41

ACTIVE REGIONS FROM THE PHOTOSPHERE TO THE CHROMOSPHERE *Gary A. Chapman*
3.1 Introduction 43
3.2 Horizontally Averaged Models 43
3.3 Facular Flux Tubes and the Fine Structure of Active Regions 48
3.4 Multidimensional Radiative Transfer 57
3.5 Radiative Cooling of the Quiet Sun and Plage Chromospheres 60
3.6 Upper Chromosphere and Transition Zone 73
3.7 Summary 79
References 81

OBSERVATIONS OF MASS MOTIONS IN ACTIVE REGIONS *R. Grant Athay*
4.1 Introduction 83
4.2 Steady Flows 85
4.3 Quasisteady Flow 87
4.4 Transient Flows 91
4.5 Oscillations 99
4.6 Nonthermal Line Widths 102
References 111

THE DYNAMICS OF ACTIVE REGIONS *Guenter E. Brueckner*
5.1 Introduction 113
5.2 The Spectra 113
5.3 Flows and Explosive Events 114
5.4 Velocities in the Corona above Plages and Sunspots 121
5.5 Influence of Dynamic Effects on the Formation of Spectral Lines 123
5.6 He 1640 Emission and the Separation of Elements in Downflows 124
5.7 The Mass and Energy Balance of Active Regions 125
References 127

SPECTROSCOPIC DIAGNOSTICS OF THE ACTIVE REGION: TRANSITION ZONE AND CORONA *Kenneth P. Dere and Helen Mason*
6.1 Introduction 129
6.2 Temperature and Emission Measure Diagnostics 129
6.3 Electron Density Diagnostics 140
6.4 Atomic Physics 148
6.5 Line Profiles 156
6.6 Remarks 159
References 160

ACTIVE REGION STRUCTURES IN THE TRANSITION REGION AND CORONA *David F. Webb*
7.1 Introduction 165
7.2 What is the Basic Structure of an Active Region? 166
7.3 Thermal Structure of Loops 176
7.4 Temporal Behavior of Loops 184
7.5 Conclusion 194
References 197

PHYSICS OF STATIC LOOP STRUCTURES AND SCALING LAWS *George L. Withbroe*
8.1 Introduction 199
8.2 Theory 199
8.3 Comparison with Observation 206
8.4 Conclusions 210
References 212

THEORY OF LOOP FLOWS AND INSTABILITY *Eric R. Priest*
9.1 Introduction 213
9.2 Siphon Flow 218
9.3 Evaporation and Draining 227
9.4 Thermal Stability of Static Loops 237
9.5 Downflow and Upflow in Intense Flux Tubes 251
9.6 MHD Instability of Loops 263
9.7 Conclusion 272
References 273

MECHANISMS OF ENERGY SUPPLY *Joseph V. Hollweg*
10.1 Introduction 277
10.2 Perspective 277
10.3 Anomalous Dissipation of Parallel Currents in the Corona 287
10.4 Minireconnections in Loops 296
10.5 Surface Waves on Loops 299
10.6 Coronal Fast Waves 304
10.7 Alfven Waves above the Convective Zone 307
10.8 Dynamics of Filament Formation 315
References 316

FUTURE ACTIVE REGION OBSERVATIONS *Randolph H. Levine*
11.1 Introduction 319
11.2 Limits to Interpretation 321
11.3 Better Morphological Systematization 325
11.4 Temporal Variations 329
11.5 Flows 330
11.6 Can We Distinguish Among Possible Heating Mechanisms? 332
11.7 The Role of the Magnetic Field 335
11.8 Summary 339
References 340

APPENDIX: LIST OF PARTICIPANTS 343

INDEX 347

PREFACE

This monograph is the offspring of the Third Skylab Workshop, which was dedicated to the study of solar active regions. Our intent has been to summarize the advances of the workshop in the broader context of current research in the physics of active regions. A detailed, comprehensive record of scientific results is to be found in the numerous journal articles now in print or in press generated by the participants.

As Director of the Workshop, it is my pleasure to thank the participants for their sustained enthusiasm and generous cooperation. Their personal and communal scientific labor forms the substance of this volume. I was fortunate to have as Workshop Advisors, R. Grant Athay, Robert Noyes, and Gerard Van Hoven, who helped in organizing the workshop and in setting its initial scientific direction, and who throughout provided scientific and practical counsel. The brunt of responsibility during the meetings fell to the Working Group Leaders: Gary Chapman, Joseph Hollweg, Alan Krieger, Eric Priest, and George Withbroe. They presided over the spirited group meetings where the basic scientific problems were identified and confronted, and prepared the written and oral summaries of progress that made the workshop a concerted effort. Randolph Levine organized and chaired a committee that documented the crucial observational and theoretical future needs of active region research that were frequently uncovered by the working groups.

The Skylab Solar Workshop Series owes its conception, organization, and overall direction to Gordon Newkirk, Jr. He has been a continual source of support, encouragement, and guidance. The very successful format for conducting the workshops evolved in the first and second in the series, directed by Jack Zirker and Peter Sturrock, respectively. Karen Ellis served as Administrative Coordinator of the series and handled all of the logistics, including arrangements for the meetings. Keith Watson was my mentor in fiscal and contractual matters. I am grateful to them both, for their helpfulness and efficiency.

In my role as editor of this book I have the pleasure of thanking the authors of the individual chapters that comprise it, not only for their careful labor, but also for their cheerful patience with me. They include R. Grant Athay, Guenter Brueckner, Gary Chapman, Kenneth Dere, Joseph Hollweg, Randolph Levine, Helen Mason, Eric Priest, Neil Sheely, David Webb, and George Withbroe. The authors had the final decision on the form and contents of their chapters, but had the benefit of suggestions and constructive criticism from a number of participants and colleagues. In this regard, I would like especially to thank Spiro Antiochos, R. Grant Athay, Thomas Ayres, Claudio Chiuderi, George Doschek, Alan Krieger, Randolph Levine, O. Richard White, and Harold Zirin.

The Technical Editor of this volume was Kenneth Zweibel. His impeccable accuracy and judgment, and his tactful guidance made it both an education and a pleasure to work with him. The comprehensive subject index was the careful work of Sabine H. Schaffner.

PREFACE

The Skylab Workshop Series was occasioned by the magnificent success of the Skylab mission. In particular, the ATM solar instrument package and the vigorous efforts of the Astronauts and the ATM Experiment Teams produced an incomparable and seemingly inexhaustible set of observations of the sun's outer atmosphere. The ATM Principal Investigators, Robert MacQueen, James Milligan, Edmund Reeves, Richard Tousey, James Underwood, and Guiseppi Vaiana, and their teams, have generously shared their data and experience with the workshop participants. A number of ATM team members have been, moreover, participants in the Workshop Series. Other solar data from the ground and from space required by the participants were supplied by other astronomers. Helen Coffey and Virginia Lincoln of the World Data Center, and Charles Wende of the National Space Science Data Center were helpful in making these data collections easily available to the participants.

The Solar Skylab Workshop Series was sponsored by the National Aeronautics and Space Administration (NASA) under Order H-18204B, and by the National Center for Atmospheric Research (NCAR) sponsored by the National Science Foundation (NSF). The foresight and effort of a number of administrators in these organizations have made the Workshop Series possible. These include David Bohlin, Noel Hinners, Harold Glaser, Charles Lunquist, William Snoddy, and Adrienne Timothy of NASA; Francis Bretherton and John Firor of NCAR; and Alan Grobecker, John Slaughter, and Georgio Tesi of NSF. Of course the opinions expressed in this book are those of its authors, and are not necessarily those of the National Science Foundation, the National Center for Atmospheric Research, or the National Aeronautics and Space Administration.

I am personally grateful for the hospitality shown to me by Robert MacQueen and the High Altitude Observatory of NCAR.

The participants of the workshop have dedicated this volume to the memory of Stefan Smerd, a fellow-participant who died suddenly in December 1978. He was present at the first meeting, and as always, was a vigorous and enthusiastic contributor. We remember him with admiration and friendship. We recall the numerous scientific meetings that he and his high-spirited Australian colleagues enlivened, and at which they presented fundamental contributions to the physics of the sun.

Frank Q. Orrall
Institute for Astronomy
University of Hawaii
Honolulu
4 December 1980

1. INTRODUCTION

Frank Q. Orrall

1.1 SOLAR ACTIVE REGIONS

Compact ropes of magnetic flux from time-to-time rise from below and break through the photosphere of the sun to produce a bipolar magnetic region (BMR) with compact, locally closed loops of strong magnetic field extending into the corona. Small, short-lived BMRs may occur at any solar latitude, and a few hundred of these ephemeral regions, each having a typical lifetime of one-third of a day, are distributed over the solar surface at any one time. Larger, longer-lived BMRs develop within the latitudes of the sunspot zone and give rise to sunspot groups having lifetimes ranging from a few days to a few months. The term *active region* is now most commonly used to designate the entire time-varying structure generated in the sun's atmosphere by the emergence of a sunspot-producing BMR. The term *ephemeral active region* is used to designate a simpler structure generated by a small, short-lived BMR of the type that occurs in any latitude and that does not produce sunspots. In either case, the emergence of the magnetic field restructures and transforms all regions of the overlying atmosphere where it intrudes.

This book is concerned with the physical consequences of this intrusive magnetic field—the effect that it has on structure, dynamics, and energetics. Solar flares are one dramatic consequence of this magnetic field, but they were the subject of Workshop II and its comprehensive monograph (Sturrock, 1980); this volume does not consider them. Although large solar flares may temporarily disrupt and transform the active region in which they arise, they do not on the average appear to dominate the energetics (Orrall and Kopp, 1976); thus, it is reasonable and useful to speak of "quiet," or "nonflaring," active regions.

The emergent magnetic field that expands into the corona is not uniformly filled with plasma. Instead, the coronal plasma is concentrated into loops, which to first order follow certain magnetic field lines. The existence of these loops had long been known from ground-based eclipse and coronagraph observations, but it was the Skylab soft X-ray and extreme ultraviolet (EUV) multispectral images that first revealed them in all of their splendor and variety, and that have provided the basic data needed for their first interpretation. Plasma loops give rise to most of the sun's soft X-ray emission as well as to much of the EUV emitted by the active region corona. It seems likely that these plasma loops hold the key to an understanding of coronal heating. On the other hand, at the level of the photosphere and low chromosphere, most of the excess emission of active region plages comes from intense magnetic flux tubes of very small scale. It is thus not surprising that observational and theoretical studies of coronal plasma loops and intense magnetic flux tubes occupy a major place in this monograph devoted to the physics of solar active regions.

1.2 THE THIRD SKYLAB SOLAR WORKSHOP

The ATM (Apollo Telescope Mount) cluster of solar instruments aboard Skylab was extraordinarily successful. In its nine months of manned and unmanned operation (May 1973 to February 1974), it returned a mass of both detailed and synoptic data that has transformed our conception of the sun's outer atmosphere. An illustrated overview has been given by Eddy (1979) of the impact of ATM, including a description of the instrument package and the conduct of the mission. Detailed descriptions of the individual Skylab experiments and their calibration occupy a large part of the April 1977 issue of *Applied Optics* (Vol. 16, No. 4).

The ATM experiment teams began a vigorous program of data interpretation that bore many important results even before the end of the mission. But the very wealth and importance of the ATM data made a broader involvement of the solar physics community both desirable and necessary. This was foreseen by Dr. Gordon A. Newkirk, Jr., then Director of the High Altitude Observatory of the National Center for Atmospheric Research (NCAR). He proposed a series of Skylab Solar Workshops that would focus on specific pertinent topics and that would establish and carry out concrete research aims to be completed within the time frame of a workshop—about one year and a half. This proposal received the enthusiastic support of the National Aeronautics and Space Administration (NASA) and the ATM experiment teams. The result was a series of workshops sponsored jointly by NASA and by the National Science Foundation, through NCAR, and conducted under Gordon Newkirk's overall direction.

Workshop I, directed by Jack B. Zirker, then of the University of Hawaii, was devoted to the study of coronal holes and high-speed solar wind streams. It had 76 participants and held three plenary meetings in Boulder between October 1975 and August 1976. Workshop II addressed the complex subject of solar flares under the direction of Peter A. Sturrock of Stanford. It had 91 participants and held four plenary sessions in Boulder between November 1976 and December 1977. Monographs produced by these two workshops are now in print, edited by their directors (Zirker, 1977; Sturrock, 1980). At the second workshop, possible topics for a third and final workshop were discussed; the subject of solar active regions was considered most promising for the next detailed study, both because of its intrinsic interest and because of the large amount of unique and pertinent ATM active region data. Moreover, a wealth of new data on active region dynamics was becoming available from the OSO-8 spacecraft and from the NRL high-resolution rocket-borne spectrograph, strongly complementing the Skylab observations.

I later agreed to serve as director of the Workshop III; R. Grant Athay, Gerard Van Hoven, and Robert W. Noyes agreed to serve as Workshop Advisors. Because the subject of active regions is very extensive, and because the resources remaining for a third workshop were considerably less than for the first two, it was essential to narrow the focus. This was done, first, by emphasizing physical processes; second, by considering only the "quiet," or "nonflaring," active region; and third, by explicitly excluding sunspots as a subject of detailed study. The decision to exclude sunspots was a difficult one, justified by the fact that there are few ATM sunspot observations and by the realization that the subject deserves a workshop of its own.

An advertisement inviting science proposals for participation in Workshop III was published in the United States and abroad. Chosen participants were to be supplied with travel support, ATM data, and computer time at NCAR, as required. A committee consisting of the advisors, Gordon Newkirk, and myself reviewed the proposals received and tentatively assigned each of the participants to one of five working groups on the basis of their proposed science. Where important pertinent science was not sufficiently represented, some additional participants were recruited or pressed into service. Gary A. Chapman, Joseph V. Hollweg, Allen S. Krieger, Eric R. Priest, and George Withbroe agreed to take on the demanding task of leading the working groups.

The three formal meetings of Workshop III were held in Boulder on 6 through 8 July 1978, 30 October through 1 November 1978, and 12 through 14 March 1979. The general format of these meetings was similar to that which proved so successful in the previous two Skylab Workshops. However, because fewer meeting days could be budgeted for the third workshop, it was essential that it get off to a very fast start. For this reason, the general aims and structure of the working groups, their membership, and their leaders were established before the workshop started. At the first meeting, following a brief plenary session, the working groups met separately to set concrete goals, identify difficulties, and discuss the basic science. Before the end of the first meeting, this was achieved, and the working groups presented their goals for Workshop III and their preliminary conclusions. The individual participants left the meeting with commitments for personal research to be carried out if possible before the next meeting. At the second and third meetings, the working groups spent much of their time hammering out the basic scientific problems. Joint sessions of two or more working groups were held to discuss related problems. Summaries by each group of work completed and in progress were presented and discussed in plenary session at the end of each meeting, and written summaries by the team leaders were prepared and circulated. Work on a number of projects generated by the workshop continued for several months after the final meeting.

The working groups, their formal memberships, and their leaders, are listed below. The number of participants present at any one meeting averaged about forty, but the total number was considerably larger: members of consortia of collaborators took turns attending the meetings. A number of participants were active in the work of more than one group or were engaged in collaborative research with members of other groups. This attests to the wide range of participants' interests, the informality of the workshop, and the interdependence and cohesiveness of the science. Unfortunately, only an imperfect list of these affiliations exists, and therefore only the formal group memberships are presented here:

- WORKING GROUP 1. *Modeling, Diagnostics, and Physical Conditions in the Transition Region and Above.* LEADER: George Withbroe. MEMBERS: Chung Chieh Cheng, Ian Craig, Kenneth Dere, Alan Hood, Bernard Jackson, Kenneth Krall, Massimo Landini, Randolph Levine, Helen Mason, Brunella Monsignori-Fossi, John Parkinson, Kenneth Phillips, Stefan Smerd, Keith Strong, James Underwood, and John Vesecky.

- WORKING GROUP 2. *Modeling, Diagnostics, and Physical Conditions in the Transition Region and Below.* LEADER: Gary Chapman. MEMBERS: Lawrence Auer, Thomas Ayres, Oddbjorn Engvold, James Heasley, Philippe Lemaire, Jeffrey Linsky, Marcos Machado, Olav Kjeldseth Moe, Stanley Owocki, Henk Spruit, Einar Tandberg-Hanssen, Jorge Vernazza.

- WORKING GROUP 3. *Mechanisms of Energy Supply.* LEADER: Joseph Hollweg. MEMBERS: Claudio Chiuderi, Gordon Emslie, Shadia Habbal, James Ionson, Max Kuperus, Dennis Papdopoulos, Robert Rosner, Dean Smith, Daniel Spicer, Gerard Van Hoven, and Loukas Vlahos.

- WORKING GROUP 4. *Dynamics and Mass Balance.* LEADER: Eric Priest. MEMBERS: Spiro Antiochos, R. Grant Athay, Guenter Brueckner, John Cook, Roger Kopp, Bruce Lites, Giancarlo Noci, Robert Noyes, and Bernard Roberts.

- WORKING GROUP 5. *Structure and Evolution.* LEADER: Allen Krieger. MEMBERS: Leon Golub, James Mosher, Neil Sheeley, David Webb, and Harold Zirin.

In confronting the outstanding problems, the working groups soon identified a number of crucial data and techniques, essential for understanding or for testing theory, that did not exist and could not reasonably be hoped for in the time frame of Workshop III. These included new observations, diagnostics, and theoretical or numerical studies. At the second workshop meeting, several participants suggested that it would be useful to collect and document a selective list of future needs. Accordingly, Dr. Randolph Levine agreed to organize this effort. He received suggestions from the participants and convened a committee of representatives of the working groups consisting of R. H. Levine (Chair), R. G. Athay, T. R. Ayres, C. Chiuderi, L. Golub, and R. Rosner. Dr. Levine later agreed to prepare an extended discussion of future active region observations in the context of the workshop. This appears as Chapter 11 of the present volume.

1.3 ACHIEVEMENTS OF THE WORKSHOP—AN OVERVIEW

This monograph is an outline of the scientific results of the workshop presented within a self-contained review of the physics of solar active regions. The remainder of this chapter looks briefly at some of the achievements of the workshop and at some of the topics that were the subject of intense discussion within the five working groups. Reference to specific authors or papers has been avoided in Chapter 1 except where needed to identify contrasting studies. Documentation can be found in the reviews that comprise the body of the book.

1.3.1 Working Group 1

Working Group 1, led by George Withbroe, studied *Modeling, Diagnostics, and Physical Conditions in the Transition Region and Above.* The group's major objectives were (1) to derive improved models of static coronal loops; (2) to study the stability of these loops and their response to a change in energy input; (3) to carry out a number of empirical studies of active regions and to compare the results with theory; and (4) to improve plasma diagnostic techniques.

Major advances were made during the workshop in understanding quasistatic loop models in energy balance. Improved models were constructed based on both analytical and numerical techniques. Such models now seem capable of approximating basic properties of the stable, higher-temperature coronal loops that constitute much of the observable active region corona. In these idealized models, energy is deposited by some process in a symmetric magnetic flux tube filled with plasma in hydrostatic equilibrium and rooted at both ends in the chromosphere or transition region. This energy is redistributed by thermal conduction and lost by radiation, although some fraction may flow out of the footpoints as a heat flux F_o. A few parameters define the gross properties of these loops—the length L, the pressure P, and the maximum temperature T_m at the loop vertex—and are related by simple "scaling laws" of the form $T_m \approx C(PL)^{1/3}$. Here C depends weakly on the form of the heating function; on the geometrical expansion of the field; and on F_o, which is limited to moderate values by the ability of the underlying atmosphere to dispose of it in the steady state. These scaling laws provide important insight into the behavior of loops and provide a useful comparison of theory with observation.

The active region corona appears to consist primarily of systems of quasistatic loops. The hypothesis is not inconsistent with the observation that many of the individual loops pass from one state of quasistatic energy equilibrium to another as the geometry or as the physical conditions at the footpoints slowly evolve. Insight into this process is given by the scaling laws. These suggest that if either the loop length or the energy deposition rate increase, the density and pressure in the loop must increase to maintain energy equilibrium, and conversely. This implies a net inflow or outflow of mass through the footpoints. There is an upper limit to the thermal conductive flux that can flow from the loop back into the chromosphere, given by the ability of the chromosphere to dispose of it by radiation. Thus, if the rate of energy input to the loop is increased, it must be matched by an increase in radiative loss and hence by an increase in pressure. This can be provided by evaporation of new mass from the chromosphere at the footpoints, which will continue until the energy balance is again restored at some new set of T_m, P, and L. On the other hand, if the input is decreased, the material in the loop will become thermally unstable and nonbuoyant, and mass will flow back into the chromosphere, reducing the pressure and thereby the radiative losses until the energy balance is reestablished. This process was studied by numerical modeling during the workshop. The greatest uncertainty in the modeling was at the footpoints, where the temperature gradients are steep and where the response of the underlying transition region and upper chromosphere is uncertain. Direct observations of the emission and velocity in the footpoints would provide guidance, but loop footpoints are surprisingly difficult to locate unambiguously on existing observations. Nevertheless, the overall structure and evolution of symmetrical quasistatic loops in energy balance and the response of these loops to changes in heating or geometry seem well understood. A comprehensive discussion of quasistatic models is given by George Withbroe in Chapter 8.

Considerable insight into the problem of comparing theoretical models and observable loop parameters was developed in the workshop. Scaling laws offer the simplest comparison, and these were explored in detail. The parameters T_m, P and L measured in plasma loops in both flares and quiet active regions can be fitted reasonably well to a common theoretical scaling law. This supports, but does not prove, the validity of the quasistatic models. The considerable scatter of the points of this comparison may represent real differences between the secondary properties of the individual loops, such as in their geometrical expansion, the temperature sensitivity of the assumed heating function, and the heat flux F_0. The effects of all of these were explored, but none alter the scaling laws in a way that presently provides a useful diagnostic for the identification and characterization of these secondary properties. The predicted differential emission measure offers another comparison with observation. Calculations show that it is altered by loop geometry and by basal heat flux but not by the temperature sensitivity of the heating function. It is an important and surprising result that quasistatic loops are not very sensitive to the form of the steady-state heating function and hence have thus far set no strong constraint on proposed heating mechanisms. Such constraints will have to be sought in the detailed structure of individual loops or in their dynamic properties. In a study discussed below and in Chapter 9, Habbal and Rosner found that even the thermal stability of quasistatic loops does not depend strongly on the heating mechanism assumed.

Several studies of individual loop systems carried out in the workshop are discussed in Chapter 7. There is little observational evidence that hot static loops are other than isobaric and isothermal except near their feet. While it has not yet been possible to derive the differential emission measure for individual loops, very careful emission measure analyses for the integrated emission of active regions were completed in the workshop, using Skylab data together with rocket X-ray data. The problem of inverting spectrophotometric data to derive stable determinations of the differential emission measure was one of the diagnostic problems that received major attention in Group 1. These are discussed as part of an extended review of active region diagnostics by Kenneth Dere and Helen Mason in Chapter 6.

1.3.2 Working Group 2

Working Group 2, led by Gary Chapman, was concerned with *Modeling, Diagnostics, and Physical Conditions in the Transition Region and Below.* The group's primary goal was to derive active region atmospheric models applicable from the photosphere through the transition zone; and as a secondary goal, to calculate radiative losses as a function of height for these models. A detailed discussion of their results is given in Chapter 3.

Group 2 concluded early in the first meeting that the small-scale inhomogeneous structure of active regions associated with concentrations of magnetic field would have to be incorporated at all levels in any realistic model, and that its study would constitute a major part of their effort. Nevertheless, they found it necessary to construct a hybrid mean-quiet-sun model from critical examination of existing one-dimensional models, since in practice the quiet sun forms the background reference for active region studies. They also constructed a homogeneous model of a plage, from the photosphere through the upper chromosphere, for comparison with observations of low spatial resolution, and as a guide and an intermediate step in constructing more-realistic models. Although these homogeneous models represent reasonably well several different types of observations, discrepancies remain that can probably only be resolved by including inhomogeneities, not only in plages, but also in the quiet sun.

The strongest evidence for inhomogeneous structure at the photospheric level in active regions of course comes from high-resolution observations of brightness enhancements (faculae and filigree) and of strong (1–1.6 kG) magnetic field concentrations, both often of sub-arcsecond scale. There is also strong, although incomplete, evidence that these brightness and magnetic structures are cospatial. Group 2 explored in detail a nonuniform model of a plage in which the inhomogeneities at the photospheric level are strong magnetic flux tubes manifest as faculae, or filigree elements, and embedded in a surrounding field-free region that has a thermodynamic structure similar to the field-free quiet sun. These flux tubes expand with height until they merge with each other—this merging height being typically 800 km in active regions. A major effort was devoted to the study of flux tubes and to the empirical or semiempirical determination of the *filling factors,* which define the fractional area occupied by the magnetic flux tube, or its emitting material, as a function of height. A number of different types of observational data and diagnostic techniques were brought to bear on this crucial problem. Another question crucial in modeling the inhomogeneous plage atmosphere is the extent to which horizontal radiative transfer between the flux tube and its field-free or quiet-sun surroundings must be taken into account. Calculations presented in the workshop indicate that for diagnostic studies using the Ca II K-line wings, horizontal radiative transport can be safely neglected even for horizontal scales as small as 70 km. In the Mg II k-line, however, due to longer thermalization lengths, contrast will be lost for flux tubes or horizontal scales as large as 350 km.

The second major goal of Group 2 was to determine the net radiative power loss per unit volume as a function of height, to constrain theories of mechanical heating. After clarifying some basic physics of radiative loss, cooling curves for both the quiet-sun and average-plage models were derived by numerical simulation and compared with observation. In the chromosphere, H I, Ca II, and Mg II were considered. In the middle chromosphere, the Ca II and Mg II resonance lines and the Ca II infrared triplet are the most effective coolants. At $T \approx 8000$ K, Hα and the Balmer continuum dominate, whereas near the top of the chromosphere ($T \approx 10^4$ K) Lα and the Lyman continuum are the most important coolants. The Paschen continuum and Lα are a source of heating in the middle chromosphere. In the plage, the region dominated by Ca II and Mg II cooling occurs at a higher temperature than in the quiet sun. Below the region dominated by Ca II–Mg II cooling (6000 K in the quiet sun and 7000 K in the plage), the cooling curves are uncertain. The controversial question of whether H^- heats or cools the low chromosphere was readdressed. The effect of hot magnetic flux tubes on the empirical estimates of radiative cooling were then considered. This effect is greatest below 6000 K, where the homogeneous model overestimates the heating requirements by a factor of 10 compared with the more realistic inhomogeneous model. In the photosphere, the principal uncertainty in determining the photospheric flux excess from empirical data is in the center-to-limb contrast change of the faculae.

A model for the mean–active region transition zone was constructed, based on Skylab, Calroc, HRTS, and OSO-8 data. The radiative losses as a function of height and the conductive flux deposited in the underlying atmosphere were calculated. A loop model for the transition region was also considered.

1.3.2 Working Group 3

Working Group 3, led by Joseph Hollweg, was concerned with *Mechanisms for Energy Balance*. They focused their major effort on looking critically at specific proposed active region heating mechanisms as well as for possible "signatures" that might allow the reality or effectiveness of the mechanisms to be assessed. For a number of mechanisms, new theoretical work had to be initiated. Since almost everywhere on the sun magnetic regions are regions of enhanced heating, the most promising mechanisms were considered those that involved, directly or indirectly, the magnetic field. Most effort was expended on processes that deposit energy on coronal loops. Major attention was also given to the transport of energy from the convection zone to the corona; the effects of sheared magnetic fields on thermal instability; and to some other topics. A number of ideas were explored and debated in the group meetings; only a fraction could be studied closely within the time frame of the workshop. A detailed account of some of the results within a broader perspective is given by Joseph Hollweg in Chapter 10. Among the topics considered by Group 3 were the following:

Coronal magnetic fields must be very nearly force free but not necessarily potential. Although the shapes of observed coronal loops often conform closely to calculated potential fields, anomalous dissipation of currents associated with a small undetected nonpotential component might heat the loop if confined to a thin sheath. This mechanism had been considered prior to the workshop; a principal effort was to consider it critically and to extend the earlier work. It was found, for example, that electrostatic ion-cyclotron waves may be the important source of anomalous resistivity (at least on short active region loops) rather than the ion-acoustic waves considered earlier. The hot sheath is expected to be far too thin to observe directly, but anomalous current dissipation should produce numerous energetic runaway electrons and hence nonthermal radio emission and X-rays. In particular, it was predicted that the radio emission should certainly be detectable, although it would not be a *unique* signature of anomalous dissipation. Turbulent line broadening may in the future prove a useful signature and diagnostic. Several other effects that were considered seem presently below detection.

Large-scale magnetic reconnection may be an important source of heating in rapidly developing coronal loops, but probably not for the high-temperature loop systems that are usually observed to develop slowly. But small-scale (mini-) reconnections distributed throughout the loop in thin current sheets might serve as a local dissipation mechanism. Workshop studies suggest that, provided sufficiently thin current sheets can form, the process may be important, and semicollisional tearing is more likely than collisional tearing. Nonthermal radio emission from runaway electrons and turbulent line broadening, again, are possible signatures, but detailed predictions are needed.

The role of surface waves in heating coronal loops was explored, extending work done just prior to the workshop. In particular, a mechanism was examined in which wave energy, propagating up the loop from below as an Alfvenic surface wave, is dissipated in a very thin sheath less than 1-km thick. Waves of 300-second period would be suitable for loops of large scale (10^9 cm). Waves of shorter period (perhaps excited by microturbulence) would be required for loops of smaller scale. A convective layer 30-km thick or less might form next to the dissipation sheath and be driven by it. Plasma convected to the top of the loop could cross into the loop interior by means of the electrostatic Rayleigh-Taylor instability, cool by radiation, and thus fall and produce a loop with a cool core. Such a convective layer could also form in other sheath heating processes, such as in the anomalous dissipation mechanism discussed earlier. Nonthermal emission from runaway electrons, again, are expected; but suitable estimates were not completed during the workshop. Observed nonthermal line widths are not inconsistent with this mechanism but do not uniquely support it.

Fast-mode MHD waves, which have some important advantages for heating a structured corona, were also considered. They can cross magnetic field lines and can be refracted (in this context, slow-mode waves seem unable to provide the needed energy in the corona). Hence, fast-mode waves of rather small energy flux density might be focused to deposit suitably large amounts of energy locally at special places. Thus, even if fast-mode waves entered the corona uniformly, they might nonetheless provide heat nonuniformly, enhancing and producing distinct coronal features. The damping mechanism considered was Landau/transit-time damping. With this mechanism in the isothermal model studied, fast-mode waves tend to be refracted toward the region where damping is largest; moreover, local heating results in an increase of the heating rate, in positive feedback. Observed nonthermal line widths are not inconsistent with this picture, but further development of the model and more detailed knowledge of coronal structure and temporal fluctuations are required before it can be tested.

It seems likely that Alfven waves are implicated in coronal heating and acceleration even though they are difficult to dissipate by any known process. It was noted that Alfven waves observed in the solar wind do show evidence of nonlinear damping, and there is also evidence that these waves originate in the corona. Alfven waves can only propagate along the magnetic field, which is where the most intense heating is observed. Since the waves are noncompressive, the flow of wave energy can be followed in detail from the convective zone into the corona. For these reasons, major attention was given to their study during the workshop. In particular, the propagation of axisymmetric Alfvenic twists on a vertical axisymmetric potential flux tube was investigated numerically in a linearized self-consistent treatment. Modeling was carried out both for the open field case and for a closed "loop" (without curvature). In the open case, for periods between 10 and 300 seconds, the predicted fluxes could heat the chromosphere and corona. The problem of damping remains, although a number of promising mechanisms were considered. An important result of the closed-loop model is that loop resonances may be excited (as a result of reflections from steep gradients in Alfven speed at the ends of the loop) and could provide the necessary loop energy.

1.3.4 Working Group 4

Working Group 4, led by Eric Priest, addressed problems of *Dynamics and Mass Balance*. The group and its effort was divided into two interacting and complementary sections, one concerned with empirical studies of active region dynamics, and the other with theoretical models of dynamic structures and processes.

The section concerned with observations could not base their study of mass motions on Skylab data alone. The ATM instrument package on Skylab obtained numerous observations of dynamic phenomena associated with active regions. However, it had only one instrument—the ultraviolet spectrograph SO82-B of the Naval Research Laboratory—with sufficient spectral resolution to make direct Doppler measurements of systematic and random velocities. SO82-B had rather low spatial resolution and a limited film supply, making it inadequate for many needed dynamic studies. Ground-based observations provide detailed information on the photosphere and chromosphere, and on the corona above the limb; but EUV observations are required to study the dynamics of the transition region and corona against the disk, and to make use of the emission lines of the middle and upper chromosphere. Fortunately, these could be provided by the French and the Colorado high-resolution spectrometers on the OSO-8 spacecraft launched in 1975, and by the Naval Research Laboratory's rocket-borne High-Resolution Telescope and Spectrograph (HRTS), which was flown in 1975 and 1978. The OSO-8 experiments had the advantage of higher spatial and spectral resolution (2 × 20 arcsec and 0.02 Å, respectively) but, also, the ability to obtain time-series observations of mass motions over a two-dimensional raster. The HRTS instrument has very high spatial resolution (one arcsec) in the range λ1175–λ1719 at all points on a 100-arcsec-long slit.

Observational studies of active region dynamics carried out or initiated in the workshop based on OSO-8, HRTS, and Sacramento Peak diode array data include the following: (1) a determination of the properties of short-lived EUV bursts as observed in C IV; (2) studies of systematic flows based on OSO-8 data in both transition region and chromospheric lines; (3) studies of velocity and intensity fluctuations at the photospheric and chromospheric levels in plages and spots using the diode array to look for evidence of Alfven wave propagations; (4) high-resolution studies of fast, small-scale downflows in spots as well as of systematic velocities in plages and spots using HRTS data. No simple characterization of active region flows is yet possible, except that they are of large amplitude and complex. A comprehensive and systematic review of mass motions in active regions that includes both ground-based results and the recent results from OSO-8 was prepared for the workshop by R. Grant Athay and appears as Chapter 4 of this volume. The unique high-resolution observations from HRTS are discussed by Guenter Brueckner in Chapter 5.

The theoretical studies carried out in the working group centered on the dynamics of coronal loops and intense magnetic flux tubes in the photosphere. A brief overview is given below; a detailed account by Eric Priest is to be found in Chapter 9.

As described earlier, symmetric quasistatic models in energy equilibrium are capable of approximating several basic properties of the plasma contained in coronal magnetic loops, including the fact that some coronal loops are observed to have lifetimes long compared to their radiative loss times. This is surprising when we consider that these magnetic loops are rooted in an atmosphere that is almost everywhere in oscillatory, transient, or systematic motion. Further, even small systematic differences in pressure at the two footpoints of a loop can drive siphon flow, and indeed flows and rapid changes in loops are also observed. The questions of thermal instability and MHD instability also arise. Major advances in these areas were made in the workshop. Studies of the processes of evaporation and draining of loops associated with changes in the heating rate, done jointly with Working Group 1, have been described above.

Study of the thermal stability of static loops in energy balance was a major effort in the group and was explored in parallel investigations. Although the results depend to some extent on the approaches taken and the physics included, some basic conclusions have emerged that can only be sketched here. Antiochos, and Hood and Priest, carried out global, linear normal-mode analyses and concluded that thermally isolated static loops (that is, those having zero heat flux through their base) are unstable to the growth of thermal perturbations. On the other hand, both analyses show that loops can be made stable if there is a sufficient heat flux through the base of the loop. However, Antiochos points out that the growth rate for perturbations increases with decreasing base temperature—that is, that the instability is associated with the cool material at the base of the loop in the transition region, where the radiative relaxation time is short compared to that in the corona. He conjectures that the coronal portions of loops may be in static equilibrium while their lower portion in the low transition region is in an oscillatory or dynamic equilibrium, alternately condensing and expanding. Habbal and Rosner chose a different set of lower boundary conditions. They argued that the temperature gradient should vanish at a fixed base temperature, but that the height of this base should be allowed to fluctuate. The conclusion seems to be that strictly static, thermally isolated loops in energy balance are thermally unstable, but that such loops may be maintained in a dynamic or oscillatory equilibrium by some feedback between the base of the loop and the lower transition region or chromosphere in which it is embedded. A study that takes account of all of the important physics of such a feedback process is for the future, but the apparent stability of some observed loops suggests that such a process occurs. Magnetohydrodynamic instabilities are clearly important in producing dynamic effects in the corona and perhaps also for coronal heating as described earlier; but the apparent MHD stability of coronal loops and loop systems is also remarkable. Several aspects of this stability were explored in the workshop. Briefly, the fact that the footpoints of the magnetic loops are anchored in the dense photosphere (line tying) seems to be the major factor contributing to stability, provided the twist of the field is not too great. Radial pressure gradients and loop curvature may also be important.

Flows within coronal loops are both observed and expected. A sustained pressure differential between the two footpoints will drive siphon flow, and such differences are probably common, especially for those magnetic loops, one footpoint of which is rooted in a sunspot, and the other in the surrounding plage. Two studies of sustained adiabatic siphon flow in coronal loops were undertaken by Cargill and Priest and by Noci as part of the workshop. Noci's study includes calculations of the expected EUV manifestations of the flow. The magnetic field is assumed strong enough to rigidly contain the plasma. For loops that are symmetric, the critical solution becomes supersonic at the vertex and subsonic again at a standing shock on the downleg (i.e., on the low-pressure side). For asymmetrical magnetic loops that expand or converge along the direction of flow, a wider range of both sonic and subsonic flows are possible. Since the following polarity in sunspot groups is almost always more spread out at the photospheric level than the leading polarity, asymmetric magnetic loops in active regions are likely to be common. Siphon flow seems capable of explaining a number of dynamic effects in the active region corona. To the present time, however, detailed comparison with observation has not been possible.

Flux tubes of intense magnetic field (1–1.6 kG) and of small scale (sub-arcsecond) in the photosphere and chromosphere are associated with dynamic effects in these layers as well as in the transition region. Contributions to three problems in flux tube dynamics were carried out in conjunction with the workshop: first, the creation of intense flux tubes by convective instabilities with the attendant production of transient downflows; second, persistent downflows in flux tubes, which are required by downflows observed in the photosphere, chromosphere, and transition zone, in both network and plages; and third, transient up-flows produced by granular buffeting, which might explain spicules and other impulsive events.

1.3.5 Working Group 5

Working Group 5, led by Alan Krieger, studied the *Structure and Evolution* of active regions and ephemeral regions, especially as manifested in the corona. A comprehensive discussion of the overall structure and evolution of active regions in the context of the bipolar magnetic regions (BMRs) that give rise to them is given by Neil Sheeley in Chapter 2. A detailed discussion of the morphology and evolution of active region structures in the corona and transition region, including the advances made in the workshop, is given by David Webb in Chapter 7.

The major efforts of Group 5 centered around the following goals: (1) to order and classify the variety of plasma loops found in active regions; (2) to establish the temporal behavior of these loops; (3) to study the location and structure of loop footpoints; and (4) to study the global properties and integrated emission measure of the active region corona. This latter goal was shared with Working Group 1. Most of the above studies were based on multispectral images in the EUV and in the soft X-ray region obtained by Skylab.

An important discovery of the workshop is that AR coronal plasma loops vary systematically in appearance and temporal behavior with the characteristic temperature of the emission in which they are viewed. The loops are observed to become more complete, more numerous, more diffuse, and longer lived as the characteristic temperature increases toward the maximum of the active region emission measure curve near $T = 10^{6.4}$ K. Thus, in lower-temperature loops, the emission is mostly confined to the legs or feet of the loops; complete loops are rare or short lived in λ465 Ne VII ($T \approx 10^{5.7}$ K). The most conspicuous features in Ne VII are sharply defined spikes with a lifetime of ~0.5 hours, most prominent in the outer regions of BMRs. At the intermediate temperatures characteristic of λ368 Mg IX ($T \approx 10^{6}$ K), the spikes become more numerous, diffuse, and longer lived, typically ~1.5

hours. The higher-temperature plasma seen in Fe XV ($T \approx 10^{6.4}$ K) forms systems, or bundles, of diffuse loops, which form arcades above neutral lines. Often the individual loops in the system are only slightly suggested and cannot be clearly resolved. The loops fade out toward their footpoints—thus opposite in behavior to the lower-temperature loops. The lifetimes of individual higher-temperature structures is typically six hours, and for the loop systems, several days. It has proved convenient to refer to the diffuse loops with $T > 10^6$ K as "hot" loops, and loops or spikes with $T < 10^6$ K as "cool" loops. Cool loops with $10^5 < T < 10^6$ K often appear as spikes or "plumes" above sunspot umbrae, where higher-temperature emissions are usually absent. Hot and cool loops are not ordinarily cospatial, nor do hot loops ordinarily have observable cool cores. Thus, hot and cool loops seem to belong to distinct classes. Quasistatic loop models in energy balance seem to be an adequate first-order approximation to hot loops but not to cool loops, where flow is probably important.

The footpoints of plasma loops in the transition region and chromosphere are difficult to locate with precision. In the workshop, a careful comparison was made between Skylab soft X-ray images, magnetograms, and chromospheric filtergrams in a typical well-observed active region. All of the soft X-ray (hot) loops and loop systems were classified and related to the magnetic field and underlying chromospheric structures. No nonflaring hot loop was observed to end in the umbra of a stable spot. Rather, the brighter hot loops without exception were rooted at one or both ends in regions of particularly strong or changing magnetic fields, bright in Hα and transition region emissions. Workshop studies based on SO82-B XUV spectroheliograms show that patches of emission bright in Ne VII ($T \approx 10^{5.7}$ K) and Mg IX ($T \approx 10^6$K) often underlie arcades of hot loops seen in Fe XV ($T \approx 10^{6.4}$ K) and soft X-rays. Patches or structures bright in Ne VII often appear at the footpoints of the loops, and they change in time without any observable change occurring in the overlying hot loop. This suggests that the temperature gradients at the feet of hot loops are not only steep but also time varying. This is of course pertinent to the studies of the thermal stability of plasma loops discussed above and in Chapter 9. These studies also show that not only are cool loops not cospatial with hot loops, but that they connect quite different portions of the active region.

In another workshop study, Culgoora measurements of the location and polarization of Type-I and -III metric bursts were used to probe loop structure and open field regions in a Skylab active region. Among the conclusions of this study are that (1) Type-III radio bursts are well understood, and their positions and polarization are in good agreement with global potential field calculations; (2) the open field lines of the active region were associated with the strongest unipolar surface fields of the active region; and (3) the Type-I storms are located on high loops overlying X-ray loops.

REFERENCES

Eddy, J. A.: 1979, *A New Sun: The Solar Results from Skylab.* Scientific and Technical Information Branch, NASA, U.S. Government Printing Office, Washington, D.C.

Orrall, F. Q. and Kopp, R. A.: 1976, in *The Energy Balance and Hydrodynamics of the Solar Chromosphere and Corona,* I.A.U. Colloquium **36,** p. 457.

Sturrock, P. A. (editor): 1980, *Solar Flares,* Colorado Associated University Press, Boulder, Colo.

Zirker, J. B. (editor): 1977, *Coronal Holes and High Speed Wind Streams,* Colorado Associated University Press, Boulder, Colo.

2. THE OVERALL STRUCTURE AND EVOLUTION OF ACTIVE REGIONS

N. R. Sheeley, Jr.

2.1 INTRODUCTION

In 1967, during the famous Budapest Symposium, both de Jager (1968) and Kiepenheuer (1968) reminded us that the term "active region" was formerly called "center of activity"—a term that D'Azambuja (1953) used to mean "the totality of all visible phenomena accompanying the birth of sunspots." Now we know that the magnetic field underlies this activity, and we regard an active region as the totality of all phenomena accompanying the birth and early development of a solar magnetic region.

Typically, this phase includes the continuous presence of sunspots, plages, and hot coronal plasma as well as the intermittent occurrence of those energetic phenomena that give the active region its name (i.e., flares, bombs, surges, eruptive prominences, and coronal transients). However, it is interesting to note that these energetic phenomena take up only part of even the early phase of a magnetic region, and that during intervening times an active region may be relatively quiet. Indeed, the strong fields in an active region seem to repress the incessant, chaotic activity that characterizes the weak-field chromosphere (cf. Zirin, 1966; Liu, 1972; Sawyer, 1974). In this chapter, we shall concentrate on the evolutionary characteristics of the magnetic field and its atmospheric tracers and pay little attention to the transient plasma events that, in effect, are the "growing pains" of active regions.

2.2 BACKGROUND

According to our current picture, all of the sun's magnetic fields originate in balanced concentrations called bipolar magnetic regions (BMRs). The BMRs are formed as bundles of magnetic flux rise to and break through the solar surface. In the photosphere, an individual BMR experiences a relatively brief growth phase during which its opposite magnetic poles separate rapidly. Then the BMR enters a relatively long stable phase during which its flux gradually spreads over a large area of correspondingly weaker average field strength. Each polarity seems to spread out independently of the other, and a fraction of the gross flux is neutralized as opposite polarities diffuse into one another across the neutral line. In this way, the individual centroids of polarity continue to separate and eventually would reach diametrically opposite points on the solar surface if no other flux were present. Subsequently, the strength of this global dipole field would decrease steadily as opposite fields neutralized one another across its magnetic equator. (This description ignores differential rotation. In the presence of such shearing, only the north-south component of the initial dipole can contribute to a global dipole field.)

However, other BMRs do exist on the sun, and, together with our first BMR, their collective growth and diffusion accounts for the large-scale distribution of the sun's photospheric magnetic flux. Thus, at any given time, the sun's global field is the superposition of flux from all previously occurring BMRs. Variations in the occurrence rate, the location, and the flux of individual BMRs occasionally lead to complex patterns of photospheric field. However, on the average these flux sources do have systematic properties such as their location (in the sunspot belts), their orientation (slightly inclined to the equator), their magnetic polarities (Hale's Law), their occurrence rate (the sunspot cycle), and the equatorward progression of their mean latitude (Spoerer's Law and Maunder's Butterfly Diagram). These systematic properties of the sources lead to corresponding long-lived systematic properties of the global field, perhaps the most notable of which is the existence and 11-year reversal of the sun's polar magnetic fields. (A summary of the properties of large-scale magnetic fields is contained in the classic paper by Babcock, 1961, and the recent review paper by Howard, 1977a.)

In the photosphere, magnetic flux does not spread uniformly from its sources, but occurs in progressively smaller patches of strong-field concentrations. Where sufficient flux is present, the individual flux elements define a network structure whose characteristic grid size is 30,000 km and whose correlation time is on the order of one day. Evidently, the photospheric flux maps out the instantaneous boundaries of the large-scale horizontal motions called supergranules. The relatively slow formation, dissolution, and reformation of these convective cells leads to the large-scale diffusive transport of magnetic flux. The short-lived, small-scale convective motions of the ordinary granulation are superimposed on this systematic, quasi-stationary large-scale flow. These small-scale motions seem to account for the small-scale (10^3 km), short-lived (10^3 s) changes of the network structure itself. Indeed, it is tempting to suppose that if the supergranulation did not exist, the magnetic flux would still diffuse outward from its source, but with a network grid size and areal diffusion rate that are characteristic of the small-scale granulation.

As one proceeds upward into the chromosphere and corona, the gas pressure drops precipitously, and pressure gradients are no longer strong enough to confine the magnetic flux. Here, the field lines fan out from their sources in the photosphere to fill the entire coronal volume. To first order, the field behaves as the vacuum extension of its photospheric sources. Changes of the photospheric fields are propagated into the corona at the speed of light. Where oppositely directed field lines are brought together, neutral points form and transient electric fields persist as long as flux is being transferred from one field line system to another. The net result is a sequence of potential fields whose adjustment to the changing photospheric sources is delayed very slightly by the relatively short propagation time (which for the corona would be about 10^5 km/3 $\times$ 10^5 km/s = 0.3 s).

However, the sun's atmosphere is not a vacuum, and the presence of highly conducting plasma reduces the response rate of the magnetic fields from the speed of light to roughly the Alfven speed. In the lower corona, this speed (about 10^3 km/s) is still fast enough for the fields to keep pace with their photospheric footpoints (provided that the footpoints do not change significantly in 10^5 km/10^3 km/s = 10^2 s). Thus, even to second order, the coronal fields seem to evolve as a sequence of potential fields.

Of course, the mere presence of the highly conducting coronal plasma distorts the field somewhat from its vacuum configuration. A multitude of oscillations may take place and the magnetized plasma should act like a shaky bowl of jelly in its response to transients. However, steady deformation such as may occur in the higher-density regions of the chromosphere should be relatively insignificant in the strong-field, low-density regions of the corona because persistent external forces are not usually present to support magnetic stresses. (By definition, relatively weak distortions of the field will have corresponding induced currents that are small compared to the electric current sources in and below the photosphere. However, it is possible that even weak coronal currents may have a profound influence on the thermodynamic and radiative properties of the coronal plasma. Nevertheless, we shall confine our interest in the solar plasma to its use as a magnetic field tracer and leave a discussion of the plasma itself to other chapters.)

2.3 SURFACE FIELDS

2.3.1 Characteristics of a Photospheric Magnetic Region

Figures 2.1 and 2.2 show, respectively, the birth and long-term evolution of a high-latitude BMR during the rising phase of the present sunspot cycle. The figures illustrate several characteristics of photospheric magnetic regions:

(1) The flux is bipolar and emerges rapidly during an interval of a few to several days. The initial separation rate of the centroids of opposite polarity is relatively fast, but it slows down after several days. (In Figure 2.1, the phase of rapid growth apparently ended just before the BMR reached the west limb.) It seems almost certain that the flux is balanced in the sense that no net flux emerges. However, several effects sometimes contribute to apparent imbalances. Among these effects are both magnetic and spatial foreshortening of regions that emerge away from the disk center, unknown calibration differences between sunspot fields and nonsunspot fields, and

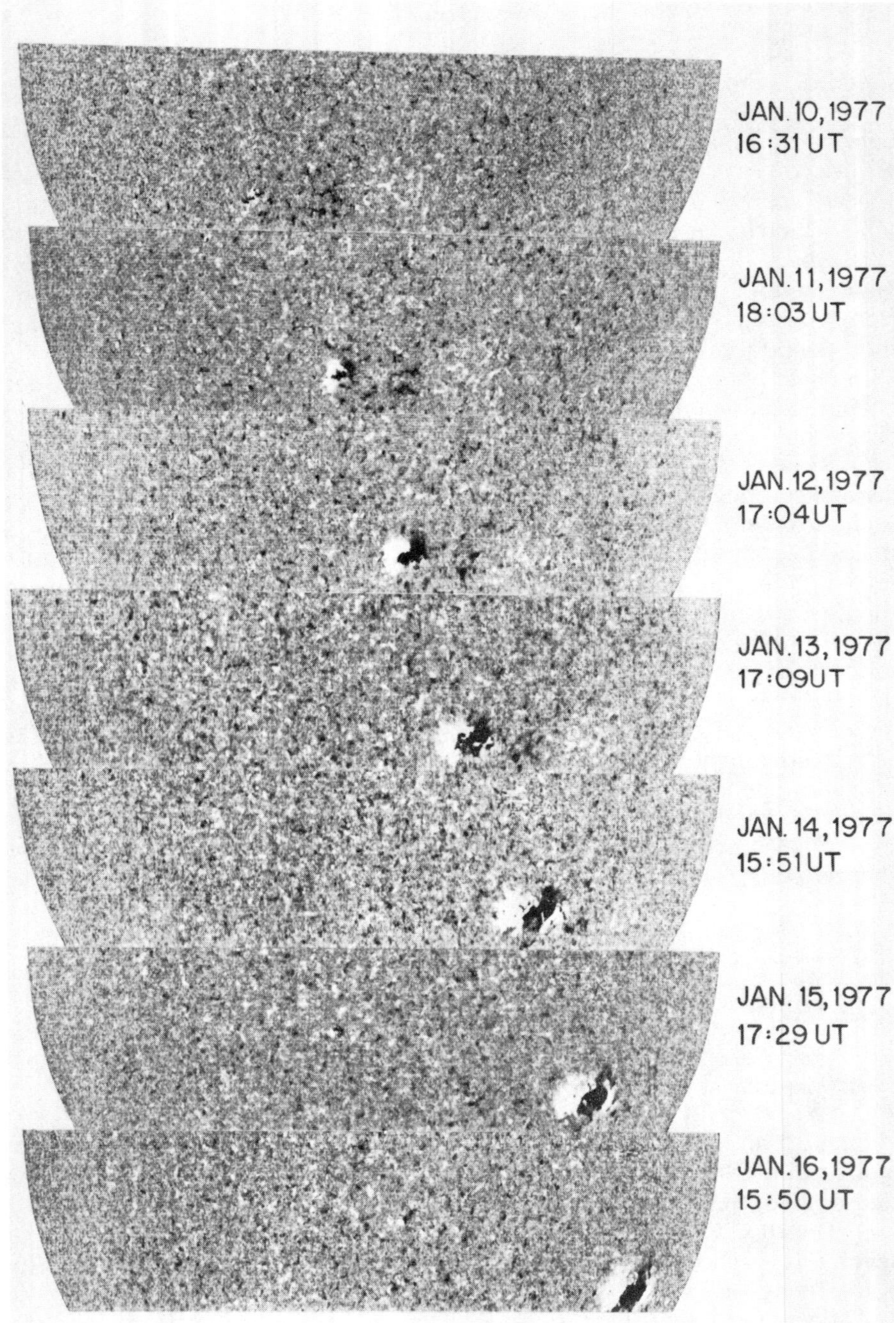

Figure 2.1. The birth and initial development of a new bipolar magnetic region during 10–17 January 1977. Solar north is up and east is to the left. Positive and negative line-of-sight field components are indicated by lighter-than-average and darker-than-average features, respectively.

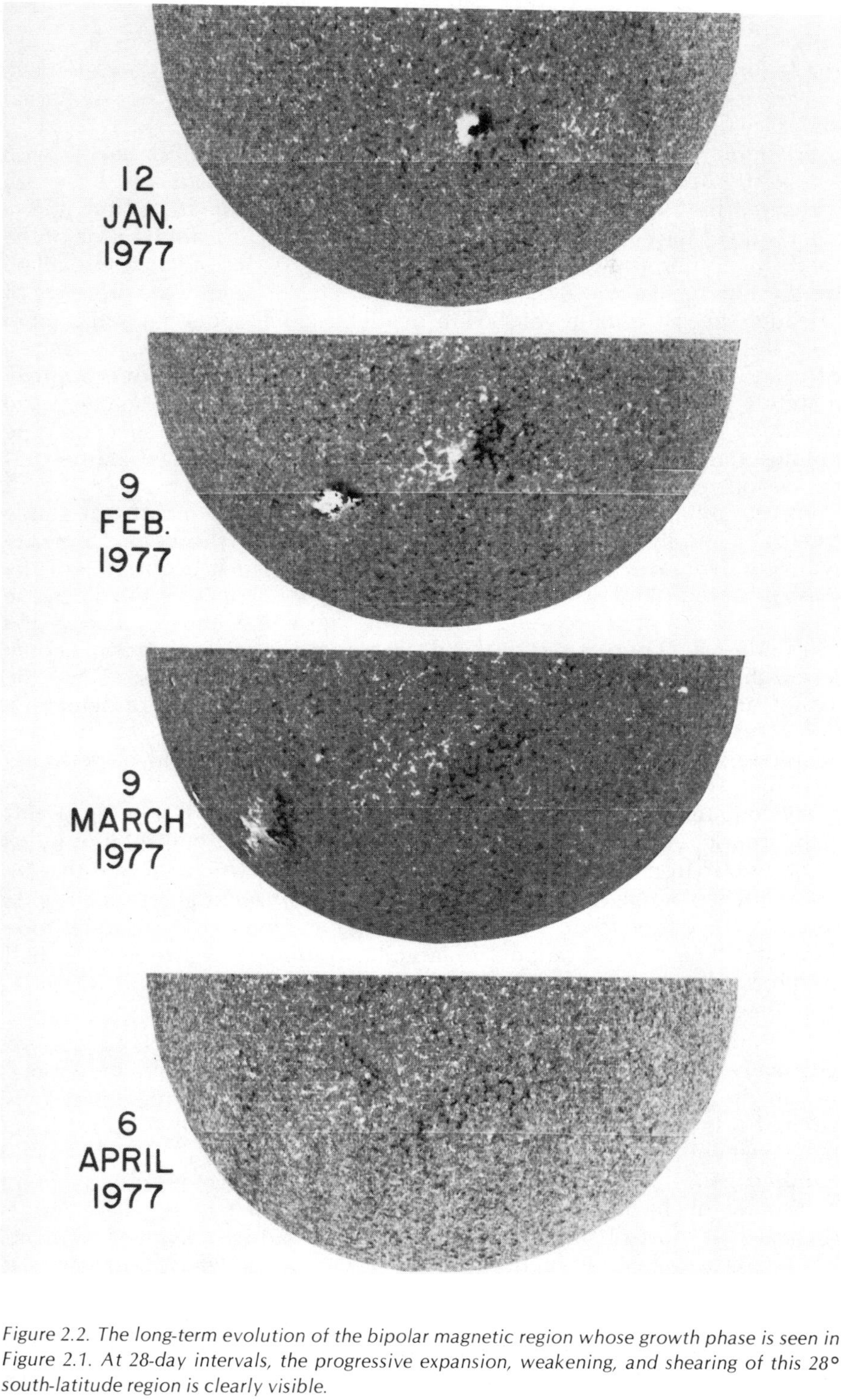

Figure 2.2. The long-term evolution of the bipolar magnetic region whose growth phase is seen in Figure 2.1. At 28-day intervals, the progressive expansion, weakening, and shearing of this 28° south-latitude region is clearly visible.

the failure to account for preexisting flux. When such effects are avoided, the measured fluxes generally balance within the typical observational accuracy of 10–20% (Sheeley, 1966; Stenflo, 1968).

(2) Bipolar magnetic regions of this size typically begin with $1\text{–}2 \times 10^{22}$ mx of each polarity (Sheeley, 1966). Although we have not measured the flux in this particular region, we can estimate it from a consideration of the sunspot flux (which constitutes a significant fraction of the total flux during the early phase of the region). As this region approached the west limb, an obvious magnetic foreshortening suggests the presence of a leader spot of roughly 40,000-km diameter. Such spots have maximum field strengths as high as 3000 gauss and average vertical strengths of perhaps one-fourth this value (Mattig, 1953). The flux in such a spot is approximately 1×10^{22} mx. Assuming that another 1×10^{22} mx may be contained in smaller leading sunspots and in magnetic elements outside sunspots, one obtains roughly 2×10^{22} mx for the total flux of leading polarity. (Of course, one expects an equal flux of negative polarity.)

(3) The initial week of growth is followed by several months of stable expansion and shearing of the region. In such cases, the flux of each polarity typically decreases by an order of magnitude before it becomes indistinguishable from the background field (Sheeley, 1966). Part of this decrease must occur as opposite polarities diffuse into one another across the neutral line. Of course, part of this decrease must also result from encounters with opposite-polarity regions of the background field as well as the loss of small flux concentrations that are below the threshold of detection. In Figures 2.1 and 2.2, the BMR lasted about four months before the combination of diffusion and differential rotation reduced it to the background level.

The long-term effect of differential rotation is particularly striking on this high-latitude (28° S) BMR. In three months, the neutral line rotated by as much as 60° from its initial nearly north-south orientation. Also in this 28-day sequence of magnetograms, the presence of differential rotation is indicated by the eastward drift of the BMR that emerged at 41° south latitude during the 9 February rotation. In a later illustration (cf. Figure 2.6), we shall examine a low-latitude (10° S) BMR whose long-term evolution was relatively free of differential rotation.

(4) As the magnetic region expands, its flux becomes fragmented into progressively smaller elements whose magnitude may range from a few times 10^{20} mx to less than 10^{19} mx. A determination of the individual field strengths is limited by spatial resolution and sensitivity. Most direct measurements give vertical field strengths ranging from a few to several hundred gauss. However, indirect studies as well as a few superb direct measurements have indicated values in excess of 1000 gauss. These results suggest that most of the sun's flux is concentrated in individual elements whose vertical field strength is on the order of 10^3 gauss. However, at present this issue is not resolved.

Relatively little is known about the quiet background flux with which an expanding BMR eventually becomes indistinguishable. However, Livingston and Harvey (1975) and Harvey (1977) have reported a mixture of polarities between the last remaining network fragments. These so-called "inner network fields" are at the limit of spatial resolution and sensitivity, but with 1500-km resolution they seem to consist of spatially discrete elements of roughly 5×10^{16} mx. In addition, they evolve significantly in a time of approximately 30 minutes.

2.3.2 Statistical Properties of Bipolar Magnetic Regions

With the aid of Figures 2.1 and 2.2, we have described the structure and evolution of a "typical" bipolar magnetic region (BMR). However, individual BMRs differ greatly in their properties, such as initial flux and size, orientation, lifetime, birthrate, and location. In this section, we shall describe the statistical distributions of these properties in order to give a background for understanding the variety of magnetic regions and complexes of magnetic regions that occur on the sun.

At its stage of maximum development, a BMR may provide an amount of flux (of each polarity) that ranges from less than 1×10^{19} mx (Golub *et al.*, 1977) for short-lived ($\sim$1 day) and physically small BMRs to more than 5×10^{22} mx for great regions with several large sunspots (Bumba, 1961; Sheeley, 1966). The lower end of this range may extend to as little as 5×10^{16} mx, which has been observed in the bipolar elements of the very short-lived and small-scale inner network field (Livingston and Harvey, 1975).

Also at this stage of maximum development, a BMR tends to be elongated along its dipole axis. The separation between the centroids of polarity ranges from roughly 5×10^3 km for the 10^{19}-mx BMRs to more than 200×10^3 km for regions in excess of 10^{22} mx. This correlation between flux and size is partly due to the fact that the flux is concentrated in elements of approximately constant areal density ($\sim 10^3$ gauss outside sunspots and two or three times this field strength in sunspots). Thus, as additional flux emerges it must cover additional area in order to maintain roughly the same field strength. Field strengths in excess of 4000 gauss have been observed only rarely (cf. Livingston, 1974, and references contained therein for discussions of measured field strengths in the range 4000–5000 gauss.)

For most BMRs with a flux (of each polarity) in excess of 10^{21} mx, the dipole axis is inclined slightly relative to the parallels of latitude, with the leading polarity at the lower latitude. Brunner (1930) found that the axial tilts of sunspot groups varied systematically from 0.6° for groups near the equator to as much as 19.0° for groups in the 30–34° latitude range. Occasionally, BMRs will occur with the following polarity at the lower latitude. However, such backward-tilted regions often undergo subsequent growth and proper motion until the normal tilt is achieved.

The very small BMRs do not seem to follow these orientation rules. Harvey *et al.* (1975) found a virtually random distribution of axial tilts for ephemeral magnetic regions. However, they also found that the axial tilts changed significantly during the one-day lifetimes of these regions, and concluded that their limited temporal resolution and observation times were not sufficient for a definitive analysis. In any case, it is not clear that the initial orientation is a meaningful parameter for such small BMRs that are quickly dominated by their surroundings.

As we have suggested in the previous paragraphs, BMRs occur with a variety of lifetimes ranging from a few hours for very small regions to hundreds of days for very large ones. The implied correlation between size (or strength) and lifetime is not perfect because the lifetime of an individual BMR is strongly affected by its surroundings. Thus, while a typical 10^{22}-mx BMR might last four solar rotations before its flux becomes indistinguishable from the background, two such BMRs might annihilate one another within a few weeks if they emerged with a suitable configuration. Indeed, the very largest BMRs often occur with complex spatial distributions that lead to their premature decay. Similarly, the remnant flux from large, old BMRs must seriously affect the lifetimes of new, relatively weak ones. However, when observations are confined to isolated BMRs near sunspot minimum, there is a definite tendency for large BMRs to live longer than small ones (Sheeley, 1966), as expected for diffusing bipolar sources of flux (Leighton, 1964). It simply takes longer for large amounts of flux to spread out to the threshold of detection.

The birthrate of bipolar magnetic regions is a strongly decreasing function of initial flux, initial size, and, thus, of lifetime. Moreover, the exact shape of this function depends on the phase of the sunspot cycle. In a previously unpublished study, Sheeley (1964) explored this relationship indirectly using sunspot records obtained daily at the Mt. Wilson Observatory during 1951–1960. Figure 2.3 shows the resulting histograms of the annual number of sunspot groups as a function of group lifetime. This figure includes a "bootstrap" correction for the observed sunspot groups whose birth and/or death occurred on the backside of the sun, but it does not include a correction for short-lived groups that may have spent their entire lifetime on the backside of the sun. Consequently, the rates in Figure 2.3 should be multiplied by a factor that varies from approximately two for 3-day groups to one for 14-day (or older) groups in order to obtain birthrates for the entire sun.

More recently, Harvey *et al.* (1975), Martin and Harvey (1979), and Golub *et al.* (1976, 1979) have analyzed the distribution of BMR birthrates with particular emphasis on the short-lifetime end of the range. As in the previous study, their objective was to combine the distribution function with a flux-lifetime relation to determine the relative flux contribution of the large number of small BMRs and the small number of large BMRs.

In an analysis of photospheric magnetograms, Harvey *et al.* (1975) estimated that during 1973 there were roughly 200 ephemeral magnetic regions on the sun at any given time. In an analysis of Skylab/ATM soft X-ray images obtained during 1973, Golub *et al.* (1976) estimated that approximately 500 X-ray bright points were on the sun at a given time. The bright points seem to be a sensitive indicator of newly emerging flux, and they may be easier to count than ephemeral regions. In any case, both studies indicated that the number was roughly a few to several hundred at any given time.

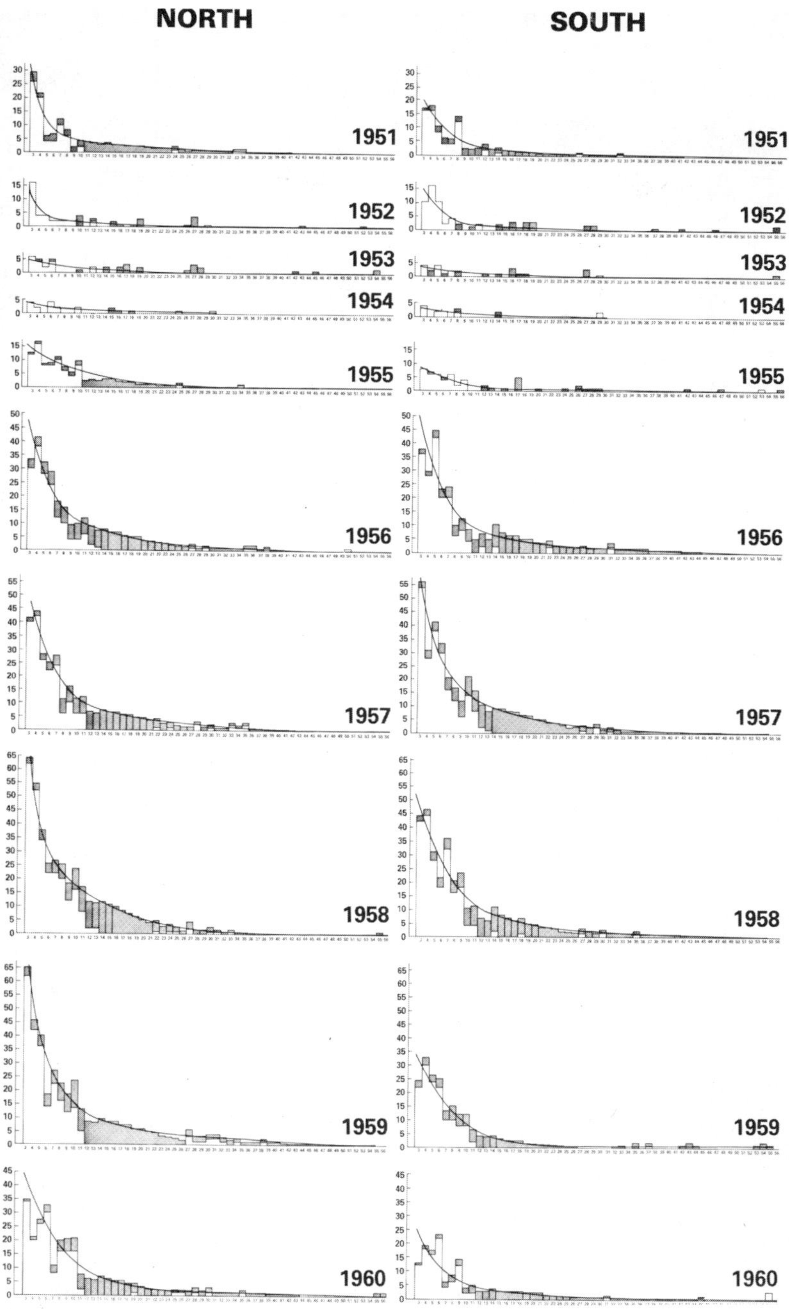

Figure 2.3. Annual histograms of Mt. Wilson sunspot groups as a function of lifetime for the northern and southern hemispheres during 1951–1960. The unshaded areas indicate sunspot groups whose birth and death were observed. The shaded areas indicate groups whose birth and/or death were unobserved on the backside of the sun and for which a "bootstrap" correction has been applied. The solid curves have been sketched by eye through the combined measurements.

Golub *et al.* (1976, 1979) converted this steady-state number of ephemeral regions to an emergence rate by noting that on the average an individual region lasted only one-third of a day. Consequently, to maintain the steady state, the birthrate must be three times the number present, or roughly 1.5×10^3 per day. Then, adopting 2×10^{19} mx as the amount of flux (of each polarity) in an ephemeral region, they concluded that the flux emergence rate was on the order of 3×10^{22} mx/day for ephemeral BMRs during 1973. By comparison, they estimated that the emergence rate for all longer-lived BMRs was approximately one-fourth of this value, or roughly 8×10^{21} mx/day during 1973.

It is possible to confirm this estimate by using the distribution in Figure 2.3 and adopting 10^{22} mx for a typical BMR whose sunspots last 10–20 days. (Of course, the nonspot flux will generally last an order of magnitude longer.) Thus, during a moderately active year such as 1956, there were about 100 BMRs with sunspot lifetimes of 10–20 days, corresponding to a flux emergence rate of roughly 10^{24} mx/yr, or 3×10^{21} mx/day. The integrated contributions of BMRs with lifetimes as short as three days may increase this figure by a factor of two to three but not much more because these short-lived BMRs have correspondingly smaller amounts of flux.

Using a limited number of rocket X-ray images, Golub *et al.* (1979) have extended this comparison to other phases of the sunspot cycle. They found that the instantaneous number of bright points varied by a factor of three, in the sense that it was largest at sunspot minimum and smallest at sunspot maximum. [Martin and Harvey's (1979) recent analysis of ephemeral regions does not support this anticorrelation with the sunspot number.] As a result, Golub (1979) concluded that ephemeral regions dominate the daily flux emergence rate at all phases of the sunspot cycle except sunspot maximum.

However, for a proper perspective on the relative importance of ephemeral regions and larger BMRs, it is important to recognize that the larger BMRs contribute the overwhelmingly greater part of the flux on the Sun at any one time. This is because the ephemeral flux dies (i.e., becomes indistinguishable from the background flux) very rapidly following its birth, so there is essentially no net contribution per day. Indeed, Golub *et al.* (1976) used this fact to derive the birthrate from the steady-state number and the one-third-day lifetime.

In contrast, the larger 10^{22}-mx BMRs may live for more than 100 days before their flux becomes indistinguishable from the background. Consequently, at an emergence rate of only 3×10^{21} mx/day, the large BMRs would contribute about 30×10^{22} mx after 100 days. In actual practice, the contribution would not be quite this large, because the presence of so much flux would enhance the decay rate of individual BMRs. However, during this same interval the ephemeral regions would have contributed essentially no additional flux and their steady-state amount would still be about 500 BMRs $\times\ 2 \times 10^{19}$ mx/BMR $= 10^{22}$ mx (or less, if their decay rate is enhanced by the presence of additional flux from large BMRs). Thus, except possibly at sunspot minimum, the relatively low rate of flux emergence from large BMRs is more than compensated by their relatively long lifetimes. Moreover, even at sunspot minimum, the ephemeral regions should not affect the large-scale field because they are distributed over a wide range of latitude with virtually random orientations of their dipole axes.

2.4 SUNSPOTS AND FACULAE

Figure 2.4 shows the initial development and interaction of several closely spaced magnetic regions and their associated tracers in the upper photosphere (or lower chromosphere). This figure compares sequences of daily magnetograms (*left*) and CN 3883 Å spectroheliograms (*right*) obtained at the Kitt Peak National Observatory during 27 June–2 July 1972 (Allen *et al.*, 1973). The CN radiation probably originates at the height of the so-called temperature minimum, and thus provides a detailed map of this level of the Sun's atmosphere. This figure illustrates several points:

(1) Allowing for differences in spatial resolution and the lack of simultaneity, sunspots and faculae (bright features) account for all of the visible magnetic elements. The relatively higher spatial resolution of the 1 July CN image suggests that the distribution of faculae consists of aggregates of unresolved small-scale elements. (These aggregates are called plages, network structures, or simply faculae, depending upon how much continuous area they cover.) Furthermore, although the direct magnetograph measurements of these aggregates usually indicate field strengths of several hundred gauss, indirect analyses (cf. Harvey, 1977; Stenflo, 1978; and references contained therein) and a few unique direct measurements (Harvey and Hall, 1975; Koutchmy and Stellmacher, 1978) indicate that the flux is concentrated in small-scale elements whose field strength is on the order of 1000 gauss. Such field strengths are only a factor of two or three less than the field strengths of sunspots.

Time-lapse sequences of high-resolution CN images show that the small-scale faculae shift their positions by roughly 10^3 km in 10^3 s, but that the small-amplitude background intensity fluctuates chaotically in only a few hundred seconds. These widely differing time scales allow one to reduce background noise by time averaging. In fact, the construction of time-averaged spectroheliograms reveals the presence of much fainter faculae than appear on an individual spectroheliogram (Sheeley, 1971a; Chapman, 1972 and 1974; Liu, 1974). Similarly, recent magnetograms with improved spatial resolution and sensitivity indicate the presence of much smaller concentrations of flux than are visible in Figure 2.4 (cf. Harvey, 1977). At present, the lower limit of this correlation between field strength and brightness is not known, but there is some evidence that all of the sun's surface area is covered by fields of some weak but finite strength even though most of the flux is contained in the strong-field elements.

(2) The initial phase of a large BMR is characterized by the growth and rapid separation of sunspots (cf. Waldmeier, 1939, as summarized by Kiepenheuer, 1953 and de Jager, 1959). In this case, the separation between the large leading-polarity spot and the large following-polarity spot had

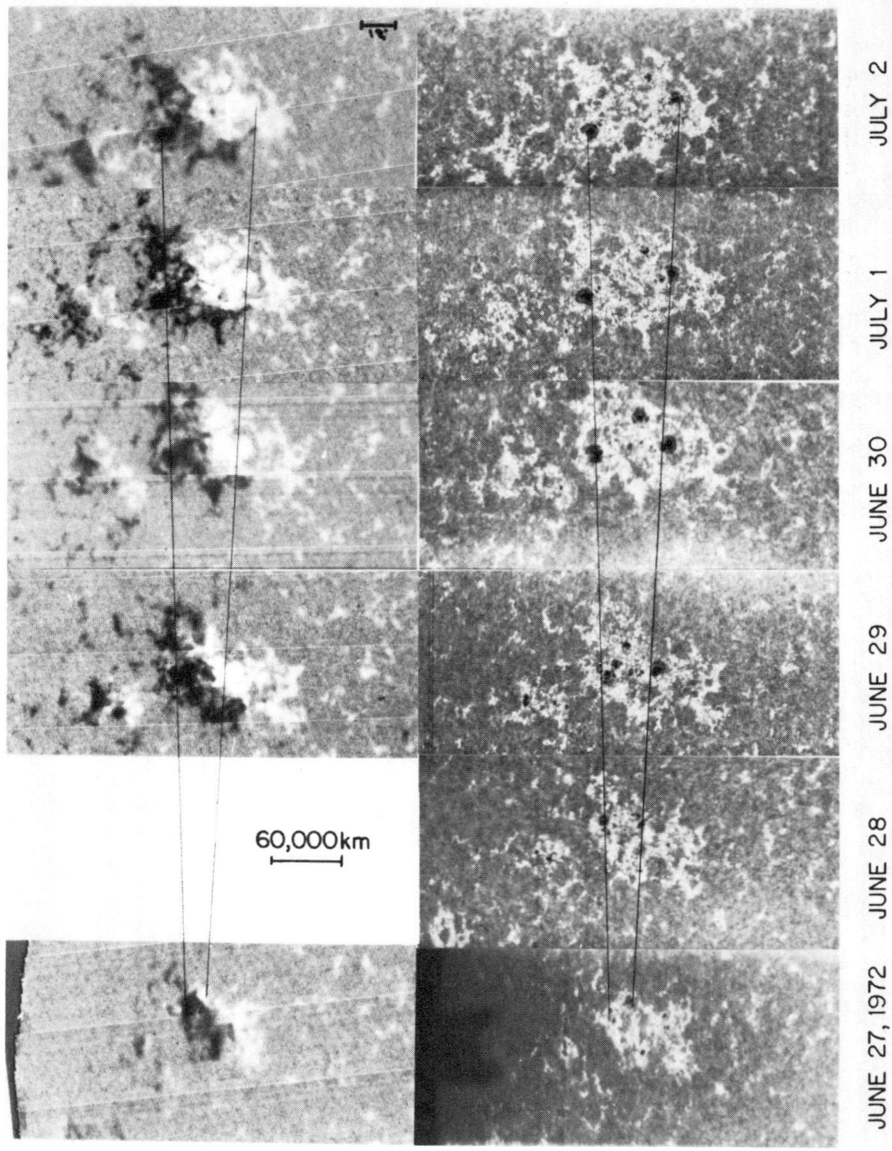

Figure 2.4. The growth phase of a photospheric magnetic region and its associated sunspots and faculae during 27 June to 2 July 1972. The solid lines have been drawn to emphasize the progressive separation of the leader and follower sunspots. Solar north is up and east is to the left. Note that time progresses upward in this figure.

reached 90,000 km within five days of the initial flux emergence. Typically, most of this 0.2 km/s separation speed resulted from the westward motion of the leader spot.

(3) The large leader sunspot developed a semicircular "moat" that was relatively free of facula aggregates (plage and network structures), but that contained a number of very tiny facula "points." As the sunspot moved westward relative to its surroundings, the moat maintained its near 50,000-km diameter and its centered location around the sunspot. Thus, almost like a snowplow or a bow wave, the moat compressed the scattered distribution of older flux to its west into a compact plage.

(4) Within the moat itself, the tiny facula points move horizontally outward from the sunspot at speeds on the order of 1 km/s until they eventually become compressed into the moat's network boundary. This movement was visible on time-lapse sequences of CN images during this interval (Allen *et al.*, 1973), and has been observed in sequences of spectroheliograms (Sheeley, 1969 and 1971b; Liu, 1973) and magnetograms (Vrabec, 1971 and 1974; Harvey and Harvey, 1973) at other times for both leader and follower sunspots.

Although elements of both magnetic polarities may originate at the boundary of a unipolar sunspot and move horizontally outward, a net flux of the "like" polarity is transported from the sunspot to the outer edge of the moat during the lifetime of the sunspot. Harvey and Harvey (1973) measured a net flux transfer rate of approximately 2×10^{19} mx/hr, which is sufficient to transfer all of the flux from a 2×10^{21} mx sunspot to the network in four days. Furthermore, two-dimensional Doppler observations of sunspots and their surroundings sometimes show large-scale, horizontally outward motions of approximately 0.4–0.8 km/s (Sheeley, 1972). Within an accuracy of a factor of two, this rate is sufficient to provide the convective transport. In any case, this magnetic outflow may be one way in which sunspot magnetic flux is converted into plage and network magnetic flux as a sunspot decays. In other cases, the spot may "dissolve" directly into plage and network structures.

Finally, although it is not illustrated in Figure 2.4, there is a tendency for a large, leading sunspot to outlive the other spots of a group. This evolution results in a characteristic "leader-spot follower-plage" configuration in which the leading magnetic flux is more strongly concentrated than the following flux. In the early studies of sunspot groups, such a configuration would have been classified "αp," where α refers to a monopolar spot (or group of spots) and p indicates that the spot (or group of spots) lies in the preceding part of the active region (cf. Hale and Nicholson, 1938, as described by Kiepenheuer, 1953).

Other configurations were designated by symbols such as βf (the β indicates a bipolar group and the f indicates that the "following" component is dominant), αf, βp, γ (complex), etc., in the Mt. Wilson classification system. On the other hand, in his revision of the Zurich system, Waldmeier (1947) used the letters A through J (excluding I) to classify sunspot groups according to their detailed spatial characteristics. Although the growth and evolution of sunspots continues to be an interesting subject of research, these classification systems seem to be pre-magnetograph anachronisms now that the magnetic flux can be mapped both outside as well as inside sunspots.

2.5 CHROMOSPHERIC TRACERS

High-resolution images in the central absorption components of strong chromospheric lines are characterized by a fibrous structure that seems to trace the chromospheric extension of the photospheric magnetic field. Most notable are the first three lines in the Balmer series of hydrogen (Hα 6563 Å, Hβ 4861 Å, Hγ 4340 Å), the ionized calcium doublet (Ca II 3933 Å, 3968 Å) and the strongest line of the triplet (Ca II 8542 Å), and the He I 10830 Å line. When comparable observations are eventually obtained in the central absorption component of H I 1216 Å (Lyα) and Mg II 2796 Å, 2803 Å, they will probably show fibrous structures, too. In this section, we shall use Hα images obtained at the Big Bear Solar Observatory (courtesy of Drs. H. Zirin and J. Mosher) to study the horizontal spreading of the field in the sun's lower atmosphere.

Figure 2.5 compares a photospheric magnetogram and a CN 3883 Å spectroheliogram with images in the core and 0.7 Å to the violet of the Hα line. These images refer to the same region on 1 July 1972 that we have examined in Figure 2.4, and they illustrate the following points.:

(1) In the small BMR at the extreme left and the large BMR at the center, arch filaments (cf. Waldmier, 1937; Bruzek, 1967 and 1969; Weart, 1970; Zirin, 1972; Frazier, 1972) and field transition arches (Zirin, 1974), respectively, join recently emerged flux of opposite polarity. As Zirin (1974) has emphasized, the confinement of these features to the neutral line means that new flux emerges on the "inside" of a BMR rather than on the "outside" where the principal leading and following sunspots are located. Thus, as new flux emerges at the neutral line, the recently emerged flux separates. Vrabec (1971) has observed this process in time-lapse movies of the photospheric field.

(2) In contrast to the arch filaments and field transition arches that join closely spaced flux elements directly across the neutral line, long chromospheric threads (Foukal, 1971) curve horizontally around the ends of the neutral line to join more distant poles of the BMR. In the off-band image, only the ends of these threads are visible, reaching toward one another from their opposite-polarity sources.

(3) As a BMR ages, its neutral-line fibril structure becomes sheared (cf. the moderately small, but well-developed BMR just to the east of the primary BMR in Figure 2.5). Eventually, the individual fibrils run nearly parallel to the neutral line and collectively become a disk filament (or prominence).

(4) In general, dark fibrils extend away from flux sources in the direction of the local magnetic field. These features are especially visible in the off-band Hα image, and they point in the direction that the field lines must go to reach the opposite polarity. Thus, fibrils from adjacent sources of oppo-

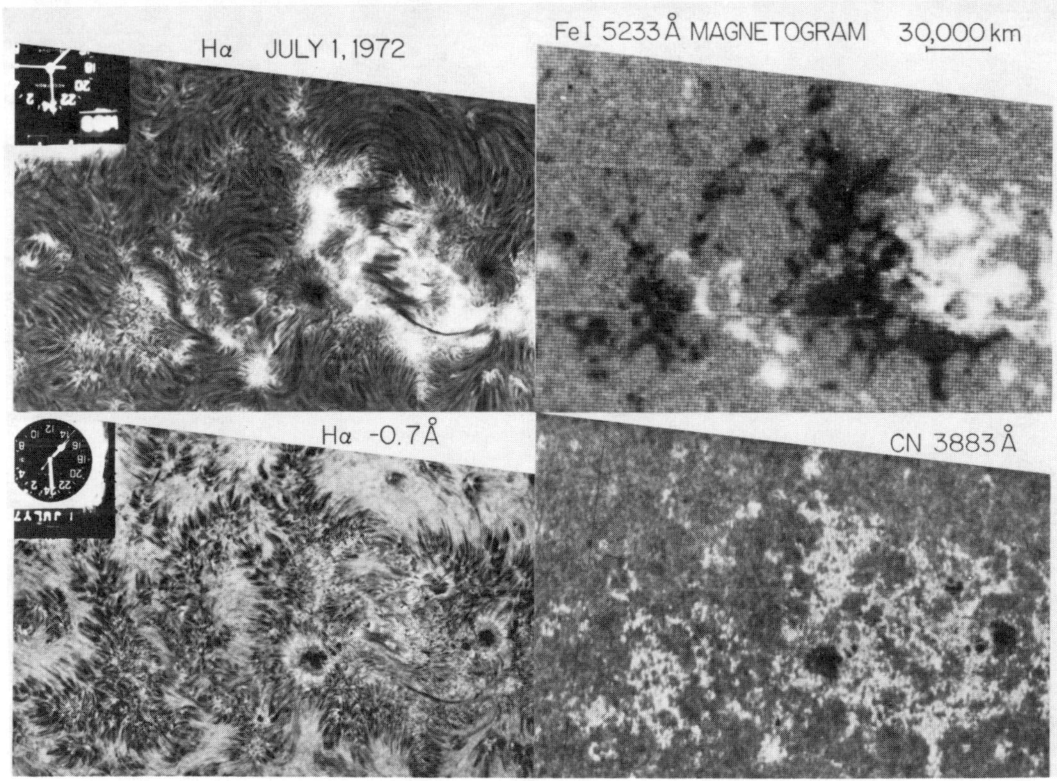

Figure 2.5. A comparison of a photospheric magnetogram with nearly simultaneous CN 3883 Å and H I 6563 Å (Hα) images illustrating the chromospheric spreading of the photospheric magnetic fields.

site polarity point toward one another, whereas fibrils from sources of like polarity all point "downwind" in the direction of the large-scale field. As we shall see in the next section, the fine structures of the transition region (Ne VII 465 Å) and lower corona (Mg IX 368 Å) are also aligned in this way.

A particularly interesting configuration is visible toward the eastern edge of Figure 2.5. In this case, a very small BMR is located "downwind" of the following-polarity end of a moderate-sized BMR. The associated fibril structure is remarkably similar to the classic field line pattern for a dipole in a uniform magnetic field.

2.6 THE TRANSITION REGION AND LOWER CORONA

In our description of the photosphere and chromosphere, we have seen that a growing BMR is characterized by an "inside" where new flux is emerging and an "outside" where accumulations of recently emerged flux are separating rapidly. Arch filaments indicate the emerging flux, and leader and follower sunspots are the vanguards of the separation. Next, in this section, we shall see that the "inside" and "outside" are also distinguished by their structure in the transition region and lower corona.

We shall use spectroheliograms in Ne VII 465 Å, Mg IX 368 Å, and Fe XV 284 Å to characterize the multithermal plasma distribution in an active region. These lines are formed at temperatures of approximately 0.5×10^6 K, 1.0×10^6 K, and 2.0×10^6 K, respectively, and represent the conditions that range from the so-called transition region to the lower corona.

2.6.1 *General Structure of the Multithermal Plasma Distribution*

Figure 2.6 compares virtually simultaneous Ne VII, Mg IX, and Fe XV negative images with photospheric magnetograms on three consecutive solar rotations, and illustrates the following general characteristics of the multithermal plasma distribution:

(1) The low-temperature plasmas are confined to the footpoints and legs of magnetic field lines, and rarely form complete loops. In contrast, the high-temperature plasma seems to originate in loops or systems of unresolved loops that cross neutral lines to join opposite-polarity regions of the photosphere. Unlike the low-temperature emission, the high-temperature emission fades out toward the footpoints themselves. (This fact, together with the diffuse character of the high-temperature emission, makes it difficult to identify individual coronal loops with extrapolated magnetic field lines. Nevertheless, there does seem to be a "first-order" agreement between the observed loop systems and the expected field line configurations.)

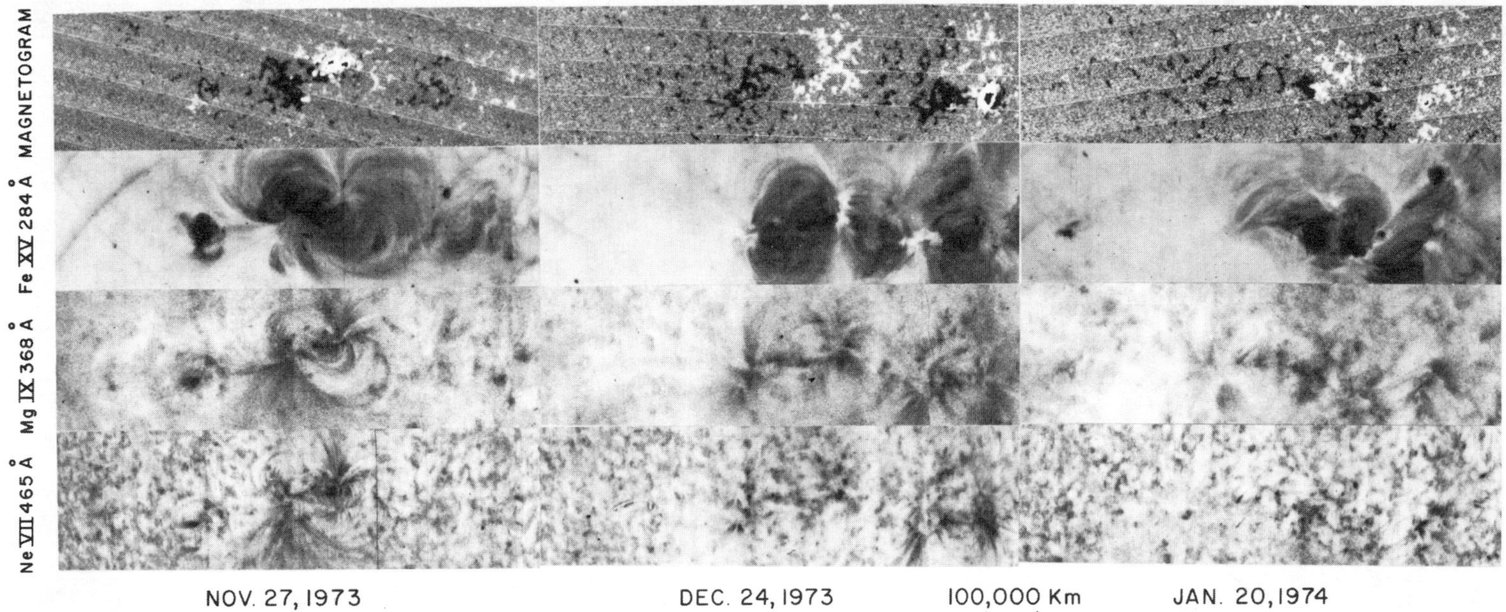

Figure 2.6. The evolution of the photospheric magnetic field and its Ne VII, Mg IX, and Fe XV atmospheric tracers at 10° south latitude, 27 November 1973 to 20 January 1974. On each day, the Ne VII and Mg IX images were obtained simultaneously, and the Fe XV image was obtained within one minute of this time. These images are negative prints. The magnetograms are printed so that positive and negative polarity are indicated by lighter-than-average and darker-than-average features, respectively.

(2) The most intense high-temperature features seem to be relatively short, low-lying loops at the center of a BMR where new flux is emerging. Less intense emission originates in long loops that join remote poles either by going high over the neutral line or around the end of the neutral line in a manner that is reminiscent of our illustration of chromospheric threads. Similar loops cross the neutral lines between adjacent BMRs.

(3) Intense low-temperature features occur in the center of a BMR at the footpoints of newly emerging flux. However, the most striking low-temperature features are long spikes that diverge from the outer ends of a BMR where the magnetic field lines have their greatest length. Thus, the "outside" of a BMR is characterized by diverging flux, long field lines, and low-temperature plasma; whereas, the "inside" of a BMR is characterized by a neutral line, short, closed loops, and high-temperature plasma.

2.6.2 Initial Development of the Multithermal Plasma Distribution

In Figure 2.7, the negative images in the second, third, and fourth columns show the nearly simultaneous evolution of the Ne VII, Mg IX, and Fe XV emission, respectively, from a new, relatively isolated BMR during 1–5 September 1973. This figure illustrates the following characteristics of a growing active region:

(1) At birth, the Ne VII and Mg IX emissions consist of intense, low-lying features that are probably at the footpoints of small, intense Fe XV loops.

(2) Within one day, larger loops are visible either over the top of or around the ends of the neutral line. At the outer ends of the BMR, Ne VII and Mg IX structures are intense but Fe XV emission is very weak. These remote magnetic poles are dominated by sunspots.

(3) As the BMR evolves, characteristic Ne VII and Mg IX "plumes" extend outward from the opposite ends of the dipole, where the Fe XV emission continues to be weak or absent. These "plumes" seem to define an outer envelope for the Fe XV loops within this active region. [However, faint Fe XV loops also extend from the negative (*black*) pole of this BMR to nearby remnants of positive flux.]

2.6.3 Reconnection and Evolution of Coronal Field Lines

Figures 2.8 and 2.9 are reprinted from Sheeley *et al.* (1975a,b) and Sheeley (1976) to illustrate the way in which a growing BMR interacts with its surroundings to produce an interconnected field line pattern. Figure 2.8 shows the evolution of the Fe XV 284 Å coronal plasma (negative prints) during the growth of a new BMR on 31 August–3 September 1973. Figure 2.9 illustrates the way that one would expect the corresponding field lines to evolve in a vacuum.

In Figure 2.8, intense Fe XV emission is first visible at 1209 UT on 1 September and grows in size by 0022 UT on 2 September. Based on our experience with Fe XV images, this emission probably consists of unresolved loops

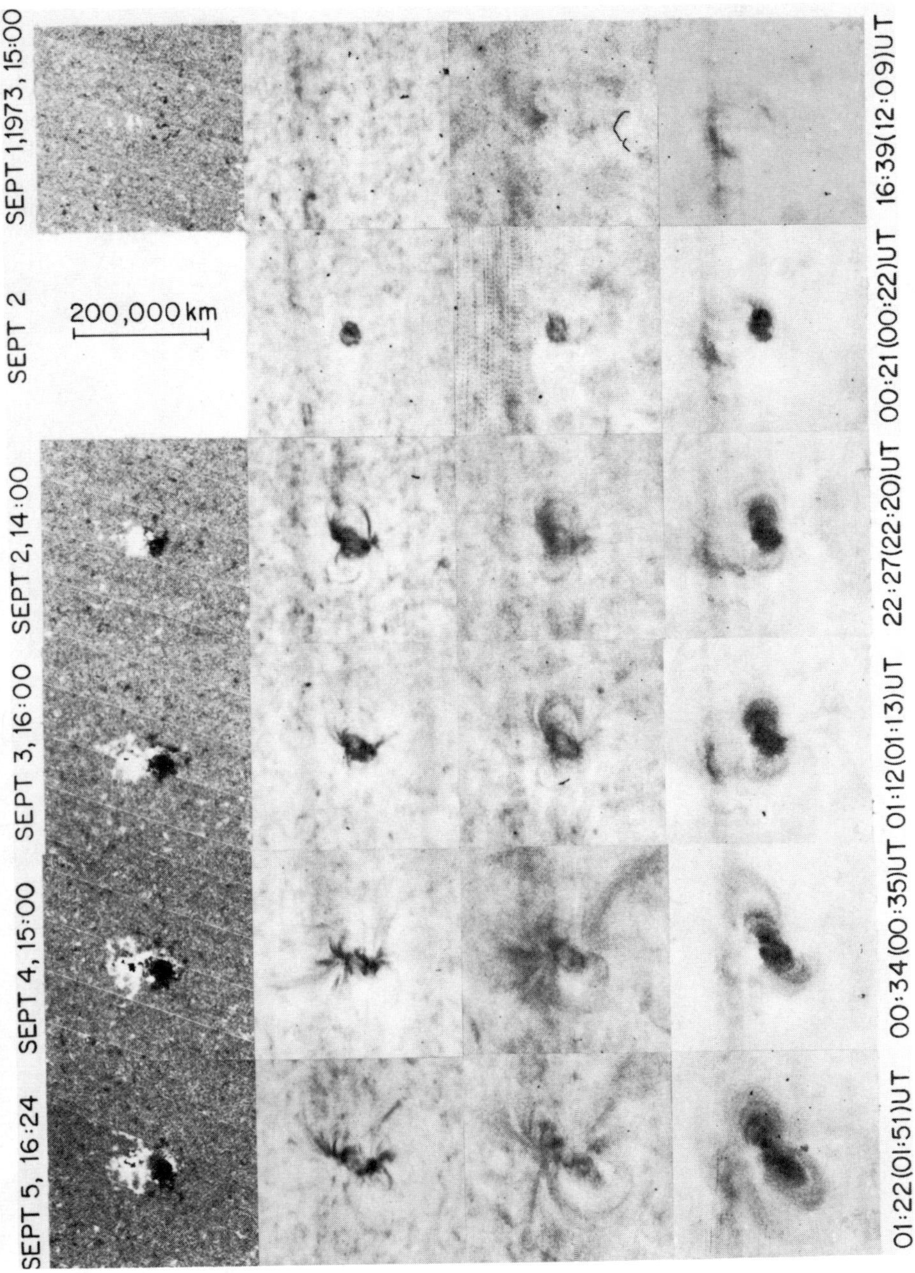

Figure 2.7. The birth and initial development of a relatively isolated bipolar magnetic region and its Ne VII, Mg IX, and Fe XV emissions, 1–5 September 1973. The magnetograms, and Ne VII, Mg IX, and Fe XV images are in columns 1 through 4 (left to right), respectively. The spectroheliograms are negative prints. The Ne VII and Mg IX images were obtained simultaneously, at the times indicated to their right. The numbers in parenthesis refer to the Fe XV images. The magnetograms were obtained at the times indicated to their left.

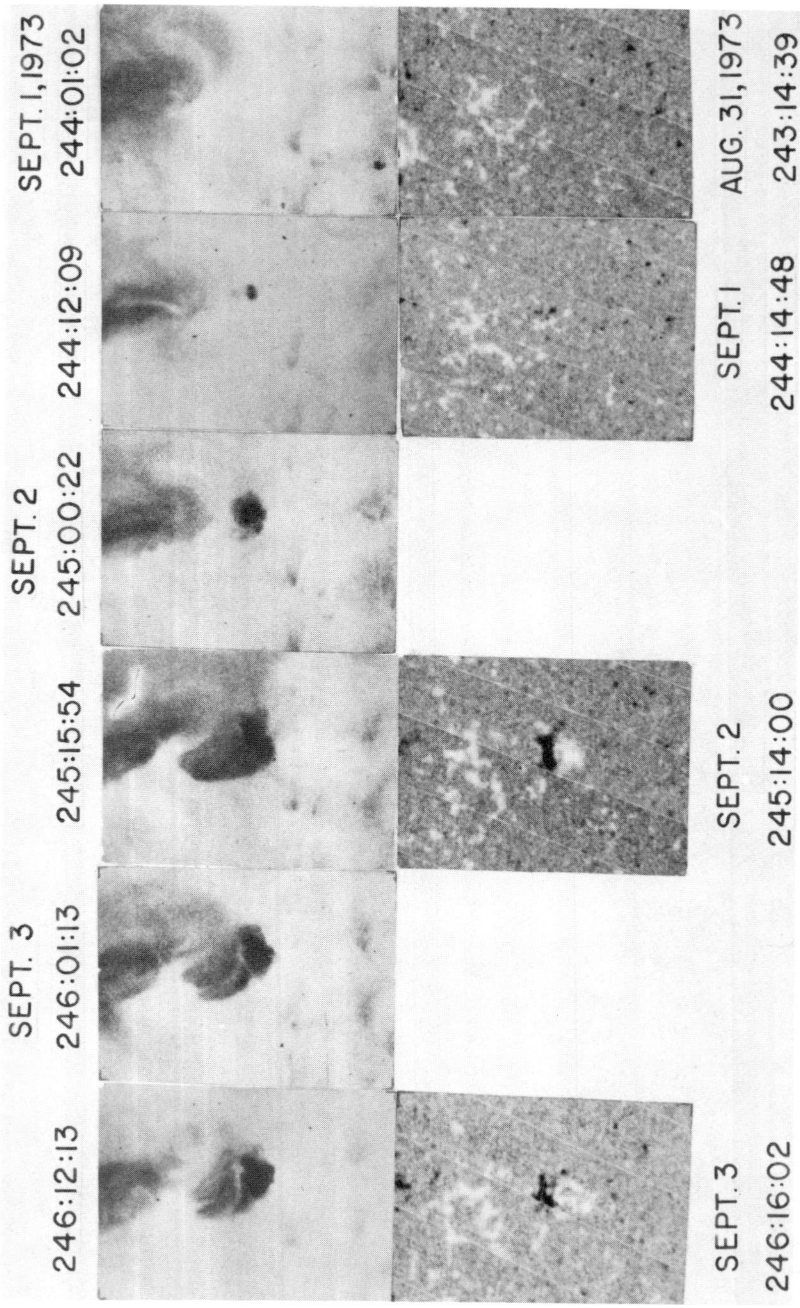

Figure 2.8. The emergence of a small bipolar magnetic region and its associated Fe XV coronal emission in a previously existing positive (white) polarity magnetic region. The Fe XV 284 Å images are negative prints. Not only is intense emission visible between the magnetic poles of the new region, but also less-intense emission loops become visible between the negative (black) pole of the new BMR and elements of previously existing positive (white) flux to its east (left).

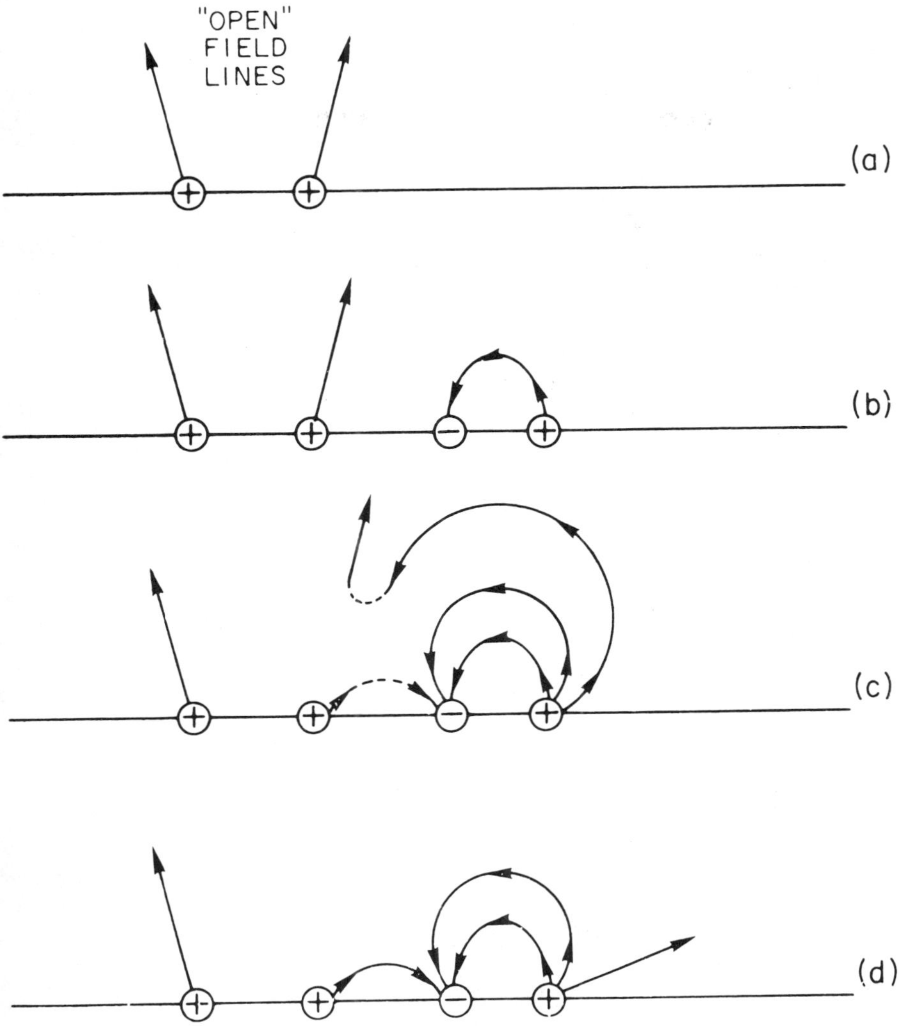

Figure 2.9. An interpretation of the magnetic field line evolution for the region in Figure 2.8: (a) *field lines extending to great distances from the previously existing sources of positive flux;* (b) *the old field lines and the field lines from the incipient bipolar magnetic region;* (c) *the changing field line connections between the old and new magnetic regions; and* (d) *the reconnected field lines.*

that join opposite poles of this newly emerging BMR. On 2 September, by 1400 UT the photospheric flux has increased significantly and by 1554 UT less-intense coronal emission joins the negative (*black*) pole of this new BMR to previously existing elements of positive (*white*) polarity nearby. Such coronal magnetic loops could not have existed here prior to the emergence of the BMR because relatively little negative flux was present to connect with the previously existing positive flux. It seems likely that this connection between the old positive flux and the negative pole of the new BMR resulted from a reconnection, as sketched in Figure 2.9. Such reconnections are not surprising in view of the fact that we have already observed them in the Hα chromosphere (cf. Figure 2.5).

As two BMRs interact, the reconnection produces two topologically different sets of field lines. An inner set of relatively short field lines joins the nearby poles of the BMRs, and an outer set of relatively long field lines joins the remote poles. We have seen that hot coronal plasma maps the shorter field lines. Depending on the magnetic environment, the longer field lines may expand into the outer corona so that their footpoints become open field regions, or these longer field lines themselves may reconnect with nearby field lines. In the former case, areas of relatively low coronal intensity (coronal holes) mark the open field regions. In the latter case, hot coronal plasma maps the connections between each remote pole and nearby sources of opposite polarity. (However, this second stage of reconnection also produces long field lines that either expand into the outer corona or reconnect with still other field lines. In this way, the process continues until flux leaves the sun.)

We have already witnessed such a process in Figure 2.6. During 27 November 1973–20 January 1974, a large BMR interacted with other BMRs to its west (*right*) to form both a pronounced, interconnecting, Fe XV loop system and a region of weak Fe XV, Mg IX, and Ne VII emission (coronal hole) at its trailing end. This example has been described in detail by Bohlin and Sheeley (1978).

2.7 EPILOGUE

We have attempted to give a "first-order" description of active regions by providing a synopsis of fundamental facts and the mental picture that they generate. In this picture, one thinks of loops of flux that break through the photosphere to form bipolar magnetic regions. Individual concentrations of flux are shifted to and fro by small-scale motions of the photospheric plasma, while at the same time they are confined to a slowly evolving pattern by quasi-stationary large-scale motions.

These concentrations are the sources of field lines that fan out with height to fill the entire chromosphere and lower corona. Displacements of these field lines are propagated as a variety of waves. Where oppositely directed field lines are brought together by motions of their footpoints or by the emergence of new flux, the field lines reconnect almost as if they were in a vacuum. The low-density, but highly conducting, coronal plasma is regarded as a field line tracer whose effect on the field itself is negligible.

Perhaps at this time it is appropriate to consider some alternate viewpoints as well as some deviations that are produced by the presence and motions of the solar plasma. (Additional, more detailed studies, can be found in other chapters.)

Howard (1977b) has argued that the Mt. Wilson synoptic magnetograms contain examples that are inconsistent with the transport of flux by the combination of random-walk diffusion and differential rotation. The most convincing examples refer to the rapid growth phase of BMRs, and probably result from the emergence and proper motion of new flux. Other examples seem to result from the emergence of new flux on the backside of the sun or from the inability of the contour-line display to adequately represent the fragmented flux distribution of very old BMRs.

In her review of prominence evolution, Martin (1973) suggested that the random-walk mechanism may be inhibited at the opposite-polarity boundary between two expanding BMRs. In particular, she noted that the encroaching Hα fibrils do not "intermingle" but instead deviate sideways to form a filament (prominence). She also cited the work of Bumba and Howard (1965), who found a similar lack of "intermingling" of opposite-polarity flux at the magnetic neutral line. More recent high-resolution observations continue to support their finding. However, in an extensive study of magnetic field evolution, Mosher (1977) showed that the observed lack of small-scale intermingling is not at all inconsistent with the random-walk concept of freely interdiffusing flux.

In his little-known, but exhaustively complete, treatment, Mosher successfully eliminated several other specific objections to the random-walk mechanism. On the other hand, he did obtain an areal diffusion rate that is from one-third to one-half of the Leighton (1964) value of 1.0×10^4 km^2/s. He concluded that the random-walk process might require a small assist from hypothetical meridional currents in order to produce the observed global magnetic field patterns. However, it is not yet clear whether his study was sufficiently accurate to justify this latter conclusion.

On the other hand, the sheared fine structure of Hα filaments does not seem to outline a potential magnetic field. Apparently, these chromospheric fields do not reconnect immediately but instead form force-free field configurations (cf. Nakagawa *et al.*, 1971 and 1973; Nakagawa and Raadu, 1972; Sheeley and Harvey, 1975). However, at greater heights the field may return to a potential configuration, because coronal loops of emission seem to lie directly across the neutral line. (Of course, it may revert back to a stressed configuration in the outer corona, where helmet streamers are visible.) If the field changes from a stressed configuration in the photosphere to a potential configuration in the lower corona, then perhaps it is not surprising to find a force-free field somewhere in-between.

Sunspot proper motions can also lead to nonpotential chromospheric configurations. Zirin and Lazareff (1975) have described fibril crossings whose eruptive untwistings seemed to be accompanied by the release of flare energy. This process appears analogous to the helical untwisting that often occurs during the eruption of a prominence.

Convincing examples of nonpotential field configurations in the lower corona are relatively rare. This is partly due to the difficulty of interpreting coronal structures uniquely in terms of field lines and partly due to the limited accuracy with which the photospheric flux can be extrapolated into the corona. However, Levine (1976) has found one or two cases for which the coronal structure can be matched by a force-free field of nonconstant α (essentially, the ratio of current density to magnetic field strength). In each case, α was distributed in such a way that the coronal field lines were compressed toward the solar surface from their potential field locations. Thus, in a BMR the remote poles would be joined by field lines that go at low heights around the ends of the neutral line rather than directly over the top of it. Such a force-free field could account for the fact that in some BMRs the large-scale loops are more prominent around the ends of the neutral line than over its top (cf. Figure 2.7). However, it seems equally plausible that this pattern could be explained as a visibility effect associated with the perspective with which the coronal emission is observed. Clearly, the existence of such force-free fields in the lower corona must be substantiated by additional work before their exciting consequences can be considered.

As a final caveat, let us note that it would have been difficult, if not impossible, to provide an accurate historical review of the very old and extensive subject of active regions. Instead, in this chapter an attempt has been made to organize the material in terms of a "first-order picture" of what is known now about the overall structure and evolution of active regions. On the one hand, this technique suffers from the fact that the first-order picture is not completely accurate as well as from the fact that several important contributions have been omitted. Notable omissions include the detailed evolution of sunspots and chromospheric fine structure and the rapid temporal variations of Doppler and intensity fields that form vast subjects in themselves. On the other hand, this technique prepares the reader for more detailed studies of the solar atmosphere, including its thermodynamic and Doppler properties as well as its variability on a short time scale. Some of these studies will be described in the remaining chapters.

Acknowledgements

In the preparation of this paper, I have benefitted from conversations with the following individuals: K. P. Dere (NRL), L. Golub (CFA), J. W. and K. L. Harvey (KPNO), R. F. Howard (Hale Obs.), A. S. Krieger (AS&E), R. H. Levine (CFA), J. T. Mariska (NRL), F. Q. Orrall (Inst. for Astronomy), A. M. Title (Lockheed), and H. Zirin (Caltech). In addition, J. W. Harvey and W. C. Livingston (KPNO) provided the magnetograms that were used in the preparation of these figures. J. M. Mosher (Lockheed) and H. Zirin (Caltech) provided the Hα images. R. F. Howard (Hale Obs.) provided the sunspot drawings that were used in the construction of Figure 2.3. All Ne VII, Mg IX, and Fe XV spectroheliograms were obtained with the NRL Skylab/ATM instrument S-082A. Financial support for this work was provided by NASA DPRs S-60404G and W-14,429. Finally, the inspiration for this paper was derived from the Skylab Workshop Series on Active Regions, which was supported jointly by NASA and NSF and managed by the High Altitude Observatory, National Center for Atmospheric Research.

REFERENCES

Allen, R., Edberg, S., LaBonte, B., and Sheeley, Jr., N. R.: 1973, *Bull. Am. Astron. Soc.* **5,** 268.
Babcock, H. W.: 1961, *Astrophys. J.* **133,** 572.
Bohlin, J. D. and Sheeley, Jr., N. R.: 1978, *Solar Phys.* **56,** 125.
Brunner, W.: 1930, *Astr. Mitt. Zurich* **124,** 67.
Bruzek, A.: 1967, *Solar Phys.* **2,** 451.
Bruzek, A.: 1969, *Solar Phys.* **8,** 29.
Bumba, V.: 1961, *Bull. Astr. Inst. Czech.* **12,** 82.
Bumba, V. and Howard, R.: 1965, *Astrophys. J.* **141,** 1502.
Chapman, G. A.: 1972, *Solar Phys.* **26,** 299.
Chapman, G. A.: 1974, *Solar Phys.* **39,** 151.
D'Azambuja, L.: 1953, *L'Astronomie* **67,** 430.
de Jager, C.: 1959, in S. Flugge (ed.), *Handbuch der Physik* **52,** 80.
de Jager, C.: 1968, in K. O. Kiepenheuer (ed.), 'Structure and Development of Solar Active Regions', *IAU Symp.* **35,** Budapest, Hungary, 4–8 September 1967, D. Reidel, Dordrecht, Holland, p. 602.
Foukal, P.: 1971, *Solar Phys.* **20,** 298.
Frazier, E. N.: 1972, *Solar Phys.* **24,** 98.
Golub, L.: 1979, in Royal Society of London discussion meeting "The Sun and the Interplanetary Medium," April 1979. (Also Center for Astrophysics preprint No. 1145.)
Golub, L., Davis, J. M., and Krieger, A. S.: 1979, *Astrophys. J. Letters* **229,** L145.
Golub, L., Krieger, A. S., Harvey, J. W., and Vaiana, G. S.: 1977, *Solar Phys.* **53,** 111.
Golub, L., Krieger, A. S., and Vaiana, G. S.: 1976, *Solar Phys.* **49,** 79.
Hale, G. E. and Nicholson, S. B.: 1938, *Magnetic Observations of Sunspots, 1917–1924,* Carnegie Institution of Washington, Part I, p. 56.
Harvey, J.: 1977, in Edith A. Muller (ed.), *Highlights of Astronomy* **4,** Part II, D. Reidel, Dordrecht, Holland, p. 223.
Harvey, J. and Hall, D.: 1975, *Bull. Am. Astron. Soc.* **7,** 459.
Harvey, K. and Harvey, J.: 1973, *Solar Phys.* **28,** 61.
Harvey, K., Harvey, J., and Martin, S. F.: 1975, *Solar Phys.* **40,** 87.
Howard, R.: 1977a, *Ann. Rev. Astron. Astrophys.* **15,** 153.
Howard, R. F.: 1977b (private communications).
Kiepenheuer, K. O.: 1953, in Gerard P. Kuiper (ed.), *The Sun*, Univ. of Chicago Press, Chicago, p. 322.
Kiepenheuer, K. O. (ed.): 1968, 'Structure and Development of Solar Active Regions', *IAU Symp.* **35,** Budapest, Hungary, 4–8 September 1967, D. Reidel, Dordrecht, Holland, p. 3.
Koutchmy, S. and Stellmacher, G.: 1978, *Astron. Astrophys.* **67,** 93.
Leighton, R. B.: 1964, *Astrophys. J.* **140,** 1547.
Levine, R. H.: 1976, *Solar Phys.* **46,** 159.
Liu, S. Y.: 1972, *Bull. Am. Astron. Soc.* **4,** 386.
Liu, S. Y.: 1973, *Solar Phys.* **31,** 127.
Liu, S. Y.: 1974, *Solar Phys.* **39,** 297.
Livingston, W. C.: 1974, *Conference on Flare-Related Magnetic Field Dynamics*, Boulder, Colorado, 23–25 September 1974, HAO/NCAR Report, December 1974, p. 269.
Livingston, W. C. and Harvey, J.: 1975, *Bull. Am. Astron. Soc.* **7,** 346.
Martin, S. F.: 1973, *Solar Phys.* **31,** 3.
Martin, S. F. and Harvey, K. L.: 1979, *Solar Phys.* (in press).
Mattig, W.: 1953, *Z. Astrophys.* **31,** 273.
Mosher, J. M.: 1977, 'The Magnetic History of Solar Active Regions', California Institute of Technology, Pasadena, California (Ph.D. Thesis; available from University Microfilms of Ann Arbor, Michigan).

Nakagawa, Y. and Raadu, M. A.: 1972, *Solar Phys.* **25,** 127.
Nakagawa, Y., Raadu, M. A., Billings, D. E., and McNamara, D.: 1971, *Solar Phys.* **19,** 72.
Nakagawa, Y., Raadu, M. A., and Harvey, J. W.: 1973, *Solar Phys.* **30,** 421.
Sawyer, C.: 1974, *Solar Phys.* **35,** 63.
Sheeley, Jr., N. R.: 1964 (unpublished).
Sheeley, Jr., N. R.: 1966, *Astrophys. J.* **144,** 723.
Sheeley, Jr., N. R.: 1969, *Solar Phys.* **9,** 347.
Sheeley, Jr., N. R.: 1971a, *Solar Phys.* **20,** 19.
Sheeley, Jr., N. R.: 1971b, in R. Howard (ed.), 'Solar Magnetic Fields', *IAU Symp.* **43,** Paris, France, 31 August–4 September 1970, D. Reidel, Dordrecht, Holland, p. 310.
Sheeley, Jr., N. R.: 1972, *Solar Phys.* **25,** 98.
Sheeley, Jr., N. R.: 1976, *Solar Phys.* **47,** 173.
Sheeley, Jr., N. R., Bohlin, J. D., Brueckner, G. E., Purcell, J. D., Scherrer, V. E., and Tousey, R.: 1975a, *Solar Phys.* **40,** 103.
Sheeley, Jr., N. R., Bohlin, J. D., Brueckner, G. E., Purcell, J. D., Scherrer, V. E., and Tousey, R.: 1975b, *Astrophys. J. Letters* **196,** L129.
Sheeley, Jr., N. R. and Harvey, J. W.: 1975, *Solar Phys.* **45,** 275.
Stenflo, J. O.: 1968, in K. O. Kiepenheuer (ed.), 'Structure and Development of Solar Active Regions', *IAU Symp.* **35,** Budapest, Hungary, 4–8 September 1967, D. Reidel, Dordrecht, Holland, p. 47.
Stenflo, J. O.: 1978, *Rep. Prog. Phys.* **41,** 865.
Vrabec, D.: 1971, in R. Howard (ed.), 'Solar Magnetic Fields', *IAU Symp.* **43,** Paris, France, 31 August–4 September 1970, D. Reidel, Dordrecht, Holland, p. 329.
Vrabec, D.: 1974, in R. Grant Athay (ed.), 'Chromospheric Fine Structure', *IAU Symp.* **56,** Surfer's Paradise, Australia, 3–7 September 1973, D. Reidel, Dordrecht, Holland, p. 201.
Waldmeier, M.: 1937, *Z. Astrophys.* **14,** 91.
Waldmeier, M.: 1939, *Astron. Mitt. Zurich* **138,** 451.
Waldmeier, M.: 1947, *Pub. Zurich Obs.* **9,** 1.
Weart, S. R.: 1970, *Astrophys. J.* **162,** 987.
Zirin, H.: 1966, *The Solar Atmosphere,* Blaisdell Pub. Co., Waltham, Mass., p. 230.
Zirin, H.: 1972, *Solar Phys.* **22,** 34.
Zirin, H.: 1974, in R. Grant Athay (ed.), 'Chromospheric Fine Structure', *IAU Symp.* **56,** Surfer's Paradise, Australia, 3–7 September 1973, D. Reidel, Dordrecht, Holland, p. 161.
Zirin, H. and Lazareff, B.: 1975, *Solar Phys.* **41,** 425.

3. ACTIVE REGIONS FROM THE PHOTOSPHERE TO THE CHROMOSPHERE

Gary A. Chapman

3.1 INTRODUCTION

The appearance of an active region is most striking at chromospheric heights (Figure 3.1). However, an active region extends throughout all levels of the solar atmosphere. Although we have learned a great deal about the chromosphere and corona of an active region, the deeper layers of the atmosphere, especially photospheric sunspots and faculae, are poorly understood.

A typical active region contains a large amount of magnetic flux, often in the form of sunspots and pores. Near the solar limb, magnetic flux not associated with spots and pores appears as white-light faculae. At greater heights, the active region is characterized by bright structures related to the complex topography of the magnetic fields.

Owing to their morphological complexity, active regions are often characterized in terms of spatially averaged properties. In principle, such averages should provide useful information, because the scale-height for emission in the quiet sun is easy to define. As we will see, however, the morphological complexity of an active region persists at all heights and greatly hinders the simple interpretation of spatially averaged observations designed to probe gross physical conditions.

Magnetic fields in photospheric faculae were first studied as a phenomenon separate from the excess brightness of active regions. We now recognize that magnetic fields and brightness enhancements are intimately related. In fact, the perturbed atmosphere in an active region confuses the interpretation of magnetic intensity data (Harvey, 1977), especially because the magnetic fields and brightness enhancements are likely concentrated into sub-arcsecond flux tubes, outside of pores and sunspots. Neglecting sunspots for the moment, the chromosphere and corona of an active region exhibit emission spectra that are greatly enhanced over the quiet sun. Neither the source of energy that drives the excess radiation of active regions, nor the energy transport mechanism, is yet established.

3.2 HORIZONTALLY AVERAGED MODELS

A variety of one-dimensional models have been developed to describe the gross physical properties of plages. Ignoring fine structure permits one to

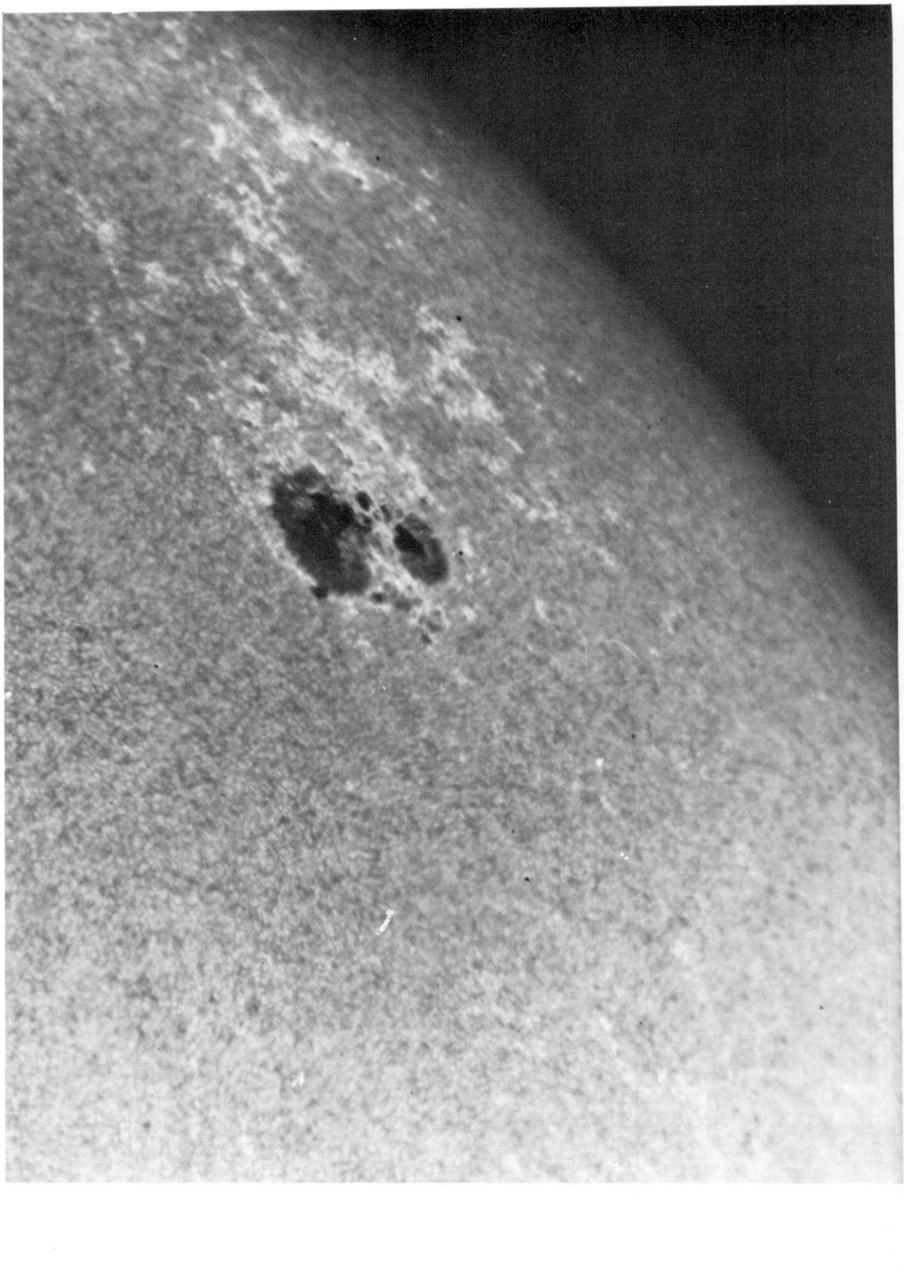

Figure 3.1. Typical active region seen in the mid-visible part of the spectrum on 17 May 1970. Towards the limb from the sunspots are several pores showing bright edges on the limb side, an indication of "bright wall" effects from the pore-photosphere interface.

concentrate on atomic physics rather than geometry. In addition, such models would seem most appropriate for comparisons with low spatial resolution observations.

A number of numerical models of solar plages are available, but those by Linsky and co-workers cover the greatest range in height. Using the damping wings of the Ca II K-line, Shine and Linsky (1974a) derived a model for the deep photosphere up to the temperature minimum. A model covering the same region was constructed by Morrison and Linsky (1978) from Skylab observations of the analogous Mg II wings. More recently, the wings of H I Lyα have been used to probe the thermal structure of the plage chromosphere from a height of about 800 km up to the base of the transition region (Basri *et al.*, 1979).

A problem frequently encountered in assessing plage and quiet sun differences is uncertainty as to what should constitute the reference quiet sun model. An example of this problem is described by Basri *et al.* (1979), who modified the VAL (Vernazza, Avrett, and Loeser, 1976) model to fit profiles of Lyα and Hα, and the Ca II K-line integrated core intensity. Lemaire has proposed a quiet sun model based on similar observational material. He finds that optically thick ultraviolet resonance lines require a temperature minimum of about 4400 K, whereas microwave continuum observations require a temperature minimum near 4200 K (e.g., VAL). Ayres and Testermann (1980) suggest, from an analysis of the first-overtone vibration-rotation transitions of carbon monoxide, that $T_{min} \lesssim 4000$ K. Gouttebroze *et al.* (1978) argue that T_{min} should be higher than 4200 K based on calibrated rocket observations of Lyα. It appears that the mean temperature distribution of the quiet photosphere is not established to better than ± 200 K. The discrepancies may be due to (1) calibration errors (especially when comparing far infrared and ultraviolet observations), (2) uncertainties in abundances or atomic parameters, (3) uncertainties in the line formation process, or, most likely, (4) a genuinely inhomogeneous photosphere and chromosphere that are not amenable to conventional mean modeling techniques.

Uncertainties in the quiet sun models affect the active region models because, in practice, active region models are based on the apparent changes in spectral diagnostics between quiet sun and plages.

3.2.1 Averaged Plage Models

In this section we will put together as best we can a model to represent the photosphere and chromosphere of a plage as seen at low spatial resolution (by low resolution, we mean observations with spatial discriminations poorer than 100 arcsec^2). Such a mean model should indicate the *directions* in which physical properties of an active region differ from those of the quiet sun. However, mean models can only be an intermediate step in describing the genuine physical character of an active region, because we already have good evidence that plages are composed of small-scale concentrations of magnetic flux.

Figure 3.2 compares our hybrid mean quiet sun model with that of an average plage. The quiet sun photosphere at and immediately below T_{min} is

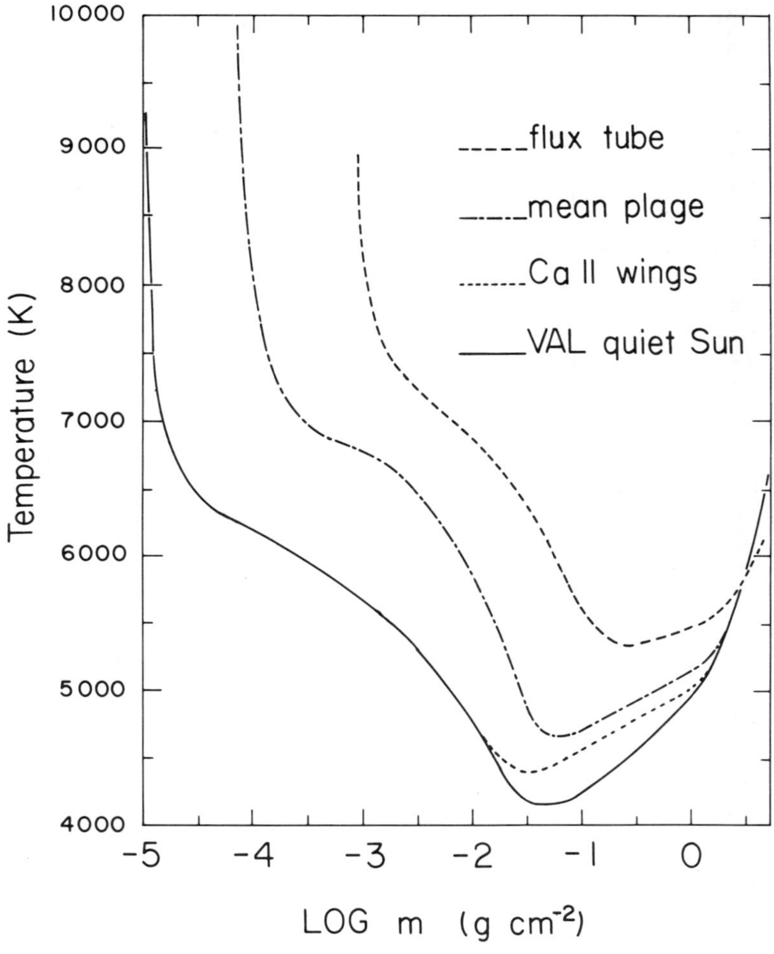

Figure 3.2. Plage, flux tube, and quiet sun models. The solid line *is the VAL quiet sun model.* The short dashed lines *(Ca II wings) represent a modification of the VAL designed to reproduce the Ca II H and K damping wings.* The dash-dot curve *is a plage model based on H I Lyα, Ca II K, and Mg II k data obtained by the French (LPSP) experiment on OSO-8.* The long dashed (higher) curve *represents a flux tube model with a chromospheric portion matching the OSO-8 plage profiles with a 20% filling factor. The photospheric portion* (m > *0.3 g cm^{-2}) is similar to the class of flux tube models advocated by Chapman (1977).*

based on the Ayres and Linsky (1976) analysis of the Ca II wings. The deep photosphere is the VAL model modified according to Ayres (1978). The photosphere of the plage is based on the models of Shine and Linsky (1974a,b) and Morrison and Linsky (1978), while the low and middle chromosphere of the plage is from an analysis of OSO-8 simultaneous measurements of the Ca II and Mg II emission cores (model J79; Lemaire, 1979). (Figure 3.3 compares the OSO-8 observations with half-profiles of Mg II k pre-

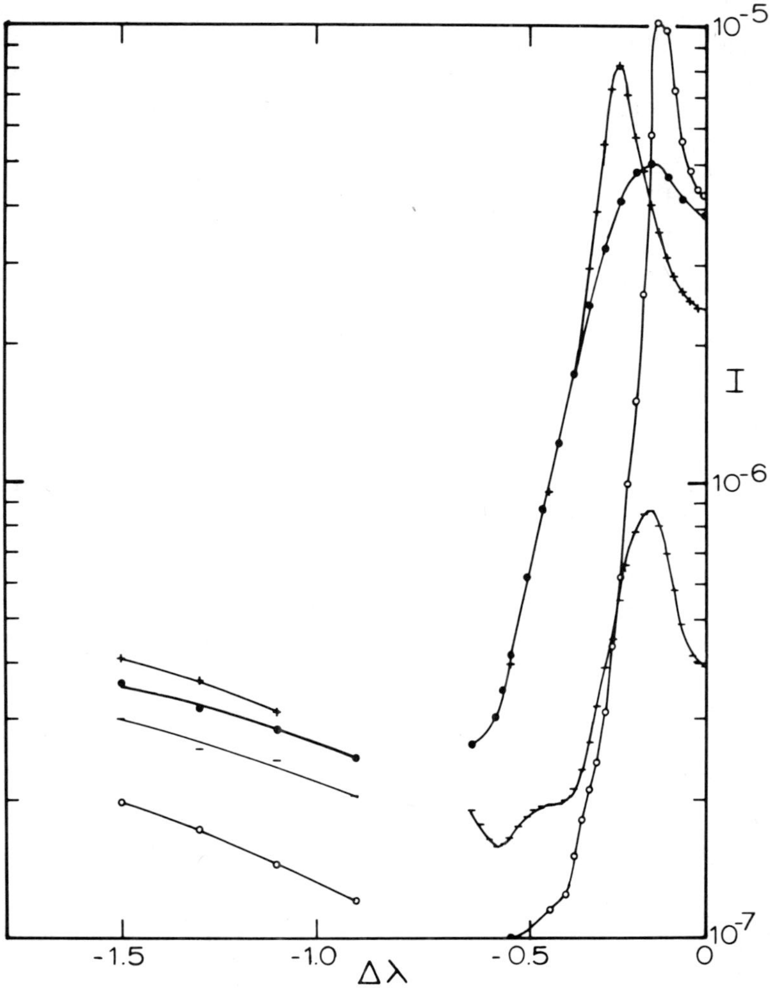

Figure 3.3. Mg II k-line half-profiles predicted by several plage models (courtesy of P. Lemaire). Filled circles *and* horizontal dashes *are from OSO-8 observations of a plage and the quiet sun, respectively. The* pluses *are from Kelch and Linsky (1978) model 5, and the* open circles *are from Basri* et al. *(1979) model P.*

dicted by several different plage models.) The upper chromosphere of the plage is based on the Lyα results of Basri *et al.* (1979).

It is clear from the comparison of mean models that plages are, in a gross sense, systematically hotter than the quiet sun at equal gas pressures. In particular, the plage chromospheric temperature inversion occurs at a larger mass column density than that of the quiet sun.

Although we feel that our composite mean plage model represents the empirical data reasonably well, we again caution that such homogeneous models probably contain serious shortcomings. For example, infrared measurements at 25 μm reveal active region contrasts that are about a factor of 2–3 smaller than those predicted from the mean plage model of Morrison and Linsky (Lindsey and Heasley, 1979). The discrepancy is probably *not* attributable to the lower spatial resolution of the infrared observations compared with the ATM S-082B observations used by Morrison and Linsky. Nevertheless, the Morrison-Linsky model fits Hawaii rocket spectra of the Mg II h and k and Mg I 2852 Å resonance lines (Heasley and Allen, 1979). The latter agreement suggests that the Morrison-Linsky plage model is probably adequate for interpreting UV spectra. However, the disagreement with measured 25-μm contrasts implies that the one-component approach does not produce a *universal* average over atmospheric structures.

We will not discuss mean models much beyond what is said above, for the simple reason that one-component models ignore the fundamentally inhomogeneous character of active regions. In particular, active regions are likely composed of filamentary concentrations of magnetic fields—the flux tubes—whose physical properties are profoundly different from the atmosphere in which they are embedded. The physical conditions inside flux tubes are *so* different from those of the external medium that it is not meaningful to define spatially averaged atmospheric properties. Consequently, our principal goal should be to isolate the physical conditions *within* the flux tubes themselves. Only then can we properly address questions concerning the material dynamics and energy flows associated with active regions.

3.3 FACULAR FLUX TUBES AND THE FINE STRUCTURE OF AN ACTIVE REGION

3.3.1 Faculae, Filigree, and Flux Tubes

A wealth of solar observations of the past decade have revealed that the magnetic fields and brightness enhancements of active regions are concentrated into structures at the arcsec level and smaller. Remarkable pictures of the filigree (resolved facular elements) obtained at Sacramento Peak (Dunn and Zirker, 1973) provided evidence for "brightness" structures as small as ½ arcsec. Evidence for sub-arcsec spatial scales of active region fields appears in studies of the flux and flux density in magnetic structures (Stenflo, 1973; Harvey and Hall, 1975; Tarbell and Title, 1977). Such observations indicate magnetic flux densities of about 1200 to 1600 G (Figure 4, Tarbell and Title) in structures comparable in size to the filigree. In fact, Ramsey *et al.* (1977) have shown that the facular brightness enhancements coincide with the magnetic field to within 0.5 arcsec. However, at chromospheric heights the correlation is much less clear. Empirical models of faculae have also indicated the need for strong magnetic fields in order to

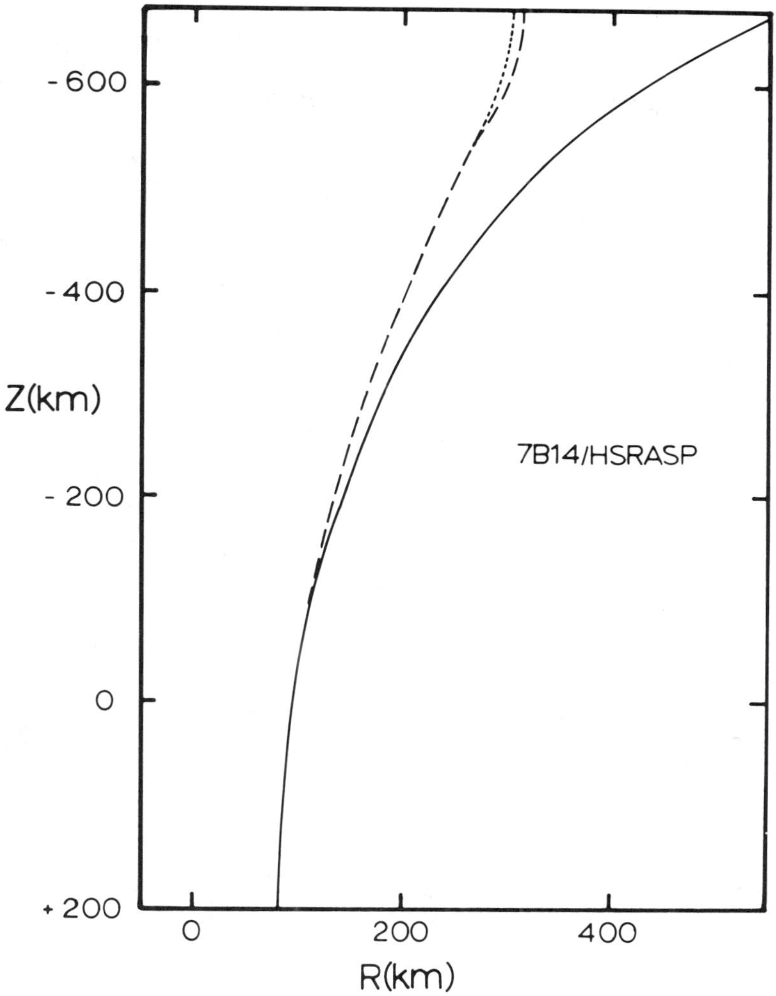

Figure 3.4. Facular flux-tube cross section (model 7B14/HSRA, Chapman, 1979). The solid curve neglects tension forces. The dashed curves represent five (short dashes) and ten (long dashes) iterations of the magnetic field to achieve equilibrium including tension for a simplified magnetic model and fixed gas pressure difference.

ensure horizontal pressure equilibrium (Chapman, 1970; p. 325). Although it has not been established conclusively that the bright filigree are cospatial with the magnetic field concentrations, we will assume here that they are the same phenomenon. Hereafter "faculae" and "filigree" will refer to the same type of physical structure.

In the current picture, a filigree element is a magnetic flux tube whose diameter increases with height. The magnetic flux density is thought to be relatively constant from one tube to another, with an amplitude of about 1500 G (150 mT). The magnetic flux of the individual elements evidently is distributed over a wide range, with 5×10^{17} Mx (5×10^{9} Wb) a representative value. Figure 3.4 depicts a cross-sectional view of such a flux tube (Chapman, 1979). The variation of field strength with height was determined according to the pressure difference between the tube interior and the external photosphere imposed by an arbitrary shift of the model height scales. Such a relative displacement is analogous to the Wilson depression in sunspots and is thought to have a value of about 150 km. The flux tube with the larger cross section was obtained directly from horizontal pressure balance, $B^2/8\pi = P_{ext} - P_{int}$, ignoring tension forces. The smaller cross section is the result of 10 iterations on the field intensity, including tension forces, assuming that $B_z \neq f(r)$ and $B_r \propto r \bullet \partial B/\partial z$. The shape of the flux tube was fitted to a polynomial, and, for a chosen flux, the equations of Dicke (Eq. 22, 1970) were used to modify the magnetic field to achieve a force-consistent flux tube. The relevant relation is

$$<\delta P> = -\frac{1}{8\pi} <B_z{}^2 + r\frac{\partial}{\partial z}(B_r B_z)> + \text{velocity terms} \ , \qquad (3.3.1)$$

where the brackets indicate averages over the flux-tube cross section. The initial guess for the field values was obtained from the horizontal pressure balance solution. The material density was obtained at each height by invoking hydrostatic equilibrium. We caution that these solutions are by no means unique, nor are they fully converged. Nevertheless, these simulations do suggest that magnetic flux tubes may expand much less dramatically with height in the upper photosphere than previously supposed.

Deep in the photosphere, the magnetic and gas pressures are comparable ($\beta \approx 1$). Observations and facular models suggest that the internal gas pressure is about one-third of the external gas pressure. Since the external pressure falls off much more rapidly with height than the internal magnetic pressure, the cross section of the flux tube must increase with height. At some level the fields of neighboring tubes begin to merge. Above the *merging height*, the field uniformly fills the available space and becomes force free. We characterize the region above the merging height as the *force-free regime*. Similarly, we designate the region below the merging height, where the field concentrations are physically distinct, as the *flux-tube regime*.

3.3.2 An Estimate of the Merging Height

An estimate of the merging height h_m is straightforward. Let B_0 be the internal tube field strength at the photospheric level ($h = 0$), say $B_0 = 1500$ G. If the cross-sectional radius of the tube as a function of height is $r(h)$, and the fraction of a horizontal plane that is penetrated by the field is $b(h)$, then

$$r^2(h)/r^2(0) = b(h)/b(0) = B_0/B(h) \ . \qquad (3.3.2)$$

At $h = h_m$, $b(h_m) = 1$ because the field fills space. Here, B is approximately equal to the spatially averaged magnetic field strength $\bar{B}$, consequently,

$$r^2(h_m)/r^2(0) = B_0/\bar{B} \quad . \qquad (3.3.3)$$

For isolated flux tubes, r depends on h approximately as

$$r \sim \exp(h/H_r) \quad , \qquad (3.3.4)$$

where the scale height H_r is about four times the pressure scale height, independent of the magnetic flux of the tube. Detailed modeling of representative tubes in the solar atmosphere (Spruit, 1976; Chapman, 1974, 1977) suggests $H_r \cong 600$ km. Hence,

$$\exp(2h_m/H_r) = B_0\bar{B} \quad . \qquad (3.3.5)$$

For $\bar{B} = 10$ G (network), we find $h_m = 1500$ km. For $\bar{B} = 100$ G (active region), we find $h_m = 800$ km. Note that these values depend on the local spatially averaged field strength, but *not on the size of the flux tubes*. [The value of 1500 km for the network is the same as that obtained by Gabriel (1976), using numerical models of the network field based on chromospheric and coronal data.] In our picture, the flux tubes, per se, cease to exist as distinct entities above h_m, although their collective field persists in higher layers as the chromospheric/coronal force-free field. However, the quiet (i.e., field-free) photosphere/chromosphere *does not extend above* h_m.

3.3.3 Filling Factors

Observations of relatively quiet areas reveal that the magnetic flux and excess emission are weak. On the other hand, in an active region the magnetic flux and excess emission are strong. One reconciles these differences in the filigree picture simply by supposing that the fractional area covered by sub-arcsec magnetic structures in the two observations is very different. The fractional area occupied by flux tubes at any given level of the atmosphere is usually called the *filling factor*. The simplest model of this sort might be characterized as follows:

a. The emission per unit volume as a function of height is the same for all flux tubes. [Alternatively, the temperature-mass column density relation, $T(m)$, (where $m = \int_0^z \varrho dz$ g cm^{-2}) is the same, independent of the size of the tube.]
b. The fraction of the tube cross section occupied by emitting matter at any height is the same for all tubes.

Present evidence favors case (a) for tube magnetic fluxes up to about 50 GWb (5×10^{18} Mx). Concerning case (b), the fraction of the horizontal surface that is covered by emitting material, $a(h)$, may be smaller than the magnetic filling factor $b(h)$ simply because the tube need not be completely filled with emitting matter. Assumption (b) requires that a/b be the same function of height for all tubes. Since $b(h)$ is approximately the same for all tubes, modulo a multiplicative factor b_0, assumption (b) ensures that the filling factor $a(h)$ itself will be more-or-less the same for all tubes. In principle, then, one could characterize all active regions on the sun by only two functions: $a(h)$ and $T(h)$.

The limits of applicability of this simple picture have yet to be explored. Nevertheless, the implied simplifications in the description of an active region are certainly appealing. For example, we need determine only a single function $T(m)$, which could be compared directly with theoretical predictions of chromospheric heating; whereas, the conventional mean atmosphere approach presents us with a *different* $T(m)$ structure for each individual active region that is modeled.

3.3.4 Empirical Determinations of Filling Factors

All observations that do not explicitly resolve flux tubes will implicitly contain a mixture of quiet and active atmosphere properties. If one knew the appropriate filling factor, one could eliminate much of the effect of the "stray" quiet sun light to derive an observation pertaining solely to the active (magnetic) elements. An obvious complication in practice is the change of the filling factor with height.

We now examine some observational constraints on filling factors. The simplest estimate of the magnetic filling factor $b(h)$ is the mean magnetic field $\bar{B}$ divided by the intrinsic flux tube strength B_0.

For example, Spruit has estimated the magnetic filling factor for the NRL ATM S-082B observations used to derive the Morrison-Linsky facular models. Figure 3.5 depicts intensity contrasts q_λ in the wings of the Mg II h and k lines plotted versus the mean magnetic field strength $\bar{B}$ determined by placing a numerical replica of the NRL slit on a Kitt Peak magnetogram of the appropriate active region. This procedure is not ideal; errors due to the positioning of the slit (*horizontal bars*) and errors due to uncertain magnetograph sensitivity will limit the precision of the mean field measurements. Despite considerable noise, there appears to be a fairly clear trend that larger values of $\bar{B}$, and consequently larger magnetic filling factors, imply larger Mg II wing contrasts. (In order of increasing contrast, the active regions are AR I, AR V, and AR VI as given by Morrison and Linsky. Their slit positions for AR II and AR III were very uncertain and could not be used, while the slit for AR IV passed through a pore.) One must examine this figure critically because the magnetic field observations refer to a level perhaps several hundred km deeper than those corresponding to the Mg II wing points. Consequently, one cannot estimate the intrinsic h and k contrasts within the tubes themselves simply by extrapolating the q_λ to $\bar{B}/B_0 \cong 1$. Nevertheless, one might expect large contrasts, approaching perhaps three to six, if one could completely resolve the flux tubes. For example, Basri *et al.* (1979) have reported plage contrasts in the Lyα core of about ten in high-spatial-resolution HRTS rocket spectra.

In fact, the filling factors determined from the KPNO magnetic field data and the S-082B spectra seem rather small compared with values quoted by Tarbell and Title (1977). They found filling factors from about 0.19 to as high as 0.44. The OSO-8 chromospheric data used to derive the mean plage

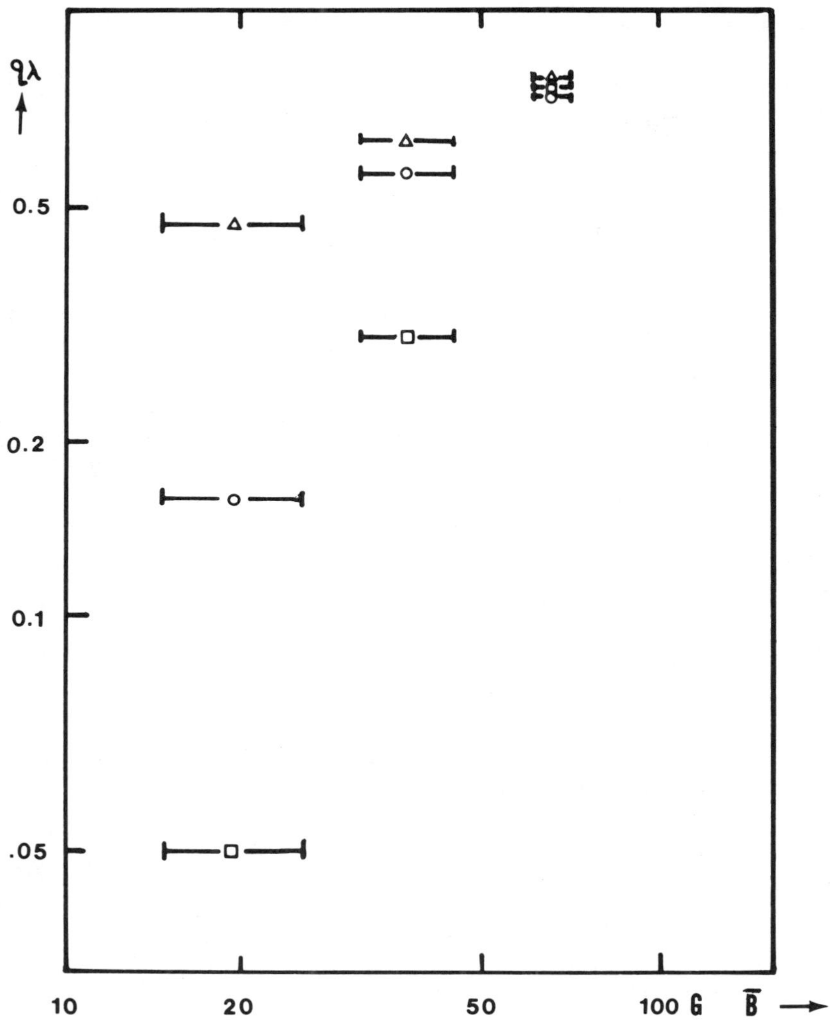

Figure 3.5. Contrast, q_λ at three wavelengths in the Mg II h and k profiles as a function of the average magnetic field strength for three active regions, from Morrison and Linsky (1978). Symbols are □ for λ = 2785.8, ○ for 2790.1, and Φ for 2800.2 Å. These wavelengths represent a sequence of increasing height of formation. Error bars correspond to 2″ uncertainty in slit position. $\bar{B}$ is from KPNO magnetograph field averaged over slit (courtesy of Spruit).

model illustrated in Figure 3.2 had a mean magnetic field strength of 230 G, corresponding to a filling factor of about 0.15 at photospheric levels.

Another technique for deducing filling factors is to compare low spatial-resolution calibrated line profile observations with synthetic line shapes computed using theoretical or empirical flux tube models. Heasley and

Allen (1979) have compared observations of the Mg II wings in plages with a weighted contribution of a flux tube model (Chapman, 1979) and a quiet sun model. Reasonable matches to the h and k wing shapes were obtained for filling factors of 0.20–0.30. The wings of the Mg I 2852 Å resonance line indicated somewhat smaller filling factors than the Mg II data.

Lindsey and Heasley (1979) have applied an analogous approach to infrared continuum observations of faculae in the 12–25 μm region obtained with the University of Hawaii 2.2-m telescope on Mauna Kea. In this wavelength range, the maximum observed contrast is 0.02 at 25 μm, while conventional one-component plage models, such as those by Morrison and Linsky, predict facular contrasts of 0.05–0.10. Using the same two-component model adopted for the Mg II comparison described above, Lindsey and Heasley find that a filling factor of order 0.10 would yield the measured IR contrasts. However, it is not clear whether the different filling factors inferred from the infrared and ultraviolet data represent a genuine expansion of the flux tubes with height, or instead are the result of instrumental effects or intrinsic variations between individual solar active regions.

Using photographs in the wing of the Mg I b_1 line, Spruit (1977) has determined a mean filling factor for an active region of 0.07. He divides the filling factor into three classes: faculae, magnetic knots, and pores. The small value of the filling factor is reasonable, because the b_1 wings are formed deep in the photosphere.

Chapman (1977) deduced filling factors, in terms of "parasitic light," of about 0.5 for a high spatial resolution spectrogram and about 0.25 for lower resolution photoelectric spectra.

The relation between facular contrast and apparent magnetic field strength appears to change at observed fluxes of 50 GWb (Frazier, 1978). At greater fluxes, the "facular" features exhibit continuum darkening and are probably pores. Such behavior suggests a pronounced change in heating. However, even pores exhibit a line weakening, when magnetic line profiles are normalized to their local continuum (Title, private communication). Consequently, significant changes in the character of the heating may not occur until fluxes comparable to those of sunspots are reached. A detailed study at high spatial resolution would be useful for probing the differences between pores and faculae.

Another way to estimate filling factors empirically is suggested by a comparison of the slope and intensity of the Si I continuum in the 1400–1700 Å region. Figure 3.6 illustrates representative HRTS rocket spectra of that region in the quiet sun (scan #13) and an active area (scan #7). A simple analysis of such data (Moe and Engvold, 1979, private communication) revealed typical enhancements in the active region Si I *brightness temperature* of only about 200 K over the quiet sun. However, the enhancement in *color temperature* is about 1200 K. One possible interpretation is that the color temperature difference is independent of the filling factor, whereas the brightness temperature difference is not. Unfortunately, the straightforward interpretation of the Si I continuum is complicated by substantial departures from LTE (Machado, 1979, private communication).

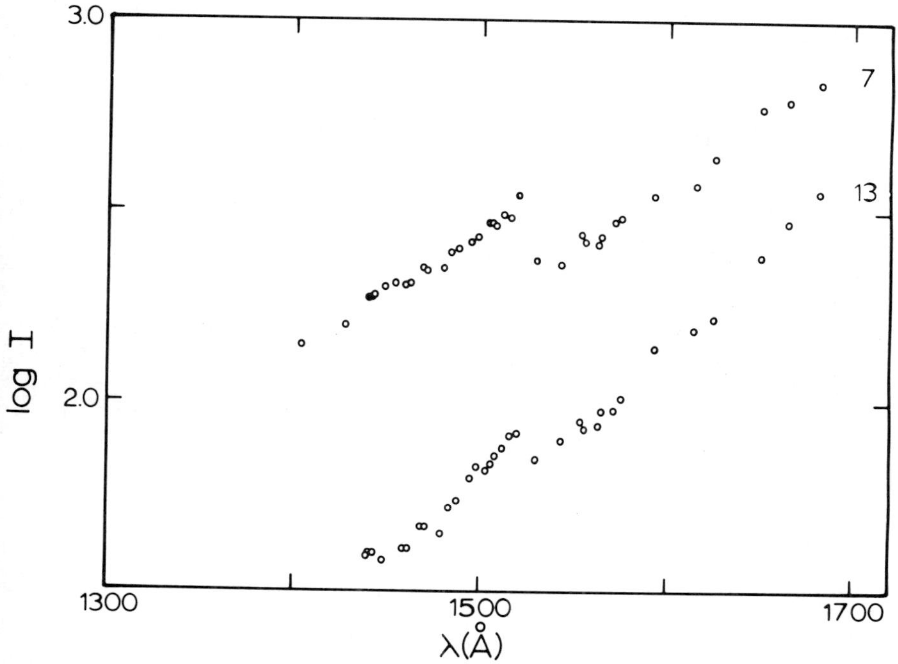

Figure 3.6. Si I continuum spectra, quiet and active sun from HRTS. The slope of the continuum between 1400 and 1700 Å for the quiet sun (scan #13) and one position in an active region (#7). The intensity is in erg s^{-1} cm^{-2} Hz^{-1} Sr^{-1} (courtesy of Moe and Engvold).

3.3.5 Height Dependence of the Filling Factor

Empirical determinations of the height dependence of the filling factor are fraught with difficulties. In particular, one must contend with at least *two* height dependences, namely those of the magnetic and brightness filling factors. Since high resolution observations of active regions in chromospheric lines reveal considerable spatial structure, the brightness filling factor for the entire active region is much less than unity. It is quite possible that the magnetic filling factor is larger at chromospheric heights than the brightness filling factor because magnetostatic pressure balancing requires a rapid falloff of field strength with height (cf. Section 3.3.2). However, owing to the presence of dynamical phenomena in the chromosphere and transition zone, the magnetic field may not spread as rapidly as the pressure scale height would seem to suggest. The spreading of the flux tubes with height may also depend on the packing density.

NRL rocket observations have detected the presence of ultraviolet continuum bright points in the photosphere that are comparable in size to the resolution of the telescope, about 1 arcsec (Brueckner, this volume). If such bright points correspond to the typical spatial scales of the flux tubes near the temperature minimum, where the UV continuum is formed, then the tube diameter must increase by roughly a factor of three between height zero and T_{min}. Of course, the actual increase may be less if the UV bright points are not in fact resolved. Basri *et al.* (1979) found that bright points in the core of Lyα were resolved at 1.8 arcsec, whereas bright elements in the Lyα wing ($\Delta\lambda = 2.5$ Å) were probably unresolved at 0.8 arcsec. However, the authors suggest that the 1.8 arcsec image size in the Lyα core probably is *not* due to flux tube spreading but rather to photon diffusion.

Koutchmy and Stellmacher (1978) have proposed a flux-tube divergence with height of the form $r = 0.12\,h + 87$, where r is the tube radius and h is the height in km. However, this relation is based on an analysis of Mg I b_1 filtergrams by Spruit (1977; p. 124–125), who claimed to find no statistically significant variation of diameter with height.

Spruit has offered the following heuristic height-dependent brightness filling factor $a(h)$ for lack of reliable empirical information:

$$a(h) = \begin{cases} a_0 \exp(h/H) & h \leq h_x \\ a_0\left(\dfrac{h - h_x}{2H} + 1\right)^2 \exp(h_x/H) & h \geq h_x \end{cases} \tag{3.3.6}$$

Here, $a_0 = a(h = 0)$, where h is the height above $\tau_{5000} = 1$, and the spreading scale height H is 290 km. Below h_x, the filling factor follows the change in size of a flux tube in pressure equilibrium. Above h_x, the filling factor corresponds to a linear increase of diameter with height (mimicking the behavior of monopole field lines). Table 3.1 provides some sample cases.

Table 3.1
Variation with Height of Hypothetical Filling Factors

H(km)	$h = 0$	$h_x = 350$ km	$h_x = 700$ km	7B14/HSRA	7B14/HSRA with tension[a]
0	1	1	1	1	1
200	1.81	1.99	1.99	2.28	2.0
500	3.47	5.30	5.30	10.9	6.8
700	4.87	8.6	11.2	42.4	11.2
1000	7.4	15.	26.		
1500	13.	30.	63.		
2000	20.	49.	117.		

[a]Results including tension are not fully converged to a stable solution.

The increase of diameter with height for the Spruit model is rather rapid compared with the albeit sparse empirical evidence cited above. Alternatively, one might invoke a linear increase in diameter of the form

$$a = a_0(1 + h/L)^2 \ , \tag{3.3.7}$$

with $L \gtrsim 580$ km.

The magnetic filling factor probably increases more rapidly with height than the brightness filling factor described above. However, tension and hydrodynamic forces may significantly modify the field strength in the upper layers. For example, Table 3.1 shows that the magnetic filling factor for facular model 7B14/HSRA increases very rapidly without tension forces, but much less rapidly when tension forces are included.

3.4 MULTIDIMENSIONAL RADIATIVE TRANSFER

An important area of concern with empirical modeling of flux tubes has been the applicability of conventional one-dimensional (1 D) radiative transfer simulations. In particular, owing to reduced internal gas pressure and because they are so narrow (100–200 km) at photospheric depths, flux tubes are nearly transparent in the horizontal plane. Facular models have been constructed, taking into account the two-dimensional (2 D) geometry, by Rogerson (1961), Chapman (1970), Wilson (1971), Muller (1975), Spruit (1976), and Hirayama (1978). Only the work of Wilson included the effects of horizontal radiative transfer. The others involved the assumption of no horizontal interaction and calculated the emergent intensity by adding the contribution of the facular and quiet sun atmospheres along the desired ray path. This approach is referred to below as 1.5 dimensional (1.5 D) radiative transfer. Recently, the effects of 2 D radiative transfer on the emergent intensity have been investigated by Stenholm and Stenflo (1977) and Mihalas, Auer, and Mihalas (1978). None of these investigations allowed the effects of 2 D radiative transfer to alter the thermal structure of the embedded atmosphere. Even more recently, Owocki and Auer (1979) have investigated the effects of multidimensional transfer on conventional spectral diagnostics, in particular the inner wings of Ca II K and Mg II k ($\Delta\lambda$ = 1–10 Å). They considered three cases: (1) 2 D, an explicit treatment of the 2 D transfer of radiation for "narrow" tubes embedded in an undisturbed atmosphere; (2) *modified two dimensional* (1.5 D), in which the tube and quiet sun radiation fields do not explicitly interact, but the specific intensities along given rays are properly computed from the individually determined component source functions; and (3) *one dimensional* (1 D), for which a laterally homogeneous model is concocted from an average of facular and quiet physical variables at common depths. A facular tube model was derived with and without a Wilson depression from the plage model of Shine and Linsky (1974a) by uniformly enhancing the temperatures in the outer layers, as illustrated in Figure 3.7 for the case without a Wilson depression. The line source function was calculated using a "partial coherent scattering" formalism. Owocki and Auer (Figure 3.8 and Table 3.2) found that for the Ca II K-line, the core thermalization lengths were small, and consequently the mean intensity $\bar{J}$ saturated to B_ν (local) on the scale of 150 km or less. The resulting line profiles given by two-dimensional and 1.5 D calculations were nearly identical (within a few percent). The differ-

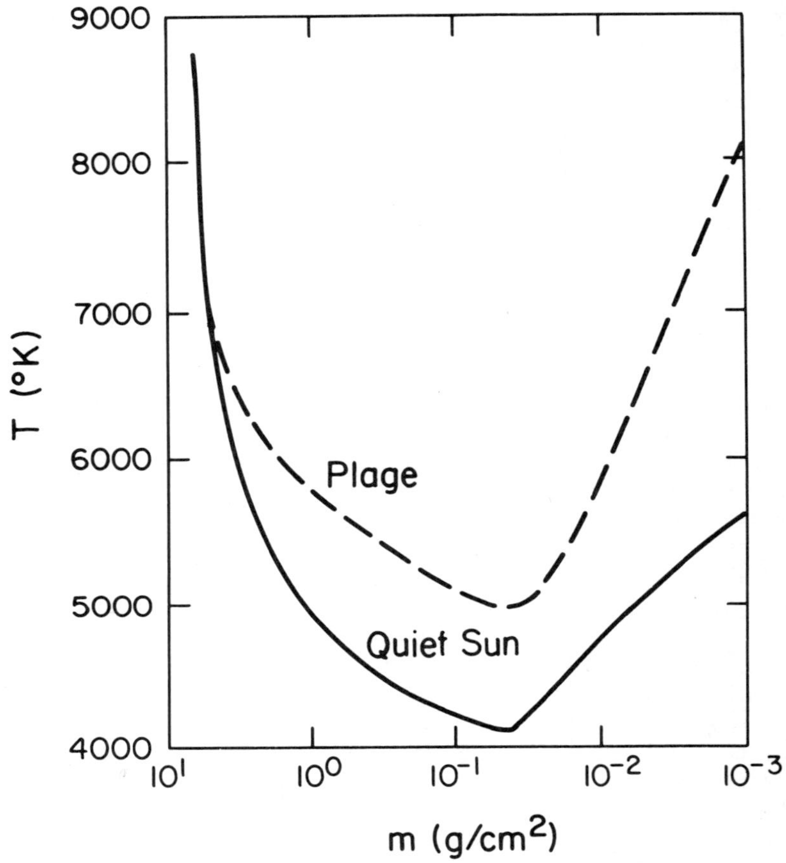

Figure 3.7. Temperature models of flux tube and quiet sun (HSRA) for two-dimensional radiative transfer simulations. The tube model is an extreme Shine-Linsky facular model (courtesy of Auer and Owocki).

Table 3.2
Summary of Parameters for Multidimensional Radiative Transfer

Case	Horizontal Period, x (km)	Plage Component Filling Factor, *f*	Plage Component Wilson Depression (km)	Plage Component Size, *fx* (km)
a	700	0.5	0	350
b	140	0.5	0	70
c	700	0.1	0	70
d	700	0.5	165	350

ences were small regardless of whether the spectra were calculated for spatially averaged or spatially resolved plage components. In the case of a flux tube with a Wilson depression of 165 km, the averaged plage contrast from 1 D calculations could differ by as much as 35% from the 1.5 D and 2 D calculations.

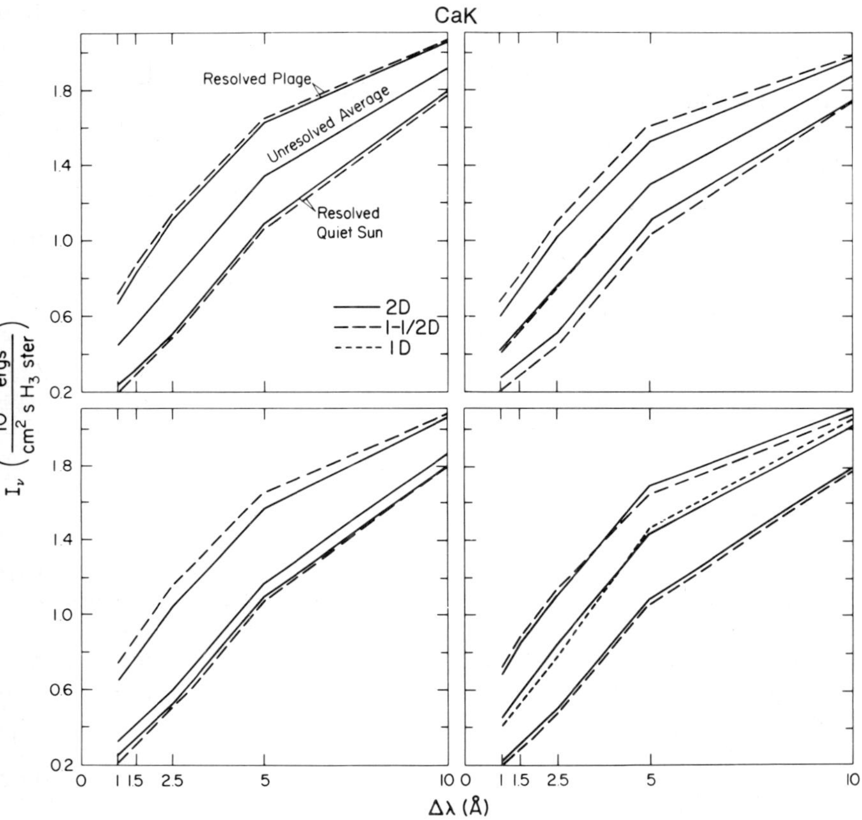

Figure 3.8. Wing profiles of disk center emergent intensity for Ca II K versus displacement from line center for four cases. The upper and lower pairs of curves are the spatially resolved plage and quiet sun components, respectively. The dashed lines represent the 1.5D solution and the solid lines represent the 2D solution. The central line or lines represent the 1.5D (dashed lines) *and 2D* (solid lines) *solution for the unresolved emergent intensity. The lower right* (d.) *shows the effect of a 165-km "Wilson depression" on the emergent profiles. (Refer to Table 3.2 for a complete description of the cases.)*

For the Mg II k-line, the thermalization lengths were much larger than for the Ca II K-line. Therefore, horizontal transfer effects were more important. The increased thermalization lengths will lower the contrasts predicted from 2 D calculations compared to 1.5 dimensional for flux tube diameters of about 350 km or less.

Stenholm and Stenflo (1977) demonstrated that for a line source function that is strongly out of LTE, the emergent intensity from the flux tube can be greatly influenced by the "hot wall" effect. They considered the HSRA model for the quiet sun and the same model, but shifted to produce a Wilson depression, for the flux tube. They found for small values of ϵ, the ratio of collisional deexcitation to spontaneous decays, that the absorption line exhibited considerable brightening (weakening). If that brightening were interpreted as an increase in an LTE source function, then false temperature enhancements of up to 400 K would be inferred for the flux tube. The importance of the "photon channeling" effect should be explored further in future studies.

In summary, the wings of the Ca II K-line have been found to be a reliable diagnostic for obtaining the physical properties of flux tubes down to horizontal scales of the order of a scale height (70 km). Only slight horizontal transfer effects exist, so that observations can be interpreted using 1.5 D radiative transfer. For the wings of the Mg II k-line, serious errors will be made in interpreting the contrast of high-resolution plage observations if lateral transfer (2 D) effects are ignored.

3.5 RADIATIVE COOLING OF THE QUIET SUN AND PLAGE CHROMOSPHERES

3.5.1 Introduction

At present, the source of chromospheric heating has not been identified conclusively. Acoustic waves have been proposed as the primary heating mechanism at the base of the chromosphere (see Ulmschneider, 1979; and references to earlier work therein). Nevertheless, studies of direct and indirect diagnostics of wave propagation in the middle and upper chromosphere have failed to find adequate acoustic fluxes to explain the radiative losses of those and higher layers (Athay and White, 1978; Bruner, 1978).

Both the claims of sufficient acoustic heating at the base of the chromosphere, and of deficiency of the wave flux in higher layers, are based on an important argument: The excess radiative cooling at various levels of the atmosphere is quantitatively estimated, and the net cooling at each level is equated to the mechanical heating by invoking steady-state energy balance. That argument is the principal way in which proposed chromospheric heating mechanisms can be tested in terms of total heating requirements as well as the depth dependence of the energy dissipation.

Solar active regions provide a particularly valuable application of the energy balance argument. For example, one can examine plage radiative losses relative to those of the quiet sun, thereby removing some of the ambiguities associated with determining *absolute* cooling rates.

We address here the following questions in that context: (1) What is the proper way to calculate radiative losses? (2) How do the chromospheric radiative losses in plages compare with those of the quiet sun? (3) What effect might inhomogeneities, for example magnetic flux tubes, have on empirical estimates of radiative cooling in active regions?

3.5.2 Radiative Losses

A plasma can transport energy by a variety of means, for example, conduction and convection. In addition, the plasma can exchange energy with the embedded radiation field. Under chromospheric conditions, conduction and convection are inefficient transport mechanisms, and virtually the entire energy flow must be carried by radiation. Therefore, the cooling of the chromospheric plasma occurs by a net transfer of energy from the kinetic and internal (excitation and ionization) reservoirs of the gas to the radiation field. In thermal equilibrium, the energy drain from the plasma is replaced by mechanical heating.

One must recognize that while some radiative processes can cool the plasma, others can be a source of heating. In fact, in the absence of mechanical heating, it is a balance between the radiative heating and radiative cooling that determines the plasma thermal state; i.e., the condition of radiative equilibrium. We must therefore carefully consider the possible radiative heating mechanisms, as well as the sources of radiative cooling, in order to properly assess the *net* cooling of the plasma.

The important radiative coolants in the solar chromosphere have been identified in a number of previous studies (see Athay, 1976; Linsky and Ayres, 1978; Giovanelli, 1978). The middle and upper chromosphere cooling is dominated by collisionally excited resonance and subordinate lines of H I, Mg II, and Ca II, and by hydrogen continua, particularly the Balmer continuum. The lower-chromosphere and temperature-minimum-region cooling is dominated by H^- recombination and a variety of effectively thin lines, particularly from Fe I and Ti II (Athay, 1970).

3.5.2.1 Line Cooling in the Middle and Upper Chromosphere

Empirical models. The state of the chromospheric plasma is usually characterized by an empirical model. The model specifies distributions of temperature, density, and velocity in the gas, as well as the plasma geometry and temporal evolution.

The most familiar empirical models, for example, the Vernazza, Avrett, and Loeser (1976) quiet sun chromosphere, usually are constructed with simple underlying assumptions, namely a plane-parallel atmosphere in hydrostatic equilibrium. Although these assumptions are perhaps oversimplified, the resulting models have significant advantages for numerical simulations of spectral line and continuum formation, if only because the number of free parameters is obviously reduced from, say, an explicitly time-dependent case.

The temperature-density stratification of an empirical model is adjusted to produce agreement between computed and observed line shapes or continuum intensities. The derived model is in some sense a "mean" atmosphere, to the extent that the observed profiles and intensities are spatially and temporally averaged.

Statistical equilibrium. Given the temperature-density structure and the geometry of a model, one can calculate the statistical equilibrium level populations in a particular model atom. The steady-state populations are established by a balance between collisional and radiative excitation and deexcitation rates. Frequently, the radiation fields in strong transitions within the test atom or ion drive the level populations far away from the values otherwise expected in LTE.

Net cooling. The net radiative cooling (or heating) by line transitions in a particular atom or ion can be derived straightforwardly from the steady-state populations and radiation fields. The cooling rate in a transition connecting lower level ℓ to upper level u is (Gebbie and Thomas, 1971):

$$\epsilon_{\ell u} = n_u A_{u\ell} \frac{hc}{\lambda_{\ell u}} (1 - \bar{J}/S_\ell) \ , \tag{3.5.1}$$

where n_u is the population of the upper level, $A_{u\ell}$ is the spontaneous decay rate, $\bar{J}$ is an average of the mean intensity J_λ in the line weighted over the absorption profile,

$$\bar{J} = \int_0^\infty \phi_\lambda J_\lambda d\lambda \ , \tag{3.5.2}$$

and S_ℓ is the integrated line source function (neglecting stimulated emissions),

$$S_\ell = \frac{n_u A_{u\ell}}{n_\ell B_{\ell u}} \ . \tag{3.5.3}$$

The quantity in parentheses in Eq. 3.5.1 is often called the "net radiative bracket."

Physically, the cooling ($\epsilon > 0$) or heating ($\epsilon < 0$) of the gas occurs only by a net exchange of photons with the local radiation field. Such exchanges are mediated by electron collisions, which can either create photons at the expense of the thermal energy reservoir of the plasma or extract energy from the radiation field by deexciting photoexcited atoms.

In fact, for a strict two-level atom, Eq. 3.5.1 can be rewritten in a form that explicitly illustrates the role of collisions (Gebbie and Thomas, 1970),

$$\epsilon_{\ell u} = n_\ell C_{\ell u} \frac{hc}{\lambda_{\ell u}} \left(1 - \frac{b_u}{b_\ell}\right) \ . \tag{3.5.4}$$

Here

$$C_{\ell u} = n_e \Omega_{\ell u}(T) \exp\left(-\frac{hc}{\lambda_{\ell u} k T}\right) \tag{3.5.5}$$

is the upward collisional rate coefficient (the collision strength parameter $\Omega_{\ell u}$ is typically a weak function of temperature), the *b*s are departure coefficients, and the quantity in parentheses in Eq. 3.5.4 is the "net collisional bracket."

For collision-dominated resonance lines (cf. Thomas and Athay, 1961), such as Ca II K and Mg II k, Eq. 3.5.4 is a reasonable representation of the net cooling in the chromosphere. In practice, it is convenient to divide the net cooling rate by the quantity $n_e n_H$, because that product contains what is typically the dominant density dependence of $\epsilon_{\ell u}$. For example, Eq. 3.5.4 reduces to,

$$\phi_{\ell u}^{e\ell} \equiv \frac{\epsilon_{\ell u}}{n_e n_H} = A_{e\ell} \frac{n_\ell}{n_{e\ell}} \Omega_{\ell u}(T) \times \exp - \left(\frac{hc}{\lambda_{\ell u} kT}\right) \frac{hc}{\lambda_{\ell u}} \left(1 - \frac{b_u}{b_\ell}\right), \tag{3.5.6}$$

where $A_{e\ell}$ is the element abundance relative to hydrogen, and $n_\ell/n_{e\ell}$ is the population of level ℓ relative to the total concentration of that element in all ionization stages. Since $n_\ell/n_{e\ell} \cong 1$ for the resonance lines of majority species in the chromosphere, and typically $b_u \ll b_\ell$, $\phi_{\ell u}^{e\ell}$ is essentially only a function of temperature.

Several important conclusions can be drawn from Eq. 3.5.6.

First, in strict LTE, $b_u \equiv b_\ell \equiv 1$, and there can be no *net* exchange of energy between the plasma and radiation field.

Second, a transition $\ell \rightarrow$ u can be an important coolant only when the lower level is overpopulated relative to the upper level compared with the LTE population ratio (i.e., $b_\ell > b_u$).

Finally, the transition ceases to participate in the net cooling (or heating) of the plasma when levels ℓ and u are in detailed balance (i.e., $b_\ell \equiv b_u$).

In short, the net exchange of energy between the plasma and radiation field is entirely a non-LTE effect. However, even when the populations n_ℓ and n_u depart significantly from LTE, there might not be any net heating or cooling in the transition connecting the levels, if the populations are in detailed balance. The latter situation is quite typical at large optical depths in resonance lines when the radiative transition has become *thermalized* (Hummer and Stewart, 1966). (Note, however, that Eq. 3.5.6 applies only to transitions that are described accurately by the two-level atom approximation. In the more general situation, especially for subordinate lines and photoionization dominated systems, Eq. 3.5.1 must be used.)

Numerical simulation of cooling curves in the quiet sun and plage chromospheres. In order to quantitatively judge how the radiative cooling curve,

$$\phi(T) \equiv \sum_{\substack{\text{all} \\ \text{species}}} \sum_{\substack{\text{all} \\ \text{transitions}}} \frac{\epsilon_{\ell u}^{\text{net}}}{n_e n_H}, \tag{3.5.7}$$

might change in active regions compared with the quiet sun, we have applied numerical simulations of H I, Mg II, and Ca II to two representative atmospheric models: quiet sun and average plage.

Ca II and Mg II radiative losses were estimated from models constructed to crudely fit mean quiet and active sun profiles of the K and k resonance lines compiled from extensive OSO-8 measurements (Lemaire, 1979). The H I simulations were done with atmospheric models constructed to fit the Lyα core and wings in high spatial resolution rocket spectra of quiet regions and plages (Basri *et al.*, 1979).

The results of these simulations can be summarized as follows:

(1) As anticipated from previous work, the dominant coolants at middle chromosphere temperatures are the Mg II and Ca II resonance lines and the Ca II subordinate infrared triplet. At somewhat hotter temperatures ($T \cong 8 \times 10^3$ K), where Ca II and Mg II have begun to ionize to higher stages, H I Balmer transitions, particularly Hα and BaC, control the plasma cooling. Finally, in the upper chromosphere ($T \cong 10^4$ K), H I Lyman transitions, particularly Lyα and LyC, are the principal radiative-loss channels.

(2) We find regions of the middle chromosphere that are *heated* by Lyα and the Paschen continuum. However, the estimated heating rates are smaller than the Ca II and Mg II cooling in those layers.

(3) The mean plage atmosphere neutral hydrogen cooling below 10^4 K is little different compared with that of the quiet sun, probably because the dominant H I coolants at those temperatures, namely Balmer transitions, are more-or-less effectively thin in both the quiet and active models. However, the temperature dependence of the Ca II and Mg II cooling changes significantly in the plage, because the resonance lines become effectively thick higher up in the dense plage chromosphere, and therefore at higher temperatures, compared with the quiet sun (cf. Shine and Linsky, 1974b). Consequently, the region of the total cooling curve $\phi(T)$ that is dominated by Mg II and Ca II shifts to hotter temperatures in the plage. [The shift extends the region of $\phi(T)$ controlled by H^- recombination (see below) from about $T \lesssim 6 \times 10^3$ K in the quiet sun to $T \lesssim 7 \times 10^3$ K in the plage.]

3.5.2.2 H^- and Line Cooling Near T_{min}

The previous discussion has centered on the middle and upper chromosphere, where the dominant sources of radiative cooling are few in number and are relatively straightforward to simulate numerically. The temperature-minimum region and the lower chromosphere ($T \lesssim 6 \times 10^3$ K) present quite a different picture. Here, the radiative cooling is complicated by a significant *heating* term in H^- and by the presence of a large number of effectively thin lines, which individually are insignificant coolants compared with the majority species resonance lines, but which collectively may well dominate the cooling function below 6×10^3 K.

It has almost universally been assumed that H^- recombination is the most important source of radiative cooling in the upper photosphere and at the base of the chromosphere (e.g., Athay, 1976). However, the proper treatment of H^- losses in empirical models has been the subject of considerable debate (Ulmschneider, 1974; Praderie and Thomas, 1976; Ulmschneider and Kalkofen, 1978), particularly with respect to departures from LTE.

The energy balance in a continuum transition is completely analogous to that in a line transition. The net cooling rate for the particular case of chromospheric H^- can be summarized as follows (Gebbie and Thomas, 1970):

$$\epsilon_{H^-}^{net} = n_{H^-}^{*}(\epsilon^{\dagger}(T) - b_{H^-}\epsilon^{(0)}) , \qquad (3.5.8)$$

where $n_{H^-}^{*}$ is the LTE concentration of H^- ions, b_{H^-} is the ratio of the actual H^- population to the LTE value,

$$\epsilon^{(0)} = 4\pi \int_0^{\lambda_0} \alpha_\lambda J_\lambda^{(0)}\, d\lambda \qquad (3.5.9)$$

is the rate of energy absorbed per H^- ion from the background photospheric radiation field (which is characterized by the surface value of the mean intensity $J_\lambda^{(0)}$), and

$$\epsilon^{\dagger}(T) = 4\pi \int_0^{\lambda_0} \alpha_\lambda B_\lambda(T)\, d\lambda \qquad (3.5.10)$$

represents the energy returned to the radiation field by radiative recombinations. (Here, stimulated emissions are treated in LTE for simplicity.) A compact expression for the departure coefficient b_{H^-} is given by Vernazza, Avrett, and Loeser (1973).

We see from Eq. 3.5.8 that, unlike the line case, H^- can be a strong source of heating ($\epsilon < 0$) even when collisions force b_{H^-} close to unity. In particular, H^- will heat the local atmosphere when $b_{H^-} \cong 1$ for temperatures below a critical value,

$$\epsilon^{\dagger}(T_{crit}) = \epsilon^{(0)} , \qquad (3.5.11)$$

where $T_{crit} \cong 4740$ K for the sun (e.g., Ayres, 1979a). (The critical temperature is essentially the LTE radiative equilibrium boundary temperature for an atmosphere having H^- as its only opacity source.) Consequently, H^- will be a net source of heating, not cooling, for $T \lesssim 4740$ K, even in LTE. In fact, when departures from LTE in H^- become significant, as they do in the low chromosphere of the sun, the critical temperature rises to the *color temperature* of the background photospheric radiation field (Gebbie and Thomas, 1970), which is roughly 5450 K for the sun (Ayres, 1979a). In the highly non-LTE situation, H^- is a source of radiative heating over an even wider range of temperatures than the LTE case.

The fact that H^- is a strong source of heating in the upper photosphere and low chromosphere is somewhat embarrassing, because virtually all of the radiative cooling estimates for these important regions of the outer atmosphere have assumed that H^- is the principal coolant of those layers (e.g., Ulmschneider, 1974; Athay, 1976; Giovanelli, 1978). How did these authors magically transform a heating source into a cooling source? The answer is straightforward. If one knows the radiative equilibrium temperature structure T_{RE} in the outer layers of the solar photosphere, one can estimate the effective cooling by any species, for example H^-, by subtracting the net cooling rate evaluated at T_{RE} from that evaluated at chromospheric temperatures in an empirical model. What results is a differential cooling curve; for example,

$$\Delta\phi_{H^-} = \phi_{H^-}(T) - \phi_{H^-}(T_{RE}) \ , \tag{3.5.12}$$

where, in this case, ϕ_{H^-} would represent Eq. 3.5.8 divided by $n_e n_{H^-}$. This procedure tends to cancel the heating terms due to background radiation fields to first order (at least in LTE) and the major residual influence of the intense photospheric $J_\lambda^{(0)}$ is on the stimulated emission terms, which are usually unimportant in the first place (Kalkofen and Ulmschneider, 1979). The procedure works, in principle, because T_{RE} is the temperature at which strict balance between the local radiative heating and cooling rates occurs. Consequently, any species that has an enhanced net cooling rate for $T > T_{RE}$, such as H^-, must be considered a genuine cooling source. Unfortunately, the practical application of the simple differential prescription is complicated by two serious problems.

First, the radiative equilibrium temperature structure for the solar outer atmosphere has by no means been established satisfactorily. Very elaborate, heavily line-blanketed RE simulations have been undertaken by Kurucz (1974) and others. However, these complex numerical models have been computed strictly in LTE. Such LTE RE simulations do not allow for possibly significant heating and cooling terms that can arise from population imbalances that frequently occur in statistical equilibrium (non-LTE; see, e.g., Gebbie and Thomas, 1971; Praderie and Thomas, 1976). For example, H^- heating can produce a pseudo "chromospheric" temperature inversion by Cayrel's (1963) nebular mechanism, without any mechanical heating (Gebbie and Thomas, 1970). Athay (1970) considered non-LTE effects in different classes of Fraunhofer lines schematically, and concluded that surface cooling by strong and medium strength lines of abundant neutrals and ions would tend to resist such a nebular temperature inversion. However, Praderie and Thomas (1976) argue that Athay neglected important heating and cooling processes in the lines, and in fact did not attempt a fully self-consistent RE simulation.

In short, we as yet do not know T_{RE} for the outer layers of the solar photosphere. Some authors, for example Ulmschneider (1974), have adopted the minimum temperature in empirical solar models as an approximation to T_{RE}. The presumption here is that T_{min} must be an upper limit to the true RE boundary temperature, owing to the likely presence of mechanical heating even at T_{min}. In reality, however, T_{min} need not be an upper limit to T_{RE} in the event that nebular heating effects in H^- and other continua drive a Cayrel-type temperature inversion.

The second problem with the differential cooling prescription is that in order for it to work, it must be applied uniformly to *all* of the important heating and cooling species, not just to H^-. In particular, even though H^- exhibits a positive net cooling trend for $T > T_{RE}$, other contributors to the overall energy balance might exhibit *negative* trends. For example, owing to the ionization equilibrium, some neutrals tend to be more effective coolants at low temperatures than at high temperatures. In addition, higher temperatures imply enhanced populations of the excited states of neutral hydrogen, which in turn implies enhanced absorption of photospheric radiation by the Balmer and Paschen continua. In short, the differential cooling by H^- is only part of the overall energy balance. The other sources of heating and cooling must also be included in the differential prescription in order to assess excess radiative losses reliably from empirical models.

We therefore have very little information concerning the detailed shape of the net cooling curve for $T \lesssim 6000$ K. We anticipate that the overall cooling must increase with increasing temperature above about 4000 K, otherwise a stable $T_{min} \cong 4200$ K could not be maintained. However, whether the cooling curve increases by a factor of 50% between 4000 and 6000 K, or by two orders of magnitude, is as yet unclear. What *is* clear is that the detailed shape of the collective cooling curve below 6000 K is of critical importance for determining the net chromospheric cooling rate and thus the mechanical energy input. In particular, the total cooling depends not only on $\phi(T)$, but also on the density product $n_e n_H$. This quantity is a rapidly increasing function of decreasing temperature in the chromosphere. Consequently, most of the chromospheric radiative cooling will occur for $T \lesssim 6000$ K, unless the plasma cooling function $\phi(T)$ decreases very rapidly with decreasing temperature below that point.

3.5.2.3 Summary of Chromospheric Radiative Cooling Functions

Schematic radiative cooling curves $\phi(T) = \epsilon_{net}/n_e n_H$ for the active and quiet solar chromosphere are depicted in Figure 3.9. As described above, the shape of the cooling curve below 6×10^3 K is very uncertain, owing to the unknown magnitude of the effectively thin line contribution. What is plotted in Figure 3.9 for $T < 6 \times 10^3$ K is the gross H^- recombination cooling rate, which may, or may not, be a reasonable representation of the total cooling at those temperatures.

Two conclusions can be drawn from the schematic cooling curves of Figure 3.9.

First, the temperature dependences of the well-established portions of the active and quiet sun cooling curves are relatively mild compared with the density dependence, $n_e n_H$. In particular, the hydrogen density $n_H = n_{H1} + n_p$ decreases several orders of magnitude from the base of the chromosphere up to the 10^4 K level. In fact, the temperature dependence of the electron fraction n_e/n_H may well play a more important role in controlling the location of the chromospheric temperature inversion than the temperature dependence of the cooling curve itself (Ayres, 1979b). The importance of the density is also illustrated in a comparison of active and quiet models (see Figure 3.10). The most striking difference between the two models is the systematically and significantly larger $n_e n_H$ values in the plage at all temperatures.

Second, the net radiative cooling in the middle and upper chromosphere can be estimated simply by considering the emission strengths of the Mg II,

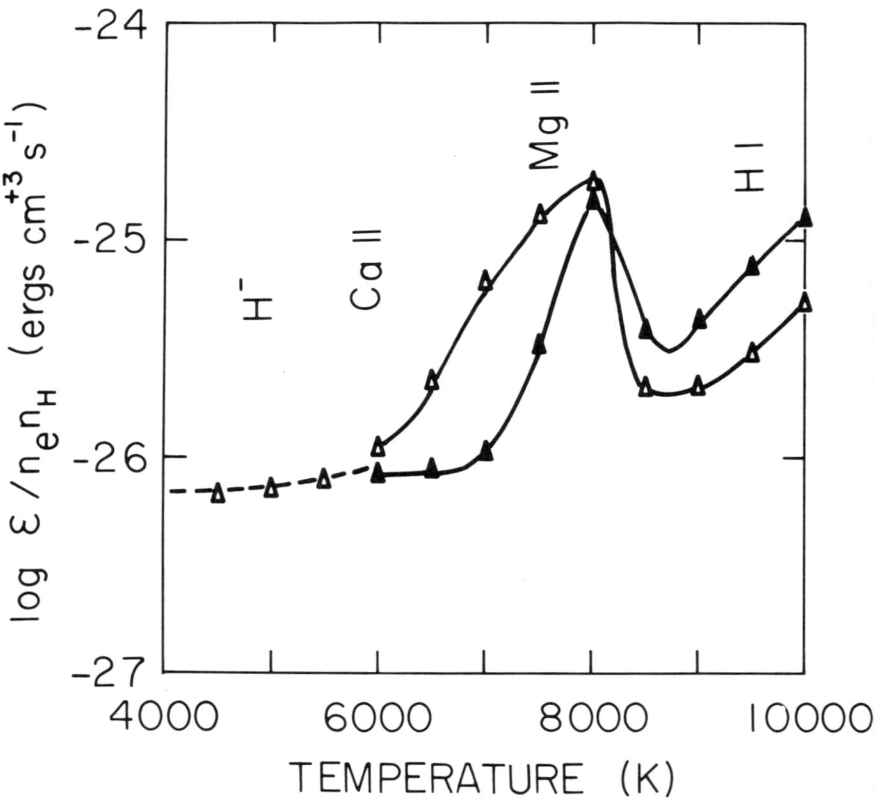

Figure 3.9. Schematic radiative cooling curves for quiet (open symbols) *and active* (solid symbols) *sun. The* dashed curve *for* T $< 6 \times 10^3$ *K is the gross H^- recombination cooling rate without correction for radiative heating. The detailed shape of the total cooling curve for* T $< 6 \times 10^3$ *K is very uncertain (see text).*

Ca II and H I resonance lines (see also Linsky and Ayres, 1978). The sum is at worst a lower limit to the total cooling above 6×10^3 K, but is likely to be reasonably close to the actual cooling owing to the importance of the few majority species emitters in that temperature range.

3.5.2.4 Empirical Radiative Losses from the Quiet Sun and an Average Plage Region

Line losses ($T > 6 \times 10^3$ K). The second consideration, above, provides a simple way to estimate the line radiative losses from the quiet sun and an average plage region. Table 3.3 summarizes core emission fluxes of Ca II H and K, Mg II h and k, and H I Lyα for both the quiet and active sun. These fluxes are based on 1 Å integrated disk center intensities from composite quiet sun and mean-plage profiles obtained with OSO-8 (Lemaire, 1979). The

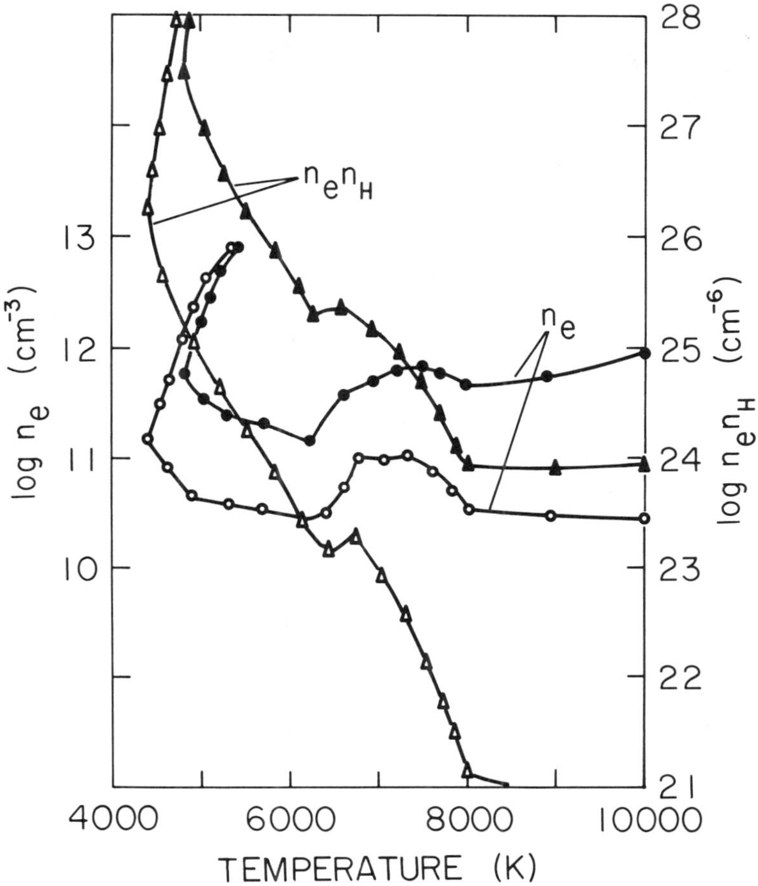

Figure 3.10. Homogeneous temperature-density models of quiet (open symbols) *and active* (solid symbols) *sun. Note the large enhancement of the* $n_e n_H$ *in the mean plage, compared with the quiet sun, at all temperatures.*

Table 3.3
Observed Chromospheric Line Radiative Losses[a]

		$F_\ell^{\pm 0.5\ \text{Å}}$ (10^5 erg cm^{-2} s^{-1}) Quiet Sun	Mean Plage
Ca II	H[b]	2.9	14
	K[b]	3.0	15
Mg II	h	4.7	23
	k	6.5	33
H I	Lyα	1.8	9
Total		19	94

[a]Based on Lemaire (1979) specific intensities.
[b]Corrected for "photospheric" contamination (see text).

integrated specific intensities I_ℓ were converted to fluxes F_ℓ using scale factors determined from the model simulations ($F/I \cong 3.0$ for all of the lines). [The 1 Å core intensities of Ca II H and K are contaminated by a large *photospheric* radiation component (see, e.g., Linsky and Ayres, 1978). The extraneous contribution was estimated by assuming that the purely chromospheric component is enhanced by a factor of five in the mean plage compared with the quiet sun. That scale factor is suggested by the behavior of the Mg II lines and Lyα, none of which are significantly affected by photospheric contamination. The Ca II H and K fluxes given in Table 3.3 have been corrected for the estimated photospheric emission contribution.] We conclude from this comparison that the quiet sun line losses are of order 2×10^6 erg cm^{-2} s^{-1}, while the mean-plage cooling is about a factor of five larger.

Low-temperature losses ($T < 6 \times 10^3$ K). The line cooling in the quiet chromosphere obtained above is comparable to the H^- radiative loss of 3×10^6 erg cm^{-2} s^{-1} estimated by Ulmschneider and Kalkofen (1978). However, as argued above, we have no way of verifying at present whether cooling estimates based solely on H^- differential cooling curves are reliable for $T < 6 \times 10^3$ K. Nevertheless, for the sake of argument (and because we have nothing better), we will assume that the approximation is close to the actual loss curve for $T < 6 \times 10^3$ K. We crudely estimate the low-temperature cooling near the base of the plage chromosphere, as follows:

$$F_{H^-} \cong \int_{Z_{T_{min}}}^{Z_{6\times10^3\,K}} \Delta\phi_{H^-}(T) n_e n_H \, dZ$$

$$\cong (1.4\, m_H)^{-1} \int_0^{m_{T_{min}}} n_e \Delta\phi(T)\, dm$$

$$\cong (1.4\, m_H)^{-1} \langle n_e \rangle \Delta\phi_0 m_{T_{min}} \,, \tag{3.5.13}$$

where $\Delta\phi_0 \cong 4 \times 10^{-27}$ erg cm^3 s^{-1} is the mean differential H^- cooling for $T_{RE} = T_{min}$ (Ayres, 1979a) and $m_{T_{min}}$ is the mass column density (g cm^{-2}) at T_{min}. In obtaining the final line of Eq. 3.5.13, we have exploited the facts that the H^- differential cooling is relatively independent of temperature, the chromospheric electron density is relatively constant with height in both active and quiet models (Shine and Linsky, 1974b; see also Figure 3.10), and most of the chromospheric mass column density is between T_{min} and 6×10^3 K. The plage model of Figure 3.10 has $\langle n_e \rangle \cong 3 \times 10^{11}$ cm^{-3} and $m_{T_{min}} \cong 0.16$ g cm^{-2}. We therefore estimate an "H^-" loss of roughly 8×10^7 erg cm^{-2} s^{-1} for the plage. The corresponding model parameters for the quiet sun are $\langle n_e \rangle \cong 5 \times 10^{10}$ cm^{-3} and $m_{T_{min}} \cong 0.032$ g cm^{-2}, which imply an H^- loss $\cong 3 \times 10^6$ erg cm^{-2} s^{-1}. This latter estimate is comparable to that obtained by Ulmschneider and Kalkofen (1978).

We again caution the reader that the "H^-" losses derived above are intended to be purely *illustrative*, and should not be construed as realistic radiative-cooling estimates. The important point is that homogeneous plage models will produce much larger low-temperature cooling rates than homogeneous quiet models, owing to the substantially increased $m_{T_{min}}$ and $<n_e>$ of the former, more-or-less independently of the detailed shape of the low-temperature cooling curve.

3.5.2.5 Geometrical Considerations

The discussion above has implicitly assumed the so-called homogeneous representation of the plage chromosphere. In particular, the active region is considered to be a spatially uniform entity characterized by a single temperature-pressure relation. However, a large body of evidence suggests that both the quiet sun and plages are not homogeneous, but instead consist of discrete, "hot" magnetic elements—the flux tubes—embedded in an otherwise uniform, "cool" photosphere. The difference between quiet and active sun would then be attributed to differences in the spatial number density of the flux tubes, rather than, say, to changes in the gross physical properties of the plage and quiet sun magnetic elements themselves.

If the flux tube picture is correct, what modifications of the radiative cooling estimates based on the homogeneous modeling of the previous section might we expect? First, we would anticipate little change in the resonance line losses, since we can identify unambiguously the emission core strength changes between the quiet sun and the active regions. Presumably, no matter what kind of flux tube model we might advocate, the fundamental confrontation between observation and theory must be the observed line shapes and integrated emission fluxes. In particular, a "realistic" model must, at the very least, reproduce the observed spatially averaged emission core intensities. Therefore, the spatially averaged excess line cooling from the plage flux tubes relative to the quiet sun is simply the difference between the line emission strengths in the active and quiet sun. (Note, however, that the actual emission strengths from the flux tubes themselves depend on the filling factor f that describes the fractional area of the plage covered by magnetic elements. For example, if the spatially averaged plage line loss is 10^7 erg cm^{-2} s^{-1} and f is only 10%, then the flux tube net surface emission is $\cong 10^8$ erg cm^{-2} s^{-1}.)

Second, we anticipate that the situation for H^- and the other important coolants below 6×10^3 K will be *very different* in the flux tube picture compared with the homogeneous models. This difference arises because we cannot directly measure lower chromosphere losses in H^- and effectively thin lines, but instead must calculate the losses from a model constructed on the basis of other diagnostics, for example, Ca II and Mg II. Since diagnostics in the ultraviolet, such as the K and k cores and wings, tend to weight exponentially towards hotter temperatures in averaging over thermal inhomogeneities, a "mean" model constructed from Ca II or Mg II observations will tend to be much closer to the thermal-pressure structure of the hot flux tubes, compared with a hypothetical mean model constructed on the basis of H^- emission. This is particularly true for inferring $m_{T_{min}}$ from plage line shapes of Ca II and Mg II. The spatially averaged intensity profile of, say, Ca II K would be much fainter than the intrinsic profile of K in a typical flux tube, since the intense emission from the fraction f of the plage area covered by magnetic structures is diluted by the perhaps pure absorption profile of K from the remaining $1 - f$ quiet component. However, the dilution should not greatly affect the important line *shape* parameters, in particular the K_1 minimum feature separation. Since the K_1 separation is used to determine $m_{T_{min}}$, and since the cooling for $T < 6 \times 10^3$ K is proportional to $m_{T_{min}}$, then the "H^-" losses estimated from the "homogeneous" plage model should more closely represent the low-temperature surface cooling *within* a typical flux tube, rather than a proper spatial average of the cooling rate. One would therefore anticipate, in the "worst" case, that the H^- losses estimated from a homogeneous plage model might overestimate the true spatially averaged cooling by a factor of $1/f$. Since f itself might only be 10–20% near T_{min} for the mean plage represented by the Lemaire (1979) profiles, the illustrative homogeneous "H^-" cooling estimate of the previous section might be a factor of 5–10 too large.

In short, the geometrical difference between homogeneous plages and flux-tube plages has a considerable effect on inferring spatially averaged cooling rates for the "unobservable" coolants that are important below 6×10^3 K. The homogeneous plage models imply a low chromosphere ($T < 6 \times 10^3$ K) energy budget up to an order-of-magnitude larger than that suggested by the flux tube picture. This difference is clearly of critical importance for understanding the heating and energy balance of active regions. The answer to the problem requires that the flux tubes, if they in fact exist, be spatially and spectrally resolved so that realistic temperature-pressure stratifications can be determined unambiguously both for the flux tubes themselves as well as for the surrounding "quiet" component. Only then can reliable cooling rates in the individual structures be estimated.

3.5.3 Empirical Measurements of Radiative Losses from the Deep Photosphere

The radiative power loss from the deep photosphere ($\tau_{5000} \gtrsim 10^{-2}$) of active regions has been studied using data from an extreme limb photometer (ELP, cf. Chapman and McGuire, 1977, and references therein). The brightness deficit of sunspots is relatively easy to measure at various positions on the solar disk, but the brightness excess of faculae is more difficult to determine. In particular, faculae are spread over a greater area of the solar surface, and their contrast varies strongly across the disk from near invisibility at sun center to maximum contrast near the limb. During 1975, four active regions were observed with the ELP, two of them on two successive days. The ELP has two apertures that were adjusted to cover an annulus of approximately 53″ across the solar disk. The area of the active regions was estimated from the *Solar Geophysical Data*.

The main uncertainty in determining the photospheric flux excess is the poorly known center-to-limb variation of the facular contrast. Several functional forms were tried that are more-or-less compatible with reported measurements of facular limb darkening. In the simplest form, we assumed that the intensity difference ΔI was independent of heliocentric angle, while the spectral variation was the same as that of the quiet sun. For that limb-darkening law, an excess flux of 2.6×10^8 erg s^{-1} cm^{-2} was determined for data from four active regions. The monochromatic flux excess was $(5.1 \pm 1.6) \times 10^8$ erg s^{-1} cm^{-2} μm^{-1}. With other limb-darkening laws, the excess flux ranged from 4×10^7 to 2×10^8 erg s^{-1} cm^{-2}. These values are not too different from that found by Skumanich *et al.* (1975), 3.8×10^8 erg s^{-1} cm^{-2}. However, their value referred only to network elements, whereas the ELP excess flux refers to an average over the entire active region. To compare the two results, one needs to know the filling factors for the individual observations. Skumanich *et al.* give a filling factor based on Ca II measurements of 0.40. The corresponding spatially averaged flux excess is 1.6×10^8 erg s^{-1} cm^{-2}, which is comparable to the ELP results. Power losses of this order are significant compared with chromospheric losses (see Section 3.5), but are only a few tenths of a percent of the photospheric bolometric radiative output.

In fact, we cannot be certain how much of the photospheric power loss represents genuine excess cooling of the plage photosphere. In particular, we don't know what fraction of the excess radiation is from heat diffusion through the thin flux tubes (Spruit, 1976), what fraction is the result of local mechanical heating of the facular atmosphere, or what fraction might simply arise from geometrical considerations, for example photon channeling or "hot wall" emission.

3.6 UPPER CHROMOSPHERE AND TRANSITION ZONE

3.6.1 Introduction

The structure and physical conditions in an active region's transition zone are of extreme importance for understanding the overall balance in solar features. In particular, the transition zone is the interface through which mass and energy flow between the chromosphere and corona. However, probing the TZ layers is hindered by a complex topography and the presence of numerous physical processes at work simultaneously. For example, we can summarize the TZ energy balance as follows (e.g., Withbroe and Noyes, 1977):

$$\nabla \cdot (\mathbf{F}_R + \mathbf{F}_c + \mathbf{F}_M + \mathbf{F}_G + \mathbf{F}_K + \mathbf{F}_E) = 0 \ , \qquad (3.6.1)$$

where the subscripts on the fluxes refer to, respectively, radiation, conduction, mechanical energy, gravitational potential energy, kinetic energy, and enthalpy. It is interesting to note that some of these terms represent net cooling at one level of the atmosphere, but *heating* at others. For example, the conductive flux $\mathbf{F}_c$ represents an important cooling channel for the 10^6 K corona, whereas the dissipation of the conductive heat flux is an important energy input to the underlying transition zone. The detailed structure of the upper chromosphere and TZ are the subject of Chapter 2 and will not be considered in great detail here. However, we will describe some results that pertain to these levels, either as horizontally averaged entities or as loop structures.

3.6.2 Observational Material

Moe has analyzed NRL HRTS data having one arcsec spatial resolution. These spectra cover the range from 1175 to 1670 Å (Brueckner *et al.*, 1978). The data were calibrated by comparing HRTS observations with Skylab CALROC measurements of quiet sun regions. Density-sensitive intensity ratios of Si II, $I(1301)/I(1296)$, and of O IV, $I(1407)/I(1401)$, were among the several considered. These transitions are formed at 3.5×10^4 K and 1.2×10^5 K, respectively. In many cases the line intensity ratios were close to the "saturation" values, consequently the derived densities are somewhat uncertain. Several lines showed contrasts of up to an order of magnitude between quiet and active areas.

The HRTS data suggest that the plage temperature gradient is steeper by a factor of two than that of quiet areas. Withbroe (1978) has described evidence for active region temperature gradients from two or three times smaller than those of the quiet sun and up to ten times larger. If the filling factor of plage material is between 0.5 and 1.0, the apparent radiative losses of the TZ can be supplied by conduction. However, if the plage filling factor is much smaller than 0.5, the TZ layers must become substantially thicker to provide the observed emission measures. Consequently, the temperature gradient is reduced at the same time that the *surface* intensity from the emitting structures is increased. In this situation, the conductive flux is no longer able to balance the large apparent radiative losses within the individual structures, and alternative heating mechanisms must be invoked. For example, Nicolas *et al.* (1979) found, from an analysis of HRTS Si II data, that electron densities in active regions range from 2×10^{10} cm^{-3} to 10^{12} cm^{-3} at 2000 km above the limb. If the filling factor for the 3.5×10^4 K material is one, then the transition zone is very thin and the divergence of the conductive flux and turbulent energy dissipation can balance the radiative losses. If the filling factor is between 0.035 and 0.001, corresponding to the base of a loop, then an energy source of from 1.5 to 390 erg s^{-1} cm^{-3} is needed to balance the radiative losses at pressures of 0.4 to 6.9 dyn cm^{-2}, respectively. The higher value is only a few times larger than the 2 dyn cm^{-2} cited by Withbroe and Noyes (1977) for a typical active region.

Lites has analyzed lines of C I–IV in OSO-8 observations of an active region within one day's rotation of the limb (that is, $\mu \approx 0.26$). He found that the resonance doublet of C IV (1548,51 Å) in the active region has self-reversed cores, whereas near disk center these lines always exhibited simple emission profiles. The active region C IV intensity was about 70 times the typical quiet sun emission. In addition, there were spatial intensity variations of up to a factor of 20 within the plage region itself. Using the absolute intensities, and the assumption of a plane parallel TZ in hydrostatic equilibrium, Lites estimated a vertical optical depth of about five in the stronger member of the C IV doublet (1548 Å), a microturbulent velocity of 13 km s^{-1}, $N_e \sim 10^{10}$ cm^{-3} at the $T \sim 8 \times 10^4$ K level, and a geometrical thickness of 400 km for the emitting region ($5 \times 10^4 \lesssim T \lesssim 10^5$ K).

3.6.3 Loop Models for the Transition Zone

Here, we summarize the analysis of EUV spectral data in terms of steady-state loop models by Emslie and Machado (1978; see also Machado and Emslie, 1979). They considered one-dimensional isobaric active region loops governed by the following simplified steady-state energy balance (cf. Eq. 3.6.1, above):

$$\nabla \cdot (\mathbf{F}_c + \mathbf{F}_M + \mathbf{F}_R) = 0 \ . \qquad (3.6.2)$$

Here, the divergences are evaluated along the loop axis. Further details are given by Emslie and Machado.

3.6.3.1 The Emission Measure Distribution

The fundamental "empirical" quantity that active region TZ models must reproduce, whether inhomogeneous or not, is the differential emission measure distribution $Q(T)$. This quantity is determined from observations of optically thin emission lines formed at a variety of temperatures. The origin of the differential emission measure is as follows:

The intensity of an EUV spectral line, per square centimeter of solar surface, is given by (Pottasch, 1967; Withbroe, 1975):

$$I(\text{erg cm}^{-2}\,\text{s}^{-1}\,\text{st}^{-1}) = 1.7 \times 10^{-16} A f \int Ne^2 g\, G(T)\, dh \ . \qquad (3.6.3)$$

The integral over height in this relation is converted to an integral over temperature by the replacement, $Q(T)\,dT = n_e^2\,dh$. Because the resonance lines of TZ ions are typically formed over only a narrow range of temperatures, the intensity of a particular emission feature is closely related to the local value of $Q(T)$. Consequently, measurements of many emission lines, spanning a large range of formation temperatures, can be used to map the detailed temperature dependence of $Q(T)$ (Withbroe, 1978).

Machado and Emslie estimated $Q(T)$ based on the composite active region spectrum given by Vernazza and Reeves (1978). Lines from the following ions were used: N VI, O VI, Ne VII, Na IX, Mg X, Si XII, Fe XVI, and one line of N I to establish the trend of $Q(T)$ at "low" temperatures. In addition, they included two observations of C IV 1548 Å from HRTS data, provided by O. K. Moe, in order to establish reliably the $Q(T)$ distribution near 10^4 K.

The resulting $Q(T)$ distribution [in its dimensionless form $q(T) = Q(T)/N_0$, where N_0 is the density at the base of the loop] is plotted in Figure 3.11, together with three of the Machado and Emslie models and a model that neglects the radiative term in the energy balance equation. The parameters α and f describe the energy dissipation and conductive heating, respectively. [Two $Q(T)$ distributions were constructed, corresponding to the different values of I(C IV). The results obtained for $I = 2500$ has a dip at 10^5 K caused by the relatively weak C IV intensity, and is probably unrealistic. The deficiency in the empirical emission measure at 10^5 causes an increment in $Q(T)$ at slightly larger temperatures to account for the N V intensities. On the other hand, the $Q(T)$ distribution obtained for I(C IV) $= 5000$ appears to be somewhat high, and we suspect that intermediate C IV intensities would probably be more representative of the composite active region spectrum given by Vernazza and Reeves.]

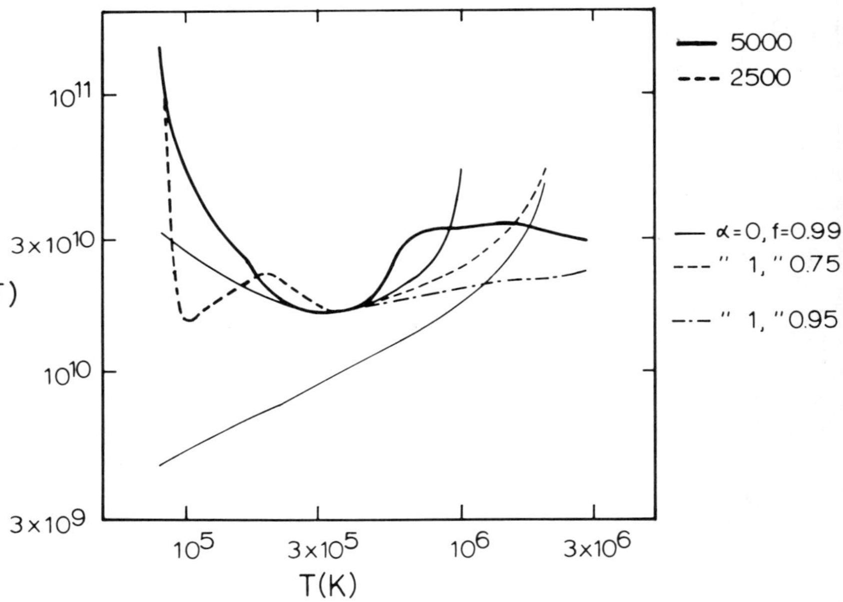

Figure 3.11. Differential emission measures (normalized to the loop base density) versus temperature for several models and data sets. The heavy curves refer to two different values of the C IV intensity (erg cm^{-2} s^{-1} Sr^{-1}) obtained with HRTS. The three medium curves are from a theoretical model, and the thin lower curve is from a model that neglects radiative losses (courtesy of Machado).

Machado and Emslie note that their results for $Q(T)$ are somewhat lower than those obtained by Withbroe (1978) for a different active region. This is not surprising, since we expect to see considerable variations not only between different active regions, but also *within* a given active region.

The approximately flat maximum in $Q(T)$ for $T > 8 \times 10^5$ K, which is not reproduced by any of the theoretical models, has several important implications:

1. The loop models are at best first order approximations to reality.
2. The instrument probably did not resolve structures in the active region with different emission measure distributions. What was observed is very likely an intensity distribution corresponding to the superposition of a wide range of physical conditions.
3. Geometrical effects, for example the expansion of magnetic flux tubes with height, may be important (see Machado and Emslie, p. 17).

Finally, discrepancies between the measured and theoretical $Q(T)$s, at low temperatures, are not surprising because at those heights the differential emission measure increases rapidly with decreasing temperature. Consequently, contributions to the emission of a given line are likely from a wide range of temperatures (see Withbroe, 1978, for further details).

3.6.3.2 A Model of the Mean Active Region Transition Zone

Given the Emslie and Machado results for $Q(T)$, one can determine the gas pressure at a given temperature. To minimize errors from the use of Li- and Na-like ions and to reduce geometrical effects, Machado and Emslie consider $3 \times 10^5 \leq T \leq 4 \times 10^5$ K. They obtain a reduced pressure, $n_e T = 2.2 \times 10^{15}$ cm^{-3} K. Their value is consistent with other independent estimates for active regions (see, e.g., Withbroe, 1978; Vernazza and Mason, 1978; Vernazza, 1977), but may be slightly on the low side, as mentioned above.

If the gas pressure is constant within the TZ, as is likely because the geometrical thickness of the TZ is much smaller than the local scale height, then we can estimate the mass column density at the top of the chromosphere, as follows:

$$m_0 = \frac{P_{\text{gas}}}{g_0} \cong 2.2 \times 10^{-5} \text{ g cm}^{-2} . \tag{3.6.4}$$

This "top mass" is somewhat lower than that given in the Basri *et al.* (1979) mean-plage model, but is within the range of the Shrine and Linsky (1974b) plage models.

A specific plage TZ model, obtained through the comparisons of empirical and theoretical emission measure distributions, is given in Table 3.4. Radiative losses calculated for the model are also included.

According to Machado and Emslie, the conductive flux at the 10^5 K level is a good measure of the conductive energy that must be radiated away by Lyα. The model in Table 3.4 implies $F_c(10^5) \sim 2.6 \times 10^5$ erg cm^{-2} s^{-1}, which is significantly smaller than the spatially averaged Lyα flux from a typical active region (Table 3.3). Additional discussion concerning the importance of conductive heating at the base of the magnetic loops is given by Emslie and Machado, and in their study of Lyα and Lyman continuum rasters.

Table 3.4
Active Region Transition Zone Model
($n_e T = 2.2 \times 10^{15}$ cm^{-3} K)

h^a(cm)	T(K)	ϕ_{RAD}(erg cm^{-3} s^{-1})
1.22 E + 05	5.00 E + 04	4.84 E - 01
3.62 E + 05	8.00 E + 04	4.83 E - 01
5.77 E + 05	1.00 E + 05	3.05 E - 01
1.14 E + 06	1.50 E + 05	1.36 E - 01
2.81 E + 06	2.00 E + 05	7.63 E - 02
4.93 E + 06	2.50 E + 05	4.89 E - 02
7.91 E + 06	3.00 E + 05	2.38 E - 02
1.20 E + 07	3.50 E + 05	1.28 E - 02
1.74 E + 07	4.00 E + 05	7.53 E - 03
2.46 E + 07	4.50 E + 05	4.70 E - 03
3.36 E + 07	5.00 E + 05	3.08 E - 03
4.49 E + 07	5.50 E + 05	2.11 E - 03
5.86 E + 07	6.00 E + 05	1.54 E - 03
7.53 E + 07	6.50 E + 05	1.32 E - 03
9.53 E + 07	7.00 E + 05	1.13 E - 03
1.18 E + 08	7.50 E + 05	9.88 E - 04
1.45 E + 08	8.00 E + 05	8.68 E - 04
1.75 E + 08	8.50 E + 05	7.69 E - 04
2.10 E + 08	9.00 E + 05	6.86 E - 04
2.49 E + 08	9.50 E + 05	6.16 E - 04
2.94 E + 08	1.00 E + 06	5.56 E - 04

[a] Above the 2×10^4 K.

3.6.3.3 Additional Remarks

Machado and Emslie have studied the characteristics of Lyman continuum emission in different solar features. Their results are summarized in Table 3.5. The tabulated color temperature was determined at the head of the continuum. Measurement errors are of order ±500 K. The apparent differences between quiet and active sun are probably genuine. If one assumes that the network is composed of the same structures as active regions, the results in Table 3.4 imply that the network filling factor is quite small. These data also suggest that upper chromospheric conditions in coronal holes are very different from those of the quiet sun and active regions.

Additional differences between networks, elements, and active regions are illustrated by the Lyα/Lyβ flux ratio. Representative values are given in Table 3.6. Unlike the LyC results described above, we do not find dramatic differences in the Lyα/Lyβ ratios between the quiet sun network and coronal holes.

Table 3.5
Lyman Continuum Observations

Feature	Color Temperature (K)	Ground State Departure Coefficient
Quiet Sun (avg)	8490[a]	200
Cell	8335	197
Network	8406	126
AR	7470[b]	4
Flare	≥8500	~1
Coronal Hole (avg)	(9607)[c]	2830
CH cell	9393	2294
CH network	9164	1102

[a]Agrees with mean quiet sun model of Vernazza *et al.* (1976).
[b]Disagrees with predictions of the Basri *et al.* (1979) model.
[c]Value uncertain.

Table 3.6
Lyman Line Ratios

Feature	$F(\mathrm{Ly}\alpha)/F(\mathrm{Ly}\beta)$
QS—Network	82.1
QS—Cell	88.5
AR	35.2
CH—Network	88.9
CH—Cell	92.6
Flare	~12

3.7 SUMMARY

We have developed a homogeneous model of a plage that is in "reasonable" agreement with several kinds of observations. In doing so, we examined the disturbingly wide range of quiet sun models. We note that at present there exist significant discrepancies between quiet sun models and certain kinds of observations, notably UV and IR spectra. Such discrepancies cast serious doubt on the reliability of homogeneous plage models for understanding the physical properties of active regions.

Observational evidence strongly suggests that active regions are composed of small-scale structures governed by concentrations of magnetic flux. We have devoted considerable discussion to the nature of these "flux tubes" and the implications of their existence for understanding observations that do not spatially resolve the fine structure. We described the concept of "filling factors" and advocate a simple, first-order flux tube theory to explain low-resolution observations. In particular, we assume that "all flux tubes are created equal"; consequently, variations in the fractional area of the active region covered by emitting structures should suffice to explain apparent changes in spectral properties. Although such a picture is likely to break down for certain parts of the chromosphere and TZ, the simplicity of the approach is certainly appealing. In effect, we retain the conventional mean-modeling approach, but restrict it to the portion of the surface that is "contaminated" by magnetic structures. Unfortunately, the practical application of our simple prescription is complicated by the uncertain height dependence of the filling factor, as well as by the possibly quite different filling factors appropriate to the magnetic field and emitting material at any given height. In addition, our simple picture ignores spicules and other dynamical phenomena. Finally, the tilting of some features away from the vertical will introduce additional complexity in interpreting high spatial resolution observations.

The effects of two-dimensional radiative transfer were explored for flux tubes of various cross-sectional sizes. Calculations were performed with and without a Wilson depression. The results indicated that neglect of explicit horizontal radiative interactions (1.5 D) is an extremely good approximation to the self-consistent two-dimensional treatment, at least for the particular problem—the Ca II K line wings—studied here. In fact, the K wings should be relatively simple, but powerful diagnostic tools for the study of facular flux tubes.

The radiative power loss from active regions was discussed in Section 3.5 for homogeneous and flux tube models. Numerical simulations suggest that the dominant coolants in the middle chromosphere are the resonance lines of Mg II and Ca II, and the Ca II IR triplet. In the upper chromosphere, the important emitters are H I Balmer transitions, especially Hα and Ba C and, at higher temperatures, Lyman transitions control the plasma cooling. In homogeneous plage models, that portion of the total cooling curve that is dominated by Mg II and Ca II shifts to higher temperatures owing to thermalization effects, thereby enhancing the importance of the low-temperature coolants, for example, H^-. In addition, a significant difference in the total heating requirements for $T < 6 \times 10^3$ K was found between homogeneous models of plages and the probably more realistic flux tube picture. The former treatment can overestimate the radiative cooling by as much as an order of magnitude.

For the case of flux tubes, no reliable estimate of the low-temperature cooling can be made without accurate knowledge of the brightness filling factors appropriate for the low chromosphere and T_{min} regions.

Observations of active regions with an ELP suggest an excess power loss at middle photospheric levels of about 2×10^8 erg cm^{-2} s^{-1}. This loss is considerably larger than the likely chromospheric power loss. Part of the excess photospheric radiation is leakage from the "hot walls" of the facular flux tubes. Perhaps half of the loss represents genuine local mechanical heating.

We confirm earlier estimates for transition zone pressures that are about an order of magnitude larger in active regions than in the quiet sun. Evidence from Lyman continuum color temperatures and departure coefficients suggests that the network filling factor is small at transition zone levels. An active region transition zone model is presented in tabular form.

The influence of magnetic loops on the heating of an active region needs further study. According to Group 1 of this workshop, coronal loops in active regions could deposit up to about 10^6 erg s^{-1} cm^{-2} through their footpoints into the transition zone and upper chromosphere.

Acknowledgments

I wish to acknowledge the efforts of various members of Group II of this Active Region Workshop both during the workshop and subsequently. I especially want to thank Dr. Marcos Machado for most of Section 3.6, Dr. Henk Spruit for his contributions to Section 3.3 and Dr. Tom Ayres for most of Section 3.5. The National Center for Atmospheric Research has provided computing time. Support for Ayres was provided by the National Aeronautics and Space Administration through NGL-06-003-057 to the University of Colorado. Finally, I wish to thank Ayres for many improvements in the manuscript.

REFERENCES

Allen, C. W.: 1973, *Astrophysical Quantities*.

Athay, R. G.: 1970, *Astrophys. J.* **161,** 713.

Athay, R. G.: 1976, *The Solar Chromosphere and Corona: Quiet Sun*. Reidel, Boston.

Athay, R. G. and White, O. R.: 1978, *Astrophys. J.* **226,** 1135.

Ayres, T. R.: 1977, *Astrophys. J.* **213,** 296.

Ayres, T. R.: 1978, *Astrophys. J.* **225,** 665.

Ayres, T. R.: 1979a, *Solar Phys.* (submitted).

Ayres, T. R.: 1979b, *Astrophys. J.* **228,** 509.

Ayres, T. R. and Linsky, J. L.: 1976, *Astrophys. J.* **205,** 874.

Ayres, T. R. and Testermann, L.: 1980 (in preparation).

Basri, G. S., Linsky, J. L., Bartoe, J. D. F., Brueckner, G., and Van Hoosier, M. G.: 1979, *Astrophys. J.* **230,** 924.

Brueckner, G. E., Bartoe, J. D. F., and Van Hoosier, M. E.: 1978, *Proceedings of the OSO-8 Workshop*, University of Colorado, Boulder, p. 380.

Bruner, E. C., Jr.: 1978, *Astrophys. J.* **226,** 1140.

Cayrel, R.: 1963, *Compt. Rend. Astron. Soc. Paris* **257,** 3309.

Chapman, G. A.: 1970, *Solar Phys.* **14,** 315.

Chapman, G. A.: 1974, *Astrophys. J.* **191,** 255.

Chapman, G. A.: 1977, *Astrophys. J. Suppl.* **33,** 35.

Chapman, G. A.: 1979, *Astrophys. J.* **232** (in press).

Chapman, G. A. and McGuire, T. E.: 1977, *Astrophys. J.* **217,** 657.

Dicke, R. H.: 1970, *Astrophys. J.* **159,** 25.

Dunn, R. B. and Zirker, J. B.: 1973, *Solar Phys.* **33,** 281.

Emslie, A. G. and Machado, M. E.: 1978, Center for Astrophysics preprint 1034.

Frazier, E. N.: 1978, *Astron. Astrophys.* **64,** 351.

Gabriel, A. H.: 1976, *Phil. Trans. Roy. Soc. London* **A 281,** 339.

Gebbie, K. B. and Thomas, R. N.: 1970, *Astrophys. J.* **161,** 229.

Gebbie, K. B. and Thomas, R. N.: 1971, *Astrophys. J.* **168,** 461.

Giovanelli, R. G.: 1978, *Solar Phys.* **59,** 293.
Gouttebroze, P., Lemaire, P., Vial, J. C., and Artzner, G.: 1978, *Astrophys. J.* **225,** 655.
Harvey, J. W.: 1977, in E. A. Müller (ed.), *Highlights of Astronomy* **4,** Reidel, Dordrecht, Holland.
Harvey, J. W. and Hall, D.: 1975, *Bull. Am. Astron. Soc.* **7,** 459.
Heasley, J. N. and Allen, M. S.: 1979, *Astrophys. J.* (submitted).
Hirayama, T.: 1978, *Pub. Astron. Soc. Japan* **30,** 337.
Hummer, D. G. and Stewart, J. C.: 1966, *Astrophys. J.* **146,** 290.
Kalkofen, W. and Ulmschneider, P.: 1979, *Astrophys. J.* **227,** 655.
Koutchmy, S. and Stellmacher, G.: 1978, *Astron. Astrophys.* **67,** 93.
Kurucz, R. L.: 1974, *Solar Phys* **34,** 17.
Lemaire, P.: 1979, private communication.
Lindsey, C. and Heasley, J. N.: 1979, *Astrophys. J.* (submitted).
Linsky, J. L. and Ayres, T. R.: 1978, *Astrophys. J.* **220,** 619.
Machado, M. E. and Emslie, A. G.: 1979, *Astrophys. J.* (in press).
Mihalas, D.: 1978, *Stellar Atmospheres,* Freeman, San Francisco, 2nd ed.
Mihalas, D., Auer, L. H., and Mihalas, B. R.: 1978, *Astrophys. J.* **220,** 1001.
Morrison, N. D. and Linsky, J. L.: 1978, *Astrophys. J.* **222,** 723.
Muller, R.: 1975, *Solar Phys.* **45,** 105.
Nicolas, K. R., Bartoe, J. D. F., Brueckner, G. E., and Van Hoosier, M. E.: 1979, *Astrophys. J.* **232** (in press).
Orrall, F. Q. and Kopp, R. A.: 1976, in Bonnet, R. M. and Delache, P. (eds.), *IAU Colloq.* **36,** 457, G. de Bussac, Clermont-Ferrand.
Owocki, S. and Auer, L. H.: 1979 (in preparation).
Pottasch, S.: 1967, *Astrophys. J.* **150,** 361.
Praderie, R. and Thomas, R. N.: 1976, *Solar Phys.* **50,** 333.
Ramsey, H. E., Schoolman, S. A., and Title, A. M.: 1977, *Astrophys. J. Lett.* **215,** L41.
Rogerson, J. B.: 1961, *Astrophys. J.* **134,** 331.
Shine, R. A. and Linsky, J. L.: 1974a, *Solar Phys.* **37,** 145.
Shine, R. A. and Linsky, J. L.: 1974b, *Solar Phys.* **39,** 49.
Skumanich, A., Smythe, C., and Frazier, E. N.: 1975, *Astrophys. J.* **200,** 747.
Spruit, H. C.: 1976, *Solar Phys.* **50,** 269.
Spruit, H. C.: 1977, Ph.D. thesis, Utrecht.
Stenflo, J. O.: 1973, *Solar Phys.* **32,** 41.
Stenholm, L. G. and Stenflo, J. O.: 1977, *Astron. Astrophys.* **58,** 273.
Tarbell, T. D. and Title, A. M.: 1977, *Solar Phys.* **52,** 13.
Ulmschneider, P.: 1974, *Solar Phys.* **39,** 324.
Ulmschneider, P.: 1979, *Space Sci. Rev.*
Ulmschneider, P. and Kalkofen, W.: 1978, *Astron. Astrophys.* **69,** 407.
Vernazza, J. E.: 1977, presentation at *IAU Colloq.* **43.**
Vernazza, J. E., Avrett, E. H., and Loeser, R.: 1976, *Astrophys. J. Suppl.* **30,** 1.
Vernazza, J. E. and Mason, H. E.: 1978, *Astrophys. J.* **226,** 720.
Vernazza, J. E. and Reeves, E. M.: 1978, *Astrophys. J. Suppl.* **37,** 485.
Wilson, P. R.: 1971, *Solar Phys.* **21,** 101.
Withbroe, G. L.: 1975, *Solar Phys.* **45,** 301.
Withbroe, G. L.: 1978, *Proceedings of the OSO-8 Workshop,* University of Colorado, Boulder, p. 2.
Withbroe, G. L. and Noyes, R. W.: 1977, *Ann. Rev. Astron. Astrophys.* **15,** 363.

4. OBSERVATIONS OF MASS MOTIONS IN ACTIVE REGIONS

R. Grant Athay

4.1 INTRODUCTION

Skylab instruments and observing routines were not designed specifically for the purpose of studying mass motions. Also, most of what has been obtained from these data relative to mass motions has been published prior to the workshop on active regions. For these reasons, most of the input to the workshop has come from the newer OSO 8 and rocket data. The objective in this chapter is to review the earlier results from Skylab together with the results from ground-based data and to summarize the newer results from OSO 8. Recent results obtained at NRL with high-resolution rocket observations are summarized by G. E. Brueckner in the following chapter.

Although this chapter is titled Observations of Mass Motions, the reader should be wary: we do not observe mass motions. We observe wavelength displacements and asymmetries in line profiles, and we observe apparent lateral translations of features identified in intensity patterns in solar images. From the wavelengths and asymmetries, we infer Doppler displacements, which we express in equivalent velocity units. Similarly, from the apparent lateral translations of features, we infer proper motions.

In many cases, there is a high probability that the wavelength shifts and lateral translations, in fact, are due to motions. However, even in such cases, we are forced to invoke a model for the motions in order to extract a numerical amplitude for the velocity. An incorrect model can lead to incorrect numerical estimates and even to incorrect signs for the velocities. These difficulties arise from the inescapable handicap of observing "integrated" effects rather than "local" effects; e.g., we always observe events integrated with their foreground and background surroundings and never in isolation.

Wavelength displacements of lines can result from several factors other than Doppler shifts, including gas pressure, gravity, isotopes, and thermal distortion of spectrographs. Even when all such factors are properly accounted for, the remaining Doppler shift is still dependent upon the velocity model. Any type of motion can produce Doppler shifts, whether it be waves, laminar flow, or turbulent eddies. In the photosphere, for example, the positive correlation between increased brightness and upward convection produces a net blue shift in low-resolution data, even though the net flow velocity is zero. Similar effects can be caused by waves. Also, downflow in the bright chromospheric network very likely produces a net red shift in low-resolution disk observation of the chromosphere.

Proper motions are a less ambiguous measure of velocity than Doppler shifts. Even so, there is still need for caution. Many solar features have lifetimes of the order of 5 minutes and proper motions of the order of 5 km s^{-1}. Thus, over the lifetime of the feature, it moves only 1500 km, or 2″. It is difficult to distinguish such motion from the apparent motion produced when two adjacent features are fluctuating out of phase in brightness, e.g., as one brightens the other fades, causing an apparent motion of the "bright" feature.

Still further difficulty arises when the scale of the motion is small compared to the angular resolution. In the penumbra of sunspots, for example, most of the Evershed flow occurs in the dark lanes separating the penumbral filaments. Thus, the average flow velocity is considerably less than the actual flow velocity. In most situations of this type, it is very difficult to discover the filamentary structure in the flow field without actually resolving it.

This latter difficulty is partially alleviated in data with high angular resolution. On the other hand, integration along the line-of-sight can never be eliminated. Such integration is most serious at the limb and under optically thin conditions. The effective path length of integration in the line-of-sight at the limb in a stratified atmosphere is $L(2\pi R_\odot/H)^{1/2}$, where H is the opacity scale height, $R_\odot$ is the solar radius, and L is the integration length at disk center. Typically, L is a few times H, and H varies from the order of 150 km in the middle and low chromosphere to 10^5 km in the corona.

An interesting consequence of the long integration path in the corona at the limb ($\approx 5 \times 10^5$ km) is that small voids in the corona are extremely difficult, if not impossible, to detect. A hot, low-density coronal loop, for example, would be inconspicuous in the electron-scattering continuum. There is not the same discrimination against bright loops, however, and one cannot readily conclude from the appearance of the EUV and X-ray corona that bright loops are more prevalent than dark loops. The same discrimination occurs in measuring Doppler shifts; i.e., it is difficult to see Doppler-shifted dark features but easy to see Doppler-shifted bright features. One should recognize, therefore, that any attempt to describe flow fields in the corona and transition region, and to some extent in the chromosphere as well, is necessarily heavily biased towards bright features.

One of the striking differences encountered in an attempt to describe the flow in active regions is that we know so little. On the other hand, the fact that we know so little may be taken itself as evidence that active regions have no universally endowed circulation pattern of substantial amplitude. Sunspots are an exception. The Evershed flow in and around large sunspots is stable in time and quite faithfully reproduced in different spots.

In the following discussion, we classify flows in five categories: (1) steady, (2) quasisteady, (3) transient, (4) oscillatory, and (5) nonthermal line widths. Steady and quasisteady flows are distinguished by the lifetimes of the flow. Thus, quasisteady flows are those that are short- or intermediate-lived, but occur often, and steady flows are those that are long-lived.

This review covers only those aspects of active region motions that are not exclusively associated with flares or energetic particles. Thus, we omit discussion of flare spectra and phenomena observed at radio frequencies. Also, we omit discussion of white-light coronal transients, which recently have been reviewed elsewhere (MacQueen, 1979).

4.2 STEADY FLOWS

4.2.1 Supergranular Flow

Supergranules are defined by photospheric horizontal flows diverging from the cell center and converging at the network-bordering adjacent cells. Magnetic knots are located at the foci of converging flow and are associated with downflow of the order of 0.1–0.2 km s^{-1}. Upflow occurs near the cell centers, but unlike the downflow it is not concentrated in small tubes, and it is lower in amplitude.

Following a lengthy and difficult program of coordinating ground-based and OSO 8 observations of supergranule flow in the photosphere and chromosphere, November, Toomre, Gebbie, and Simon (1979) established a continuity in the supergranule circulation from the photosphere to the middle chromosphere spanning some 1400 km and 11 density scale heights. Although the sense of the flow is preserved over this extended height range, the detailed pattern changes markedly. At photospheric levels, the horizontal flow velocity of about 0.8 km s^{-1} greatly exceeds the vertical flow velocity, whereas in the middle chromosphere the two are approximately equal and are increased to at least 3 km s^{-1}.

The height range included in the observations of November *et al.* (1979) includes the levels of Hα and K line plages. To the extent that the supergranules penetrate the plage regions, which appears to be quite general, one can infer therefore that the plages participate in the supergranule flow. Large plages are larger than individual supergranules, however, and a single large plage area may be subdivided into two or more supergranule cells.

4.2.2 Evershed Flow

The Evershed flow associated with large sunspots has both striking similarities and striking differences from supergranule flow. Among the similarities are the predominance of horizontal flow and the divergence from a well-defined center. Among the differences are the concentration of vertical magnetic flux at the center of divergence of horizontal flow (the sunspot umbra) rather than at the foci of convergence, and the reversal of the sense of flow between the photosphere and chromosphere.

An excellent review of the conceptual and observational developments related to the Evershed effect has been given by Schröter (1965). Up to the time of his review, the following points had been established: (1) the radial outflow at photospheric levels extends beyond the umbra and penumbra into the surrounding photosphere; (2) the maximum average outflow varies in both amplitude and location from one spot to another, but usually occurs somewhere in the penumbral region; (3) the penumbral outflow is filamented with localized channels of high flow velocity; and (4) the flow reverses direction in the low and middle chromosphere with filamented channels of inflow. Outflow velocities in the photosphere average about 1 km s^{-1} at maximum. However, local flow velocities in channels reach values of 4 to 6 km s^{-1}. At the level of Hα formation in the middle chromosphere, the inflow velocities in channels may exceed 8 km s^{-1}.

The prevailing view in recent studies is that the outflow at low levels is mainly in the dark lanes separating the penumbral filaments (cf. Mattig and Mehltretter, 1968; Bones and Maltby, 1978), which is contrary to the model proposed by Schröter (1965). The level of flow reversal appears to be in the upper photosphere, so that at and above the temperature minimum, the dominant flow is inward (Bones and Maltby, 1978). At chromospheric levels, the inflow appears to be along the Hα fibrils.

Efforts to measure vertical and azimuthal velocity components in sunspots are contradictory. However, the most careful analysis of vertical motions to date shows no residual vertical motions in sunspot umbrae in excess of 25 m s^{-1} (Beckers, 1977). Some observers (cf. Lamb, 1975; Schröter, 1965) report substantial azimuthal components of velocity. On the other hand, other observers have failed to find evidence for such motion. It seems probable that such motion is more likely to be present in complex sunspot groups than in simple groups as suggested by Schröter (1965).

Studies of proper motions of small magnetic features and small faculae near sunspots by Sheeley (1969, 1972) revealed an evident continuation of the outflow at photospheric levels to some 10,000 to 20,000 km beyond the outer boundary of the spot. This extended region of outflow, often referred to as the "moat," is clearly evident for some sunspots but not for others. Also, it is present for only a portion of the spot lifetime.

Outflow velocities in moats average about 1 km s^{-1}. When the moat is present, the network structure is swept to the outer edges of the moat; i.e., the moat appears to exclude the supergranule network structure. It is not known whether the moat flow is channeled or is more or less uniform. However, the outflow of magnetic flux appears to account for most of the decay in the flux of the parent sunspot (Vrabec, 1974).

A few observations have been made of systematic flows in the areas of embryonic active regions. Kawaguchi and Kitai (1976) report a systematic downflow in the vicinity of growing sunspots and prior to the appearance of spots. Similar results were obtained by Gopasyuk (1969) during the formative stages of sunspots. The downflow occurs in lines formed in the low chromosphere as well as in the photosphere, with the velocity amplitude increasing with height. Maximum velocities of 1 to 1.5 km s^{-1} are observed in the Na D_1 line, which is formed in the low chromosphere. The downflow ceases early in the life of the sunspot.

4.2.3 Chromosphere and Transition Region Flow

As a contribution to the workshop on active regions, Lites (1980) has studied the steady flow of chromospheric and transition region material over active regions using OSO 8 data. He finds a persistent, but not universal, blue shift directly over sunspots and a mixture of red and blue shifts over plages, with a dominance of red shifts. At lower transition region levels, where lines of C IV and Si IV are formed, the blue shift over sunspots is typically about 20 km s^{-1}. In the low and middle chromosphere, the blue shift is still apparent, but it is reduced in amplitude to 1–4 km s^{-1}. In the upper chromosphere, where lines of C II are formed, the blue shift over the sunspot is approximately 10 km s^{-1}.

Over plage areas, Lites (1980) finds a predominance of redshifts with velocities up to a few tens of km s^{-1}. Blue shifts of considerable amplitude are observed also, and the horizontal gradient of the vertical velocity is found to be large, implying large horizontal shear. Similarly, LeMaire *et al.* (1978) find both red- and blue-shifted regions over a plage with velocities of ± 10 km s^{-1} in lines of O VI.

Other observations of flow velocities over sunspots and plages are reported in the next chapter in this volume, by Brueckner. The composite picture that emerges is rather incomplete but clearly establishes that the chromosphere and transition region in active regions are characterized by large amplitude flows.

4.3 QUASISTEADY FLOW

We define this class of motions to consist of short-lived phenomena that occur more or less continually and longer-lived phenomena that occur commonly but are not known to occur universally. In the former category, we include spicules and fibril motions; in the latter, we include coronal rain and motion in coronal loops.

4.3.1 Fibrils

Hα fibril patterns around sunspots create the characteristic "iron filing" appearance of active regions. Such patterns are often stable for a substantial fraction of the lifetime of individual sunspots. There is a tendency, though perhaps not a universal one, for matter to flow along the fibrils into the sunspot areas. This flow is believed to be part of the inward Evershed flow of the chromosphere (Maltby, 1975).

Superposed on the flow along the fibril is a body motion of the fibril itself. Little is known concerning such motion. However, Maltby (1975) reports a characteristic time of 8 minutes for perturbations in the flow along individual fibrils even though the individual flow channels persist for approximately 70 minutes. Thus, the 8 minutes appears to be characteristic of the time scale for the body motions of the fibrils.

A phenomenon known as flagellant motion of fibrils is observed in time-lapse movies made in Hα during periods of exceptional seeing. Individual Hα fibrils are observed, at such times, to flutter in position. However, the more prominently developed fibrils in active regions show conspicuously less flagellant motion than do the fibrils in the quiet sun. It is clear from such observations that the intense magnetic fields in active regions stabilize the individual fibrils as well as the overall fibril pattern.

4.3.2 Spicules

Much has been written concerning spicule motions, and little if anything new has been added since the reviews by Beckers (1968) and Athay (1976). The most notable difference between spicules seen in active regions, as opposed to those seen in quiet solar regions, is the more chaotic appearance of the active region spicules. In fact, the chaotic nature of active region spicules is very likely one of appearance only. Spicules in active regions, as elsewhere, probably flow along magnetic flux tubes. The view at the limb of the magnetic field above a horizontal plane through an active region reveals a complex mixture of polarities and field orientations that provides an appearance of chaos even though the fields are locally highly ordered. Thus, the spicule flow in active regions is probably as ordered as it is elsewhere, except that the correlation distance is reduced.

To the author's knowledge, there has been no systematic study of spicule motion in active regions. Thus, we can only presume that the flow characteristics of spicules are basically the same wherever they occur. Spicules have the appearance of pencil-like jets that elongate by the injection of new material at their base; i.e., the total spicule mass increases more or less in proportion to the spicule length. The typical evolution of a spicule is an elongation phase of 5 to 10 minutes duration at velocities of about 20 km s^{-1} followed by a fading in brightness. Only a minority of spicules retract.

Lateral body motions of spicules have been reported by some observers, but quantitative studies are exceedingly difficult because of seeing effects. Such motion is probably related to the flagellant motion of fibrils.

Spectroscopic studies of spicules at the limb indicate that the elongation is due to matter flow and not to a moving excitation front. Some spicules show "tilted" line profiles believed to be caused by spiraling motion in which opposite sides of the spicule are moving in opposite directions in the line-of-sight.

Spicules rise out of the chromosphere network, which is generally assocaited with downflow. It appears, therefore, that spicules are a manifestation of a particular property of network flow and are not indicative of the general flow pattern. From the matter density in spicules, which is of the order of 10^{11} hydrogen atoms cm^{-3}, it is inferred that spicules originate in the upper chromosphere.

The driving force for spicules has been a subject of much speculation. Perhaps the most serious constraint on such models is provided by the very low gradient of gas pressure along the axis of the spicule. The effective scale height for the gas pressure is large compared to the hydrostatic scale height, which requires that the accelerating force be distributed along the axis of the spicule rather than localized at its base. This eliminates all models that utilize accelerating forces derived from pressure gradients.

4.3.3 Coronal Rain

Coronal rain is a phenomenon observed in Hα over active regions at the limb. Bright Hα condensations at coronal levels are observed to funnel into localized areas as though the matter were channeled along magnetic lines of force above one pole of a dipole. This is different from the situation for post-flare loops in which matter is seen flowing in both legs of a closed loop system, i.e., into both poles of a dipole. Also, the flow trajectories for coronal rain typically have less curvature than in post-flare loops, and the focus of the inflow is less well defined than in the case of post-flare loops. The more diffuse nature of the coronal rain has led some observers (Menzel and Evans, 1953) to classify coronal rain as not being associated with a particular sunspot.

Little is known about excitation conditions of the condensed matter in coronal rain and about its association with other types of activity. In general, however, coronal rain is not identified with energetic phenomena related to flare activity.

Material velocities for the falling condensation in coronal rain are typically of the order of 100–200 km s^{-1}. The motion appears to be subsonic in nearly all cases.

4.3.4 Coronal Loops

Although it is generally agreed that matter flows downward in the legs of coronal loops (as in post-flare loops), there actually are no observations of Doppler shifts or proper motions of condensations in such loops. The presence of motion is inferred from analogy to flows observed in Hα loops and from other indirect means.

Foukal (1976), for example, concludes that loops emerging from spots have cool, dense cores and infers that the dense condensation must fall under the influence of gravity. Similarly, Levine and Withbroe (1977) conclude that in the tops of coronal loops maximum brightness occurs later for lower-temperature ions, and they infer from this that matter is condensing into the tops of the loops from the ambient corona. In order to maintain continuity, the condensing matter must flow downward. On the other hand, Pye *et al.* (1978) postulate that loops are heated preferentially near their tops, and that such heating eventually reaches the transition region and upper chromosphere via thermal conduction. Material then evaporates from the heated chromosphere to fill the loops from below. Such a model is not necessarily in opposition to the suggestions of downflow for the condensed, low-temperature matter. However, the suggestion for filling of loops by evaporation does complicate the flow structure within the loops.

A few good observations of Doppler shifts or proper motions of condensations within coronal loops would hopefully clarify the picture. Such observations are perhaps possible with the instruments on the Solar Maximum Mission satellite.

4.4 TRANSIENT FLOWS

In this category, we include surge and spray prominences and EUV transients of the type observed with Skylab and OSO 8. The exact nature of the phenomena associated with EUV bursts is unknown. Bruner and Lites (1979) proposed the possibility that the transients are downward-propagating shock waves. However, Athay, White, Lites, and Bruner (1980) suggest that most of the transients may be associated with surges and coronal rain. There is further comment on these possibilities in the following discussion.

4.4.1 Surges and Sprays

Surge and spray prominences are associated with flare activity. Both consist of violent upheavals of major proportions. Also, both are associated closely with flare-like events at the seat of the ejection. Some authors, e.g., Bruzek (1968) and Giovanelli and McCabe (1958), consider that all true surges occur as expansions of flare material. However, in this definition many of the associated flares are small, short-lived events.

Sprays, in general, are more violent events than surges and are more typically associated with larger flares. The ejection velocity in sprays exceeds that in surges, and some authors, e.g., Smith and Smith (1963), define sprays as ejections for which the outward velocity exceeds the escape velocity. Others have defined sprays in terms of their fragmented appearance. In this definition, a spray consists of small, discrete condensation or droplets, whereas surges are more typically fibrous streams of ejected material. Still others (e.g., Tandberg-Hanssen, 1974) classify sprays as those that appear to rupture the magnetic field pattern and surges as those that appear to follow the magnetic lines of force.

Surges have typical velocities of 100 to 200 km s^{-1}. Their upward motion of 10 to 20 minutes duration, and often along a curved trajectory, is usually followed by a retraction along more or less the same path. This is consistent with the conclusion that the magnetic field pattern remains intact. If one considers sprays as those events for which the velocity exceeds the escape velocity, then spray ejecta are found to typically move in straight lines and never reverse direction. Spray velocities up to 1500 km s^{-1} are observed.

Upward-moving ejecta with constant velocity V do work against the gravitational field and have an equivalent potential energy flux given by

$$\text{P.E. flux} = \varrho g V R_{\odot} \quad , \tag{4.4.1}$$

where ϱ is the density. Their kinetic energy flux is

$$\text{K.E. flux} = \tfrac{1}{2} \varrho V^3 \quad . \tag{4.4.2}$$

The two fluxes are equal when V is the escape velocity, V_{esc};

$$V_{esc} = (2gR_{\odot})^{1/2} = 620 \text{ km s}^{-1} \quad . \tag{4.4.3}$$

Below the escape velocity, most of the energy flux is potential energy flux.

In order for the magnetic field to confine matter moving at the escape velocity, the field strength B must satisfy the inequality

$$B > V_{esc}(4\pi\varrho)^{1/2} \quad . \tag{4.4.4}$$

We adopt a mean value for ϱ of 2×10^{-13} g cm^{-3}, which requires $B \gtrsim$ 1000 G. It is not likely that B is this large at the locations of sprays. However, it is also unlikely that B is a great deal smaller than 1000 G, so that the criterion that sprays rupture the field pattern, in practice, may not be very different from the criterion that they exceed the escape velocity.

4.4.2 EUV Bursts

Observations made in active regions using EUV spectrometers with narrow fields of view show frequent flare-like bursts of intensity. Compared to flares, however, the bursts are smaller in size, shorter lived, and have larger associated Doppler shifts. Although bursts were first recognized in Skylab data from the Harvard spectrometer, these data gave no information on Doppler shifts. Subsequent observations using the University of Colorado spectrometer on OSO 8 provide Doppler information for an appreciable number of bursts. The observed Doppler shifts are large, and they appear to be essential to the interpretation of the burst phenomenon.

Studies of Skylab bursts are reported by Emslie and Noyes (1978). Bursts observed with OSO 8 have been studied by Bruner and Lites (1979), Lites and Bruner (1980), Lites and Hansen (1977), and Athay, White, Lites, and Bruner (1980). The latter study is the most comprehensive and is summarized in the following paragraphs.

All of the following data were obtained from time series of line profiles made by repetitively scanning the C IV line at λ1548 Å while the instrument remained pointed at a fixed location on the solar disk. The observations were made with a 1″ × 20″ slit; however, telescope resolution and satellite jitter give an effective resolution of 2″ × 20″. All but one of the bursts selected for this study were observed near active regions.

The bursts studied can be roughly classified in three categories: fast, slow, and exponential. The fast and slow events have more or less symmetric brightness profiles with half-lives ranging from less than 120 s (fast) to about 5 minutes (slow). A few of the fast bursts have half-lives that are shorter than the minimum resolution time of 27.5 s. Exponential bursts are characterized by a relatively rapid rise to maximum brightness followed by a more or less exponential decay. Only 3 of the 51 bursts studied were classified as exponential. For 2 of these, the e-folding decay time is near 300 s, and for the third it is near 600 s. The remaining 48 bursts are equally divided between fast and slow events.

For the 4 bursts studied by Emslie and Noyes (1978), the observed brightness increase in the O IV line at λ554 is five- to tenfold. For the 48 fast and slow bursts studied here, the median increase is fourfold; but 17 percent of the bursts increase by tenfold or more. The maximum observed is twentyfold. Although the statistics are poor, it appears that the typical brightness increases observed are not different by more than a factor of two in the two sets of data. The data are consistent with a burst diameter of the order of 3″ or less, which is smaller than the resolution size in either the Skylab or OSO 8 data.

An estimate of the frequency at which bursts occur in a given active region can be made from a control set of data. In a set of 85 orbits of data, 14 bursts were recorded. The total observing time for the 85 orbits was about 4200 minutes; so, on the average, one burst occurred for each 300 minutes of observing. If we assume that bursts occur randomly in a typical active region area of 100″ × 100″ and have a typical diameter of 3″, the OSO 8 observing aperture samples approximately 0.012 of the bursts. Thus, the average burst frequency for this hypothetical region is one burst each 3.6 minutes. We do not know, of course, how large the typical burst area is, so we cannot have much confidence in this estimate. However, the mere fact that we observe bursts relatively frequently with a 2″ × 20″ aperture means that they must be very common features of active regions, as is confirmed by the series of OSO 8 spectroheliograms reported by Lites and Hansen (1977). It is clear, also, that some active regions produce bursts much more frequently than do others. We lack quantitative data that would allow us to relate burst production rates to other properties of individual plages, however.

From the individual line profiles, we derive five parameters, defined as follows:

I_b = background intensity in the line wings;

$A = \sum_{\lambda} I_{\lambda} - I_b$ = total intensity in the line;

I_0 = intensity at line center above I_b;

$W_1 = \frac{A}{I_0} \Delta\lambda_i$ = equivalent width;

and

$$M_1 = \frac{\sum_{\lambda} \lambda (I_{\lambda} - I_b)}{\sum_{\lambda} (I_{\lambda} - I_b)} = \text{centroid of profile.}$$

The quantity $\Delta\lambda_i$ is the stepping interval of 75 mÅ. For details concerning the determination of I_b and I_0, the reader should consult White and Athay (1979a). For each of the above parameters, we construct time series 30 minutes in length that include each of the bursts to be studied.

Unfortunately, our observations do not provide reliable measures for I_b at the time of the burst itself. During many of the bursts, the C IV line broadens to such an extent that the 0.75 Å bandwidth used in the line scan is insufficient to reach outside the line. Since the background intensity near λ1548 comes mainly from scattered light at longer wavelengths radiated from the upper photosphere and temperature-minimum region (cf. Athay and White, 1979a), we assume that the bursts are not accompanied by changes in I_b. However, the bright flare-like transients near sunspots, reported by Lites and Bruner (1980), do show changes in the background. Our results for similar regions then overestimate the line variation, but the steadiness of our background estimates suggests that such strong events are a small part of our sample. To obtain I_b throughout the bursts, we use a cubic polynomial extrapolation determined from a least-squares fit to the values of I_b for the remainder of the 30-minute observing period outside of the burst. Typically, the background shows a low-amplitude oscillating brightness with a period near five minutes (Athay and White, 1979b) superimposed on longer-period drifts of similar amplitude. The cubic polynomial removes these long-period drifts but not the five-minute oscillations. However, the five-minute oscillations are not of large enough amplitude to produce a strong effect on the burst profile.

A. Intensity and Doppler characteristics. The brightness characteristics as a function of time for a number of bursts are illustrated in Figures 4.1 and 4.2. Figure 4.1 shows two fast and two slow bursts; Figure 4.2 shows two exponential bursts and two groups of bursts. The time resolution is 27.5 s in each of the examples except experiment 24394, for which the time resolution is 49.5 s.

Experiment 9770, shown in Figure 4.2, shows the brightness feature observed in C IV during the first year of operation. The observations were made in McMath region 14143 beginning at $20^h\ 7^{min}$ on 28 March 1976. At that time, two class 1B flares were in progress in their decaying phases. A later segment of this same experiment is shown in Figure 4.5.

Among the bursts shown in Figures 4.1 and 4.2, some are moderately symmetric, some have a slow rise followed by a more rapid decline, and some have the opposite asymmetry. This diversity is typical of the total sample; i.e., no preferred shape is evident.

The fine structure in the brightness curves for the bursts in experiments 9948 and 25623 is not unusual. In these cases, and in several others, the modulation in brightness about the mean burst curve is much larger than expected from counting statistics. However, we have not attempted a systematic study of the fine structure in bursts.

The centroid M_1 could provide a reliable measure of the relative Doppler shift during bursts if the proper conditions were met. These conditions require that the burst fill the slit aperture, that the line profile be symmetric, and that the wavelength scan be sufficiently wide. None of these conditions

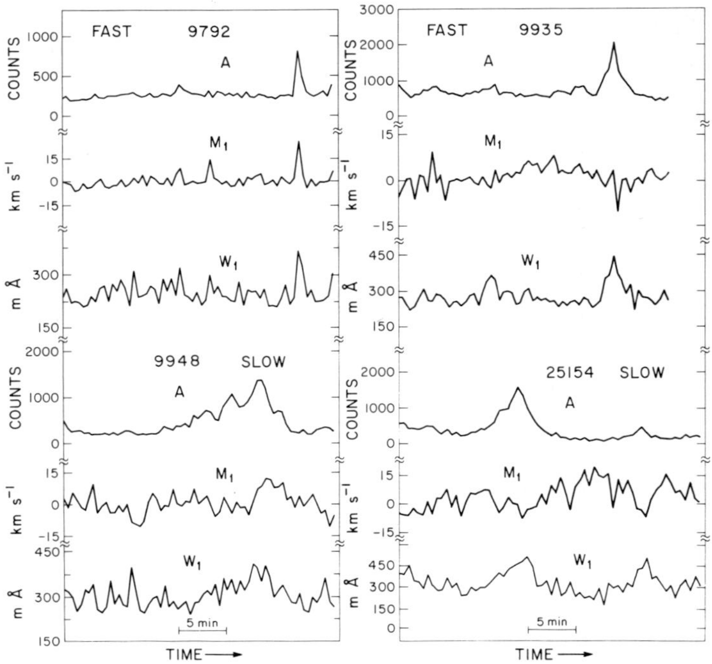

Figure 4.1. Examples of the time variation in the total line intensity (A), the Doppler shift (M_1), and the line width (W_1) for two fast and two slow bursts.

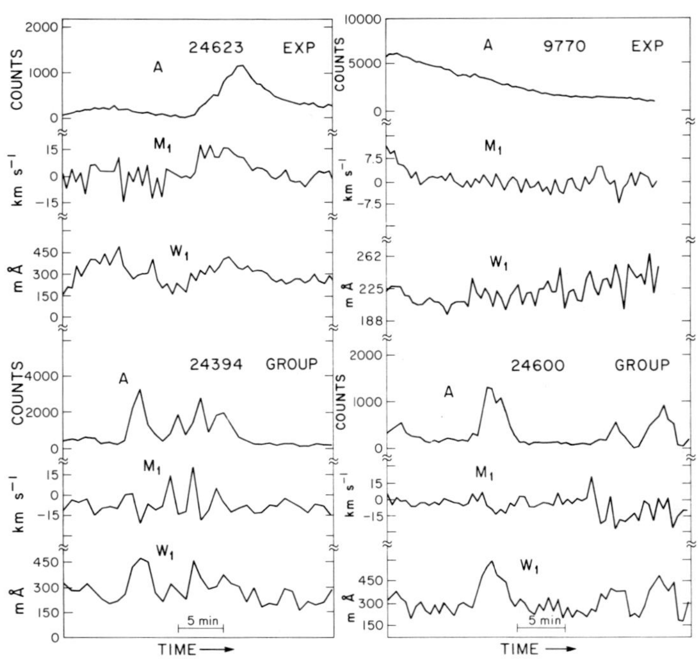

Figure 4.2. Times series similar to those in Figure 4.1, but showing examples of exponential bursts and groups of rapid bursts.

is met, however. During bursts, the maxima in the Doppler-shifted profiles are often near the edge of the scan window, which makes determination of the line shift difficult. Also, within the scan window, the profiles are often asymmetric, and, in addition, it is clear from Skylab data that the bursts are normally much smaller than the 20″ slit length used on the OSO 8 instrument. Thus, the observed profile is a weighted average of the burst profile with the undisturbed profile.

Since there is no absolute wavelength reference on OSO 8, only relative changes in M_1 can be measured. An increase in M_1 indicates a relative red shift. Relative shifts in M_1 during bursts are typically 20 km s^{-1} or less, but shifts as large as 30 km s^{-1} are sometimes observed. Asymmetries of the individual line profiles during bursts strongly indicate that the actual Doppler shifts of some of the emitting material in the burst are much larger than the changes in M_1 suggest.

Changes in line profiles associated with two fast and two slow bursts are shown in Figures 4.3 and 4.4. Each panel in these figures has five profiles, except for the upper panel in experiment 25154, which has four profiles. The top panels show the beginnings of the bursts and the lower panels show the later stages. Time resolution between profiles is 49.5 s in experiment 24434 and 27.5 s in the other three. Burst 25154 in Figure 4.3 occurs about 20

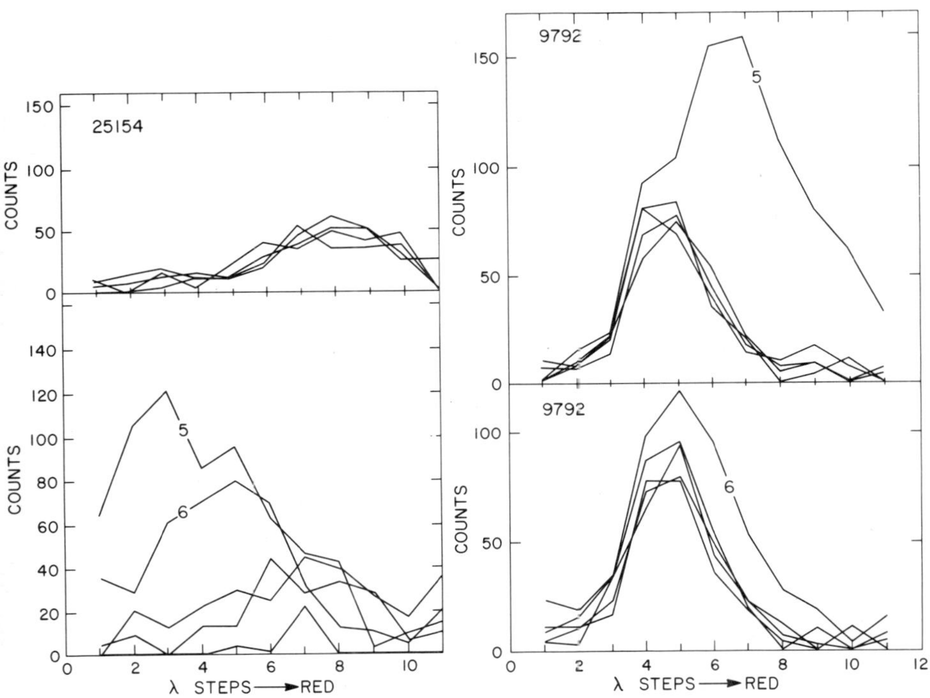

Figure 4.3. Sequences of line profiles measured during two fast bursts. The upper frame shows the profiles observed just before the burst for experiment 25154 and just before and including the strongest emission for experiment 9792. The profiles for no. 25154 show the occurrence of a large blue shift, while no. 9792 shows another event in which a short-lived red shift was observed during the burst.

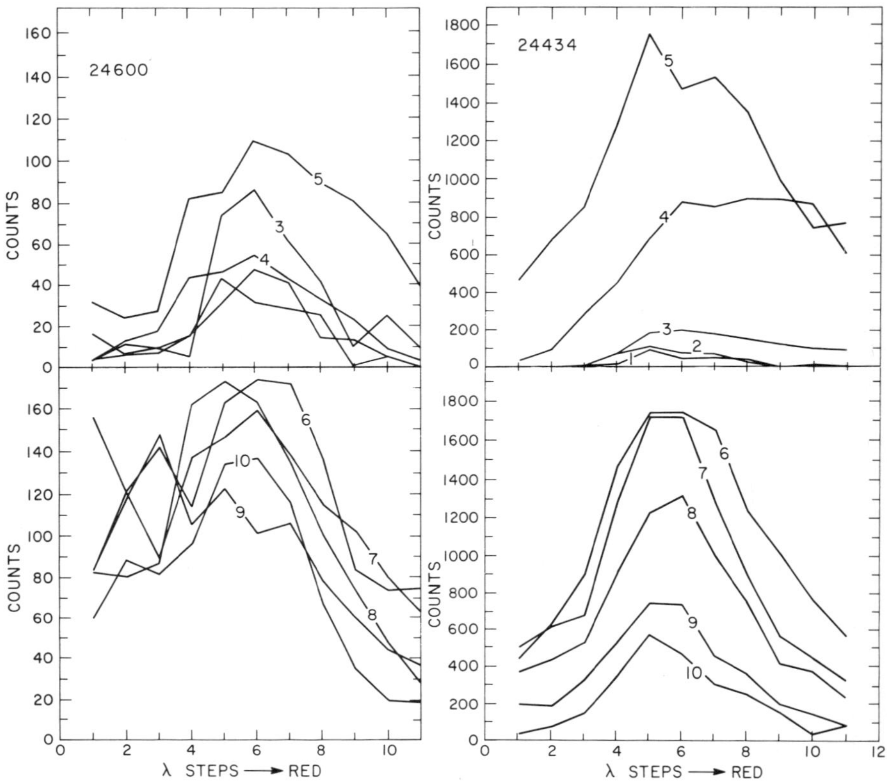

Figure 4.4. Sequences of line profiles measured during two slower-developing C IV bursts. Experiment 24600 shows a blue shift followed by a red shift; experiment 24434 shows an initial red shift that disappears before the intensity subsides.

minutes after burst 25154 in Figure 4.1. Burst 9792 in Figure 4.3 is the same as burst 9792 in Figure 4.1, and the burst in 24600 in Figure 4.4 is the same as the first large burst in 24600 in Figure 4.2.

The profiles measured during two fast bursts in Figure 4.3 were selected to show examples of strong blue (25154) and red (9792) shifts. In burst 25154, the maximum in profile five is 0.375 Å to the blue of the centroid of the preburst profiles in the upper frame, and in burst 9792 the secondary maximum in profile scan five is 0.225 Å to the red of the centroid of the preburst profiles. These wavelength offsets correspond to velocities of 73 and 44 km s^{-1}, respectively. The maximum Doppler shifts coincide with maximum brightness in both cases.

The slow burst 24600 (Figure 4.4) begins with a red shift of considerable enhancement in brightness out to about 0.375 Å from line center (profile 5). In the next four profiles, there is strong brightening in the blue wing ≈ 0.4 Å from line center. Maximum brightness occurs at profiles six and seven. Burst 24434 (Figure 4.4) begins also with a strong brightening ≈ 0.4 Å from line center in the red wing in profiles three and four. However, in profiles five and later, the entire line is enhanced with little remaining asymmetry to the red. Maximum brightness occurs near profile six. Thus, the red shift is again confined to the early part of the burst.

The cases shown in Figures 4.3 and 4.4 are not atypical. Several examples of each type are present in the data. Of the 51 bursts studied, only 3 show no evidence of Doppler shift. For the remaining 48, the correlation between the phase of maximum brightness and the major feature of the Doppler shift shows a clear tendency for red shifts to precede maximum brightness and for blue shifts to occur at or somewhat after maximum brightness. Red shifts dominate in about two thirds of the cases. There is little apparent difference between fast and slow bursts so far as these correlations are concerned. The three exponential bursts all have red precursors.

In about half of the cases dominated by blue shifts, the blue shift is immediately preceded by a red shift similar to that shown for experiment 24600 in Figure 4.4. However, no clear examples of blue shift immediately preceding red shift were observed. Thus, in the blue-shifted cases, only about half occur without a red-shifted forerunner; whereas, all of the dominant red shifts occur without a blue forerunner.

The example shown in burst 24434 in Figure 4.4, in which the wavelength of maximum intensity in the line shows very little Doppler shift during the burst, is representative of many bursts. Similarly, the disappearance of any evidence of Doppler-shifted emission before or shortly after burst maximum also is representative of many bursts. In other words, bursts may show evidence of enhanced brightness that is not Doppler shifted. These stationary sources of enhanced emission often occur at the same time as the Doppler-shifted emission and remain after the moving component disappears.

It is difficult to obtain a clear indication of the true Doppler velocity of the moving components of the bursts because of the integration over moving and stationary regions. The best indication of the Doppler shift comes from the line profiles. By roughly approximating the profiles as the sum of two displaced Gaussian profiles, we assign the burst a Doppler velocity as given by the center of the Gaussian that is farther from the center of the preburst profile. In this way, we find that 28 bursts have velocities less than 30 km s^{-1}, 12 have velocities between 30 and 60 km s^{-1}, and 11 have velocities in excess of 60 km s^{-1}. Since we have used only a crude measure of velocity and since the scanning window was not wide enough to show the complete profiles in many cases, the preceding figures probably distort the true velocity statistics. However, it seems clear that the bursts are frequently associated with large-amplitude velocities.

The group of bursts with experiment numbers greater than 24000 were observed during the period 10 April to 3 May 1978, when the sun was moderately active (Lites and Bruner, 1980). These bursts were observed in active regions that produced many small flares and subflares and it is likely that some of the bursts are flare associated. However, the proportion of blue shifts to red shifts and the magnitudes of the Doppler shifts are not significantly different from the earlier set of bursts observed in regions of low flare activity.

B. Interpretation of bursts. The following characteristics of bursts summarize the results of the preceding discussion:

1. Bursts occur frequently, at a rate of approximately 15 per hour in some large active regions.
2. Brightness increases in the C IV resonance lines exceed tenfold.
3. Lifetimes range from less than a minute to about seven minutes.
4. Large material velocities occur mainly during the rise to maximum brightness and are most often directed downward.
5. In about 20% of the observed bursts, material velocities in the moving components observed in C IV exceed the sound velocity of 48 km s^{-1} at 10^5 K, where C IV is normally formed.

These characteristics provide clues to the physical causes of bursts and help to eliminate some of the contending candidates. We consider first the question of whether the bursts are caused by "local" increases in the amount of matter in the transition region or by "nonlocal" additions of new emitting regions, such as would be produced by prominence material.

One difficulty with the "local" concept of bursts is that the material velocities are often in excess of the local sound velocity. This, in itself, contradicts the assumption of localized effects. In the average solar atmosphere, the thickness of the layer emitting C IV radiation is substantially less than 100 km, and a source of emission moving with a line-of-sight velocity component equal to the sound speed would pass through the "local" region of C IV emission in less than two seconds. Thus, the Doppler shifts are too large to be compatible (in the local hypothesis) with the observed lifetimes.

The lifetimes of bursts clearly rule out the possibility that the bursts decay radiatively. An estimate of the radiative cooling time τ can be obtained by equating the thermal energy density to the radiating power given by Cox and Tucker (1969). If we represent the latter by $Q(T_1)$, we obtain

$$Q(T_1)n_e^2\,\tau = 2n_e kT_1 \quad , \tag{4.4.5}$$

or

$$\tau = \frac{2kT_1}{Q(T_1)n_e} \tag{4.4.6}$$

Using $T_1 = 10^5$ K and $Q(T_1) = 5 \times 10^{-22}$ erg cm^3 s^{-1}, we find $\tau \approx 6 \times 10^{10}/n_e$ s. For the average sun, $n_e \approx 6 \times 10^9$ cm^{-3} at $T_1 = 10^5$, which gives $\tau \approx 10$ s. At higher densities, τ is even shorter. Thus, the burst lifetimes appear to be too long to be explained by radiative decay following a sudden infusion of additional material at temperature T_1. This means that the bursts have a sustained source of enhanced emission rather than an instantaneous one.

The interpretation of bursts as being due to additional emitting regions associated with moving prominence or coronal loop material is attractive from several standpoints. It provides a ready explanation for the material velocities, for the observed lifetimes, and for the small size of the burst region. Also, it is not inconsistent with the relatively greater intensity increases in the optically thin transition region lines.

We recognize that the velocities associated with active region prominences are usually 100 to 200 km s^{-1}, which is much larger than the velocities associated with bursts. However, because of the narrow spectral window used for the C IV line, we have strongly biased the results towards lower velocity cases. Also, we measure only the radial component of velocity. Thus, we cannot rule out the prominence possibility.

In the interpretation of bursts as prominence-like material, we interpret the upward, blue-shifted bursts as surge-type phenomena and the downward, red-shifted bursts as coronal rain or matter streaming down loops. Bruner *et al.* (1976) give a similar interpretation of more steady downward flows observed in C IV near sunspots. The bright afterglow following the Doppler-shifted portion of the burst could be the afterglow caused by the heating due to infalling material or, in the case of surges, to an afterglow in the region from which the surge is ejected. In either case, the afterglow is identified with local changes in n_e and dT/dh. This is not objectionable in this case, since the afterglow occurs after the strong Doppler shifts have subsided.

4.5 OSCILLATIONS

Short-period solar oscillations have been studied extensively at photospheric levels, moderately at chromospheric and transition region levels, and only slightly in the corona. The properties of the oscillations do not appear to change markedly in active regions, except in the sunspots themselves.

At photospheric levels, most of the power in the oscillation occurs at periods between 400 and 250 s and at spatial scales ranging upwards from about 5000 km. The dominant period in the photosphere is the familiar five-minute oscillation, but a single dominant scale is not evident (cf. Beckers and Canfield, 1976). Deubner (1976) finds convincing evidence for short-period waves with periods down to about 15 s. The velocity amplitudes of these short-period waves are difficult to determine because of observational constraints and are subject to debate. Recent work by Cram *et al.* (1980) suggests that the velocity amplitude is of the order of 0.1 to 0.2 km s^{-1}. The longer-period waves have velocity amplitudes in the photosphere of the order of 0.5 km s^{-1}.

Within sunspot umbrae at photospheric levels, Beckers and Schultz (1972) find a dominant oscillation period of approximately 180 s, but increasing back to the 300-s level in the penumbrae, with evidence of even longer periods in the extreme outer penumbrae. On the other hand, Bhatnagar, Livingston, and Harvey (1972) find dominant periods ranging from 450 to 310 s in spot umbrae, and Soltau, Schröter, and Wohl (1976) find three separate ranges of periods: 470–300 s, 196–164 s, and 123–110 s. The first two agree with the earlier results. In the latter results, all three periods occur in individual time series. However, the spatial resolution of 6″ is large enough that two or three oscillating cells could be included. The correlation length for umbral oscillations at chromospheric level is reported by some observers to be as small as 2″ to 3″. It is probably no larger than this in the photosphere.

In addition to the successful reports of velocity oscillations in sunspot umbrae, there are many reports of unsuccessful attempts. This could be due to the small correlation distance, to the difficulty of observing the sunspots without strong contamination from stray light, or to the absence of oscillation in many spots. The first two factors have certainly influenced some negative results, but we cannot rule out the possibility that the oscillations are suppressed in some spots.

The velocity amplitudes in the umbral oscillations are similar to those in the photosphere, viz., of the order of 0.5 km s^{-1} for the longer-period oscillations. For the shorter-period oscillations near 180 s or lower, the amplitude is considerably smaller.

At photospheric levels, the dominant intensity oscillations are approximately 90° out of phase with the velocity oscillations. There is little doubt that the waves are primarily evanescent, and that they are generated as the high-order, nonradial P-modes predicted by Ando and Osaki (1975).

At chromospheric levels, the oscillations are notably more diverse than at photospheric levels. Different authors have reported the primary power maximum at periods ranging from 500 s to 170 s. From the large volume of time series data acquired by OSO 8 for chromospheric lines of Si II at λ1816.93 and λ1817.45, it is clear that different regions of the solar surface oscillate with different periods, or, equivalently, that the dominant period of oscillation changes with time at a given location.

From averages of several hundred power spectra, Athay and White (1979a) conclude that oscillations occur with periods ranging from longer than 400 s down to less than 30 s. Most of the power, however, is concentrated in the band of periods between 400 and 100 s, with a distinct maximum near 300 s. Within bright plage regions, the primary band of power narrows markedly and is concentrated in the 400 to 200 s range, with a maximum at 330 s. In any one power spectrum, however, the power is concentrated in a much narrower band. Thus, it is clear that the broader bands of power in the average spectra represent the distribution functions for finding power at a particular frequency rather than the spread of power at a given time and location. This is consistent with the reports of different power maxima by different observers using limited data samples.

Although the photospheric five-minute oscillations are clearly evanescent, there is evidence that chromospheric oscillations with periods of five minutes or less are propagating. White and Athay (1979b) find a systematic time lag between oscillations in the two lines, λ1816.93 and λ1817.45, of Si II. The time lag is in the sense of upward propagation at roughly the sound speed. Within a given line, maximum intensity leads maximum blue shifts by an average of approximately 60°. However, both the time delay between oscillations in the two lines and the phase lag between intensity and Doppler shift are statistical averages subject to considerable uncertainty and to considerable variance from one realization to another. The statistical evidence favors propagation, but it could easily be the case that the propagation occurs only in selected regions.

A few observations of sunspot oscillations at chromospheric levels have been made in the Hα line. In contrast to the quiet sun, sunspot umbrae oscillations have a relatively small correlation length of the order of 2″ to 3″. Different oscillating elements within a given spot umbra tend to show somewhat different periods. Giovanelli (1972) finds periods ranging from 145 s to 185 s, and Phyllis (1975) finds periods from 145 s to 180 s. Moore and Tang (1975) report periods of 150 s and 170 s.

Umbral oscillations show predominantly vertical motion. In the penumbra, however, horizontal running waves are observed emanating from the umbra in Hα. The running waves at chromospheric levels have propagation speeds of 10 to 20 km s^{-1} and periods closer to the five-minute oscillation in the photosphere. Approximate periods found by different observers are 300 s (Zirin and Stein, 1972), 210 s (Giovanelli, 1972), and 270 s (Moore and Tang, 1975).

The pattern of vertical oscillations in the umbra and horizontally propagating waves in the penumbra is suggestive of waves moving upwards through the umbra and ducted by the diverging magnetic field into the predominantly horizontal waves in the penumbra. Giovanelli (1972) and, more recently, Lites, White, Chipman, and Athay (1979) find, in fact, that the penumbral waves can be traced well into the umbra. On the other hand, if the waves were simply being ducted along magnetic field lines, the period should be the same in the umbra and penumbra, which is clearly not the case. Apparently the shorter-period umbral oscillations excite the penumbral waves but not on a one-to-one basis.

At transition region levels, the presence of oscillations is subject to interpretation. Vernazza *et al.* (1975) concluded from Skylab data that no periodic oscillations in intensity were present. Athay and White (1979b), however, find evidence of regular oscillations in both intensity and Doppler shift in the C IV line at λ1548, with periods of 200 to 330 s in about 20% of the OSO 8 time series. Both results may be somewhat misleading. The transition region lines fluctuate in intensity and wavelength more or less steadily but without a well-defined period. Such fluctuations are present essentially all of the time, and the mean time between intensity pulses is close to five minutes. This suggests that the oscillations may be present at all times, but that their periodicity has been destroyed as a result of their propagation through the chromosphere. The expected transit time through the chromosphere for sound waves is of the order of four minutes, and changes in transit time of 10% between two wave pulses with a period of five minutes would shift the phase between the two pulses by about 30°. Such changes are plausible and are sufficient to destroy the periodicity. It may be, therefore, as suggested by Athay and White (1979b), that the main difference in the character of the oscillations at different levels in the chromosphere and transition region is in the progressive loss of periodicity as the waves propagate upwards through the complex chromosphere fine structure.

Velocity amplitudes of the chromospheric and transition region oscillations (and "quasi-oscillations") are of the order of 2 to 3 km s^{-1}. This is well above the velocity amplitude in the photosphere; but it is well below the velocity amplitude expected for adiabatic waves.

Tsubaki (1977) finds some evidence for low-amplitude periodic oscillations near 300 s in Doppler shifts of coronal lines observed in an active region at the limb. Similar evidence was not found for fluctuations in line intensity. Velocity amplitudes in the oscillations observed are of the order of 1 km s^{-1}. To the author's knowledge, no attempts have been made to study oscillations in coronal lines observed on the solar disk.

In both the photosphere and chromosphere, the velocity amplitude of the oscillations decreases from center to limb, with very little residual velocity amplitude at the limb. This indicates quite unambiguously that the oscillations are essentially radial.

4.6 NONTHERMAL LINE WIDTHS

Nonthermal line width is defined as that portion of the Doppler width of a line that is not accounted for by thermal and saturation broadening. It is assumed that the nonthermal line width is due to unresolved fluid motions in the line-forming layers, and the amplitude of the unresolved motion is derived by adopting a model for the influence of the motion on the line width. Most analyses invoke the assumption of micro- and/or macroturbulence. In the case of microturbulence, one simply broadens the absorption coefficient according to the relation

$$\Delta\lambda_D = \frac{\lambda}{c}\left(\frac{2kT}{m_A} + \zeta^2\right)^{1/2} , \qquad (4.6.1)$$

where $\Delta\lambda_D$ is the Gaussian width, m_A is the mass of the atom, and ζ is the rms velocity amplitude. Under the macro assumption, one computes a profile, using only thermal broadening, then smears the profile with a Gaussian broadening function, i.e.,

$$I'(\lambda) = \int_{-\infty}^{\infty} I(\lambda - \Delta\lambda) f(\Delta\lambda) d\Delta\lambda \quad , \qquad (4.6.2)$$

where $f(\Delta\lambda)$ is the broadening function usually assumed to be Gaussian. The macro assumption allows the line-of-sight velocity component V to be a function of x and y but not z, whereas the micro assumption requires that the average value of V_z ($<V_z>$) is independent of x and y and that V_z fluctuates about $<V_z>$ on a scale that is small compared to $\Delta\tau_\lambda = 1$, where τ_λ is the optical depth at wavelength λ. It is assumed, also, that local averages of V_z, $\overline{V}_z$, taken over intervals in z corresponding to $\Delta\tau_\lambda \approx 1$ are independent of z. In other words, all nonthermal motions occur on a scale that is small compared to $\Delta\tau_\lambda = 1$.

In the real sun, of course, the scales of motion are not small compared to $\Delta\tau_\lambda = 1$, and, in general, $\overline{V}_z$ is a function of z as well as x and y. This produces effects in the line profiles that mimic a combination of micro and macroturbulence; i.e., the lines broaden as well as Doppler shift, and, in addition, the lines are usually asymmetric.

Recent work on radiation transfer in the presence of "mesoturbulence," which allows $\overline{V}_z$ to vary with z as well as with x and y, offers a promise for improved descriptions of the velocity field. Applications to solar line profiles have not yet been made, however.

In the case of optically thin lines, the microturbulence and macroturbulence assumptions are equivalent. What is required in this case is that fluctuations in V_z occur either on a scale in the x,y plane that is small compared to the largest aperture dimension Δx or Δy, or on a scale in the z direction that is small compared to the thickness Δz of the emitting region. In a layer where Δz is small, such as the transition region, observations with sufficiently high spatial resolution should show some evidence of thermally broadened profiles. However, in geometrically thick layers, such as the photosphere and chromosphere, one should not expect to observe thermally broadened profiles even with perfect spatial resolution.

Two commonly used schemes for determining the nonthermal components of line width are spectrum synthesis of optically thick lines and analysis of optically thin lines. The latter scheme is much simpler to apply and is the most commonly used. It is based on a direct application of Eq. 4.6.1 and requires only knowledge of the temperature T where the line is formed. The biggest disadvantage is that a given spectral line may be associated with a considerable range in T, and all that one obtains from the analysis is a rather ill-defined correlation between T and ζ. All depth information in the z direction is lost except that which can be inferred from an independent relationship between T and z.

The spectrum synthesis method requires an explicit relationship between T, ζ, ϱ, and τ, therefore it contains the required depth information. However, the method is very laborious and is often nonunique, in that a family of relationships between T, ζ, ϱ, and τ fits the data. By using data for many lines, the family of solutions, in principle, can be narrowed as much as desired. However, one then encounters a new source of nonuniqueness arising from the inhomogeneous nature of the solar atmosphere. Also, so long as one is using crude descriptions of the velocity field in terms of the micro and macro concepts, there is little point in conducting elaborate analyses for their evaluation.

Before proceeding further with the analysis of nonthermal line widths, it is helpful to review briefly what is known about the nature of the macroscopic motions and to what extent they may contribute to the line width. On the scale of 1″ (725 km), the sun is highly structured. The density scale height in the chromosphere, for example, is approximately 0.2″, and chromospheric fine structure exists on scales down to less than 1″. The entire chromosphere is approximately 2″ thick, except for local extensions due to fine structure. Individual features associated with the fine structure, e.g., fibrils and spicules, have associated flows of several km s^{-1}. One can infer from this that the velocity gradients associated with fine structure are typically a few km s^{-1} per arcsecond.

Waves are present also. A wave propagating at the sound speed with a period of 100 s has a wavelength of about 1″ in the chromosphere. Such waves can also contribute velocity gradients of the order of a few km s^{-1} per arcsecond.

Most of the data used to infer nonthermal line widths has spatial resolution of several arcseconds. It is clear, therefore, that much of the nonthermal line width can be attributed to waves and to motions associated with fine structure. Furthermore, it is clear that such motions will produce an appearance of both micro- and macroturbulence. Whether any additional motion is needed to account for the observed line widths is debatable.

The outstanding features of nonthermal line width (hereafter referred to simply as microturbulence) about which there is general agreement can be summarized as follows: (1) microturbulence increases with height from the temperature minimum to the transition region; (2) microturbulence in the chromosphere and transition region is roughly isotropic; and (3) there is no pronounced change in microturbulence that is typical of active regions as compared to quiet regions. These broad features of the microturbulence account for the limb broadening of optically thick lines and the tendency for little or no limb broadening of optically thin lines.

Amplitudes of microturbulence found in the literature are often discordant, due in part to different methods of analysis and different data. Much of the discordance is due to the spatial and temporal fluctuations in line widths that persist even for appreciable areas on the sun and appreciable time intervals. The best high spatial resolution data are discussed by Brueckner in a separate chapter. In the following, we review the temporal and spatial fluctuations in width obtained from OSO 8 time series data. These data represent nearly 400 hours of observing and nearly 40,000 individual profile scans for lines of Si II at λ1816.93 and λ1817.45 and a line of C IV at λ1548. The Si II lines are formed in the middle and low chromosphere, where $T \lesssim 7000$ K, and the C IV line is formed in the lower transition region, where $T \approx 100{,}000$ K. All observations were made with fixed pointing on the solar disk, using an effective area of 2″ × 20″ and time resolution ranging from 15 s to 28 s for Si II, and from 22 s to 49 s for C IV. Details of the analysis of the time series data are contained in a paper by Athay and White (1980).

Figure 4.5 shows 30-minute time series for W_1, together with the corresponding time series for A, I_0, and M_1 for two experiments for C IV. In experiment 10378, made in a plage region, a regular oscillation is present in the time series for A and I_0, with a period somewhat over five minutes. During the oscillation, W_1 has a mean value of 400 mÅ and fluctuates through a range of approximately 300 mÅ, with a tendency for larger widths to correlate with increased intensity. Such fluctuations are not unusual. The average fluctuation range in a 30-minute interval for all of the plage data is $\Delta W_1/\langle W_1 \rangle \approx 0.3$. The extremes of $\Delta W_1/\langle W_1 \rangle$ are 0.2 and 1.0.

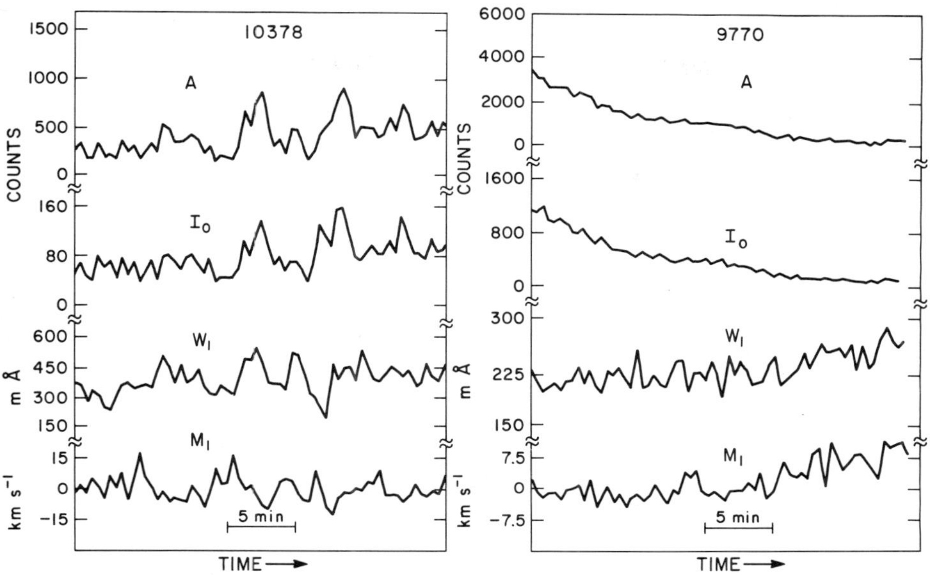

Figure 4.5. Time series (30-min) for four line parameters for C IV with time resolution of 27.5 s. Note short-term changes in line width for experiment 10378 (small-amplitude point-to-point fluctuations are due mostly to measurement noise), and long-term changes in experiment 9770. (The segment of experiment 9770 shown in Figure 4.2 occurs earlier than the segment shown here.)

The characteristic time for fluctuations in W_1 for C IV is a few minutes; i.e., most of the range in W_1 tends to occur in time intervals of five minutes or less. Experiment 9770 shown in Figure 4.5 is an exception. In this 30-minute interval, during the decay phase of a small flare, the intensity decreases by about a factor of 20, and the line width increases from about 200 mÅ to 275 mÅ. The line is unusually narrow during the flare and gradually comes back to a normal value as the flare subsides. Gradual changes of W_1 of this type are also relatively common.

Figure 4.6 shows distributions of $<W_1>$ for C IV averaged over 30-minute intervals for three classes of data: network, faint plage/bright network, and bright plage. The average width decreases somewhat as the intensity increases. The broad wings on the distributions reflect, in part, the long-period temporal changes in width at a given location as well as the differences in width from place to place in features of equal brightness. Note that the extreme range in $<W_1>$ is nearly 200 mÅ, compared to an average value for the ensemble of 285 mÅ, i.e., $\Delta<W_1>/\ll W_1\gg \approx 0.7$. Note, also, that most of the values of $<W_1>$ fall in the narrow central cores of the distributions, which reflects the fact that the dominant fluctuations have periods much less than 30 minutes.

C IV is an optically thin line formed in a shallow layer. The width of the line, therefore, is subject to fluctuations associated with any small-scale feature within the field of view. This probably accounts for the short characteristic time for the fluctuations and implies that the small-scale fluctua-

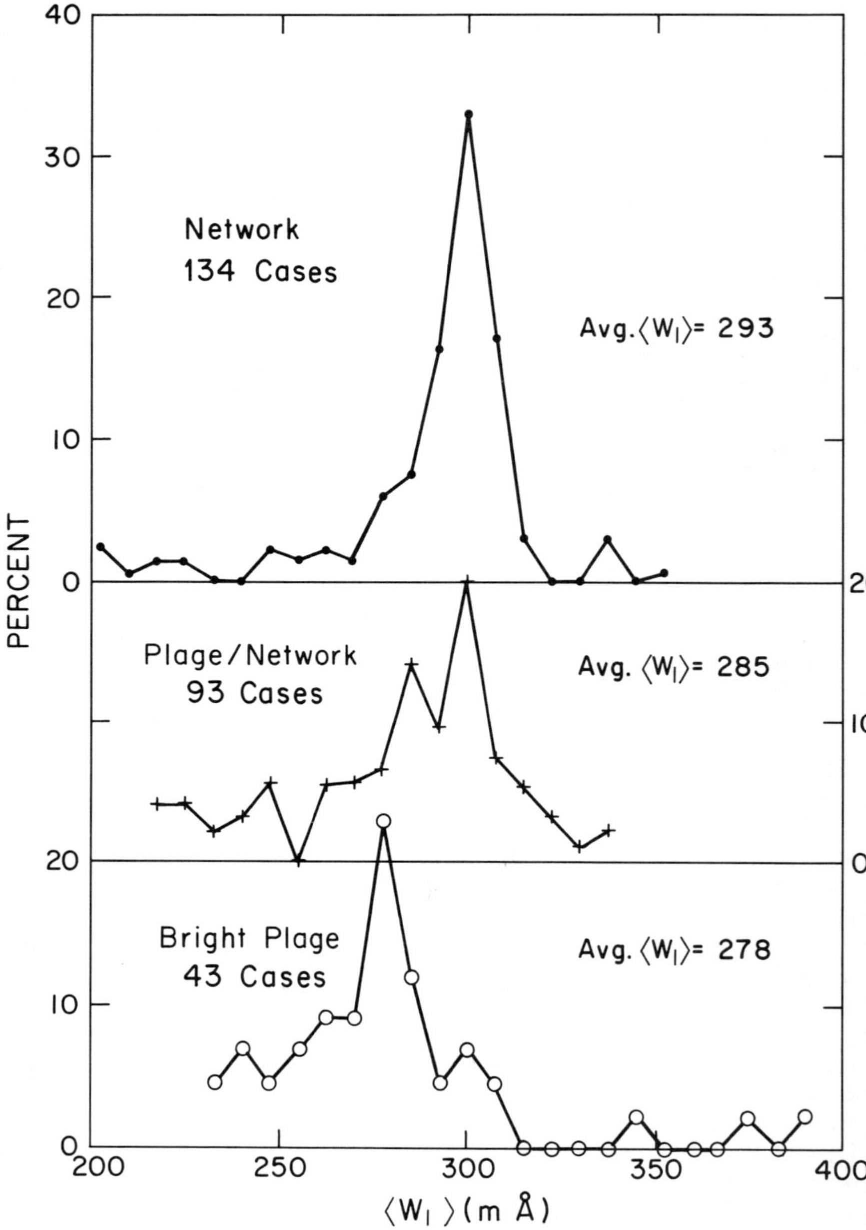

Figure 4.6. Distributions of temporal averages $<W_1>$ for C IV for network and plages.

tions are even larger than those observed. This is confirmed in the NRL results.

Figure 4.7 shows time series for the full width at half maximum, W_2, for the Si II line at λ1816.93 for given time series. The line strength *A* is plotted for one of the experiments (7713). It is readily apparent from these time series that the fluctuations in W_2 have a longer characteristic time scale than those for C IV. The relative ranges in width, $\Delta W_2/<W_2>$, during the 30-minute series, are very nearly the same as for C IV, however. In general, there is little correlation between changes in width and line strength for Si II. Also, there is only a weak correlation between changes in widths of the two lines of Si II, even though the two lines are members of the same multiplet. This lack of correlation between the two lines suggests that the changes in width are the result of local motions and not merely increased saturation. The two lines differ in optical depth by a factor of 10 and are formed approximately 300 km apart in the chromosphere (Athay and White, 1979a).

Distributions of $<W_2>$ for λ1816.93 for different classes of solar features are shown in Figure 4.8. We now find a substantial increase in average width with increasing line strength. Also, the wings of the distribution are again

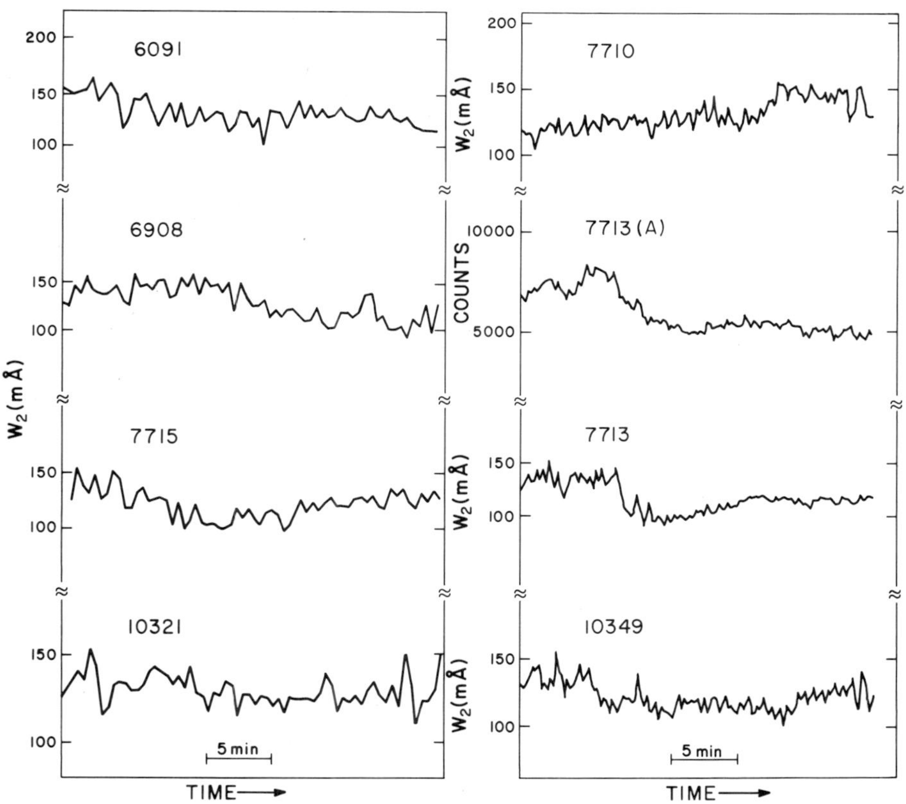

Figure 4.7. Time series (30-min) of line width for λ1816.93 of Si II (a time series for line strength A *is included for experiment 7713). Note the tendency for long-term changes in line width.*

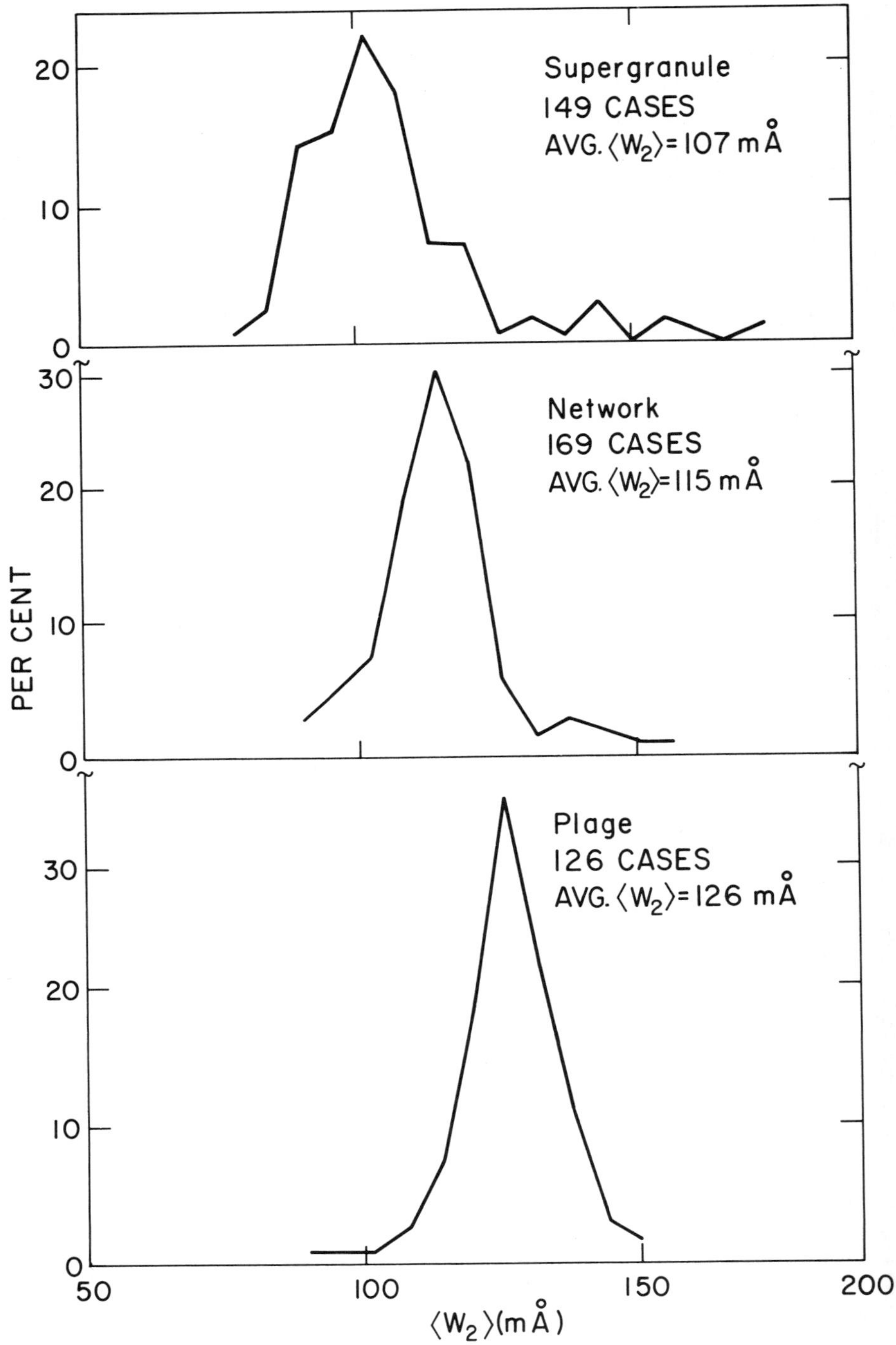

Figure 4.8. Distributions of temporal averages $<W_2>$ for λ1816.93 of Si II for supergranule cells, network, and plages.

broad, with an extreme range in $<W_2>$ of 110 mÅ, corresponding to $\Delta<W_2>/\ll W_2\gg \approx 0.95$. This is considerably larger than the value found for C IV. In addition, the central cores of the distribution for $<W_2>$ for Si II are considerably broader than those for C IV. From the half width $\Delta(<W_2>)_{1/2}$ of the distribution in Figure 4.8 for network and plages, we find $\Delta(<W_2>)_{1/2}/\ll W_2\gg \approx 0.14$, whereas the corresponding quantity for $<W_1>$ from Figure 4.6 is approximately 0.06. The broader cores in the Si II distributions confirm the impression that the characteristic time for fluctuations in W_2 is not short compared to 30 minutes.

The other Si II line at λ1817.45 does not show a pronounced increase in width with increasing line strength. In fact, $\ll W_2\gg$ is very nearly constant from quiet sun to plage. Also, the line broadens only modestly from center to limb, whereas λ1816.93 broadens quite strongly.

The Si II lines are optically thick, which means that the line widths are determined by a combination of motion and saturation. Also, they are formed in regions of the atmosphere where the nonthermal motions are increasing rapidly with height. It is most likely the increase of ξ with height that accounts for the increased width of λ1816.93 in plages and near the limb; i.e., we attribute the increases in $<W_2>$ to increases in the height at which the lines are formed.

A further consequence of the Si II lines being optically thick is that the line width is not strongly influenced by local changes associated with individual elements of the chromospheric fine structure. Instead, the width is formed by cumulative effects over a large height range, and only large-scale fluctuations are observed. This explains the relative absence of short-term fluctuations in line width compared to the optically thin line of C IV.

The picture that emerges from these data is that line width is a rather elusive parameter. Observations made at a given time and for a given location on the sun have a good probability of differing substantially from the mean. Similarly, center-to-limb variations of widths determined by sampling small areas of the disk for short time periods may give very atypical results. It is not surprising, therefore, that published line widths and their center-limb variations are sometimes discordant.

Although the nonthermal line width has an uncertain interpretation, the amplitude of the inferred velocity is so large that it commands our interest. A plot combining different determinations of the nonthermal line width in the chromosphere and transition region from a variety of sources is shown in Figure 4.9. The results from Feldman *et al.* (1976) include a quiet sun region and a coronal hole, both observed at the limb. The remaining results, except for the point from Billings, are quiet sun values extracted from disk data. In plotting the results from Feldman *et al.*, we have adopted their assigned temperatures, except for Si II, where we reduced the temperature of line formation to 20,000 K as suggested by Tripp *et al.* (1978).

The solid curve in Figure 4.9 is the sound velocity computed from the Vernazza *et al.* (1973) model of the chromosphere. Near the temperature minimum ($\log T = 3.6$), the nonthermal velocity is less than 1 km s^{-1}. However, in the middle chromosphere, some few hundred km higher, ζ has increased to approximately the sound speed V_s and remains near the sound

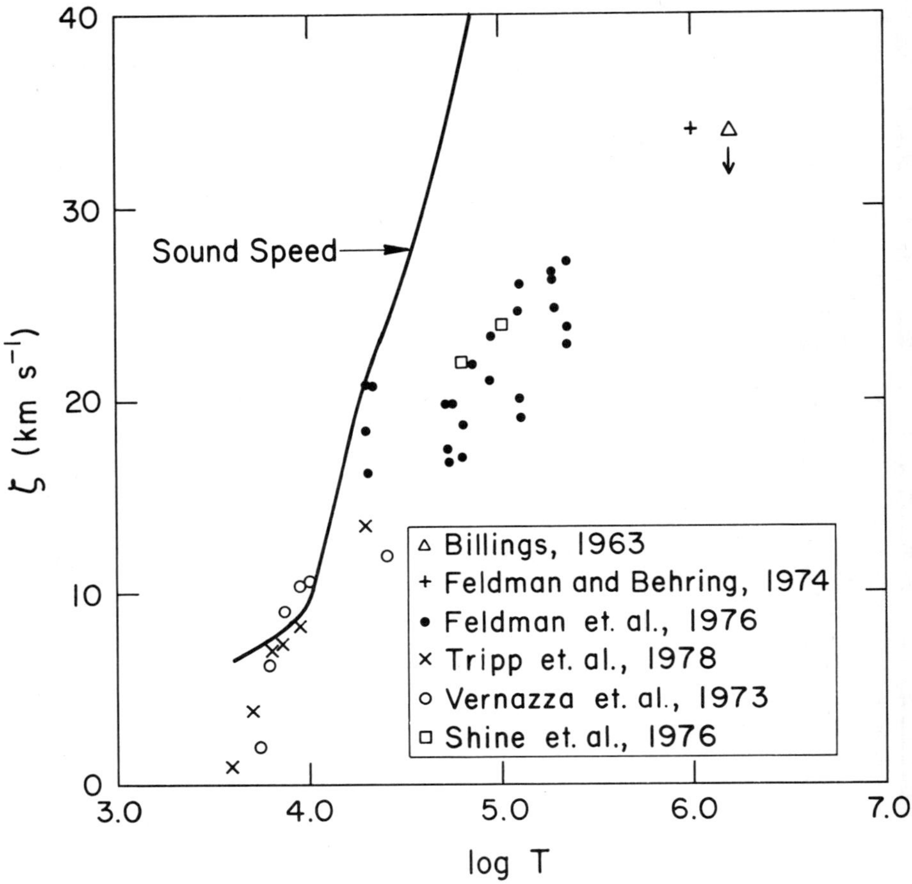

Figure 4.9. Correlation plot between nonthermal line width and temperature from several sources. (The curve is the computed sound speed.)

speed throughout most of the remainder of the chromosphere. Near the base of the transition region, V_s rises sharply due to the rapid rise in temperature and because of the rapid decrease in mean molecular weight caused by the ionization of hydrogen and helium. The nonthermal line width increases in the transition region, but not in proportion to V_s. At 10^5 K, $\zeta \approx 1/2V_s$; and at 10^6 K, $\zeta \approx 1/5V_s$. The value of ζ plotted at log $T = 6.2$ was obtained by interpreting all of the broadening of the λ5303 line of Fe XIV given by Billings (1963) as being due to microturbulence. If thermal broadening at $T = 1.6 \times 10^6$ is included, ζ is decreased to 20 km s^{-1}.

We do not yet have the required data to construct a plot corresponding to Figure 4.9 for active regions. However, from what data are available, there is little reason to conclude that the plot would differ substantially from Figure 4.9.

Clearly, there is much to learn about motions in the solar atmosphere from a proper interpretation of nonthermal line widths. The high spatial resolution data provided by the NRL rocket instruments and the time series data provided by OSO 8 point the way to still further improvements in the observations. New insights provided by these observations already suggest that at least a major portion of the line broadening is closely related to the streaming velocities of structural features and to wave motions of the type known to exist in the solar atmosphere.

REFERENCES

Ando, H. and Osaki, Y.: 1975, *Pub. Ast. Soc. Japan* **27,** 581.

Athay, R. G.: 1976, *The Solar Chromosphere and Corona: Quiet Sun*, D. Reidel, Dordrecht, Holland.

Athay, R. G. and White, O. R.: 1979a, *Astrophys. J. Suppl.* **39,** 333.

Athay, R. G. and White, O. R.: 1979b, *Astrophys. J.* **229,** 1147.

Athay, R. G. and White, O. R.: 1980, *Astrophys. J.* (in press).

Athay, R. G., White, O. R., Lites, B. W., and Bruner, E. C., Jr.: 1980, *Solar Phys.* (in press).

Beckers, J. M.: 1968, *Solar Phys.* **3,** 367.

Beckers, J. M.: 1977, *Astrophys. J.* **213,** 900.

Beckers, J. M. and Canfield, R. C.: 1976, *CNRS Colloquium,* Number 250, 207.

Beckers, J. M. and Schultz, R. B.: 1972, *Solar Phys.* **27,** 61.

Bhatnagar, A., Livingston, W. C., and Harvey, J. W.: 1972, *Solar Phys.* **27,** 80.

Billings, D. E.: 1963, *A Guide to the Solar Corona,* Academic Press, New York.

Bones, J. and Maltby, P.: 1978, *Solar Phys.* **57,** 65.

Bruner, E. C., Jr., Chipman, E. G., Lites, B. W., Rottman, G. J., Shine, R. A., Athay, R. G., and White, O. R.: 1976, *Astrophys. J. Lett.* **210,** L97.

Bruner, E. C., Jr. and Lites, B. W.: 1979, *Astrophys. J.* (in press).

Bruzek, A.: 1968, in K. O. Kiepenheuer (ed.), 'Structure and Development of Active Regions', *IAU Symp.* **35,** Budapest, Hungary, 4–8 September 1967, D. Reidel, Dordrecht, Holland.

Cox, D. P. and Tucker, W. H.: 1969, *Astrophys. J.* **157,** 1157.

Cram, L. E., Keil, S. L., and Ulmschneider, P.: 1980, *Astrophys. J.* (in press).

Deubner, F. L.: 1976, in R. M. Bonnet and P. Delache (eds.), 'Energy Balance and Hydrodynamics of the Solar Chromosphere and Corona', *IAU Symp.* **36,** G. de Bussac, Clermont-Ferrand.

Emslie, A. G. and Noyes, R. W.: 1978, *Solar Phys.* **57,** 373.

Feldman, V. and Behring, W. E.: 1974, *Astrophys. J. Lett.* **189,** L45.

Feldman, V., Doschek, G. A., van Hoosier, M. E., and Purcell, J. D.: 1976, *Astrophys. J. Suppl.* **31,** 445.

Foukal, P.: 1976, *Astrophys. J.* **210,** 575.

Giovanelli, R. G.: 1972, *Solar Phys.* **27,** 71.

Giovanelli, R. G.: 1975, *Solar Phys.* **44,** 299.

Giovanelli, R. G. and McCabe, M.: 1958, *Aust. J. Phys.* **11,** 91.

Gopasyuk, S. I.: 1969, *Izv. Krymsk Astrofiz. Obs.* **40,** 111.

Kawaguchi, I. and Kitai, R.: 1976, *Solar Phys.* **46,** 125.

Lamb, S. A.: 1975, *Mon. Not. Roy. Ast. Soc.* **172,** 205.

LeMaire, P., Skumanich, A., Artzner, G., Gouttebroze, P., Vial, J. C., Bonnet, R. M., and McWhirter, P.: 1978, *Bull. Am. Astr. Soc.* **10,** 441.

Levine, R. H. and Withbroe, G. L.: 1977, *Solar Phys.* **51,** 83.

Lites, B. W.: 1980 (submitted).

Lites, B. W. and Bruner, E. C., Jr.: 1980 (in preparation).

Lites, B. W. and Hansen, E. R.: 1977, *OSO 8 Workshop,* University of Colorado, Boulder.

Lites, B. W., White, O. R., Chipman, E. G., and Athay, R. G.: 1979 (in preparation).
MacQueen, R. M.: 1979 (in press).
Maltby, P.: 1975, *Solar Phys.* **43,** 81.
Mattig, W. and Mehltretter, J. P.: 1968, in K. O. Kiepenheuer (ed.), 'Structure and Development of Active Regions', D. Reidel, Dordrecht, Holland.
Menzel, D. H. and Evans, J. W.: 1953, *Accad. Naz. Lincci. Convegno Volta* **11,** 119.
Moore, R. L. and Tang, F.: 1975, *Solar Phys.* **41,** 21.
November, L. J., Toomre, J., Gebbie, K. B., and Simon, G. W.: 1979 (in press).
Phyllis, G. L.: 1975, *Solar Phys.* **41,** 71.
Pye, J. D., Evans, K. D., Hutcheon, R. J., Gerasemenko, M., Davis, J. M., Krieger, A. S., and Vesecky, J. F.: 1978, *Astron. and Astrophys.* **66,** 123.
Schröter, E. H.: 1965, in V. Xanthakis (ed.), *Solar Physics,* Interscience, London, p. 325.
Sheeley, N. R., Jr.: 1969, *Solar Phys.* **9,** 347.
Sheeley, N. R., Jr.: 1972, *Solar Phys.* **25,** 98.
Shine, R. A., Roussel-Dupre, D., Bruner, E. C., Jr., Chipman, E. G., Lites, B. W., Rottman, G. J., Athay, R. G., and White, O. R.: 1976, *Astrophys. J. Lett.* **210,** L107.
Smith, H. J. and Smith, E. P.: 1963, *Solar Flares,* MacMillan, New York.
Soltau, D., Schröter, and Wohl, H.: 1976, *Astron. and Astrophys.* **50,** 367.
Tandberg-Hanssen, E.: 1974, *Solar Prominences,* D. Reidel, Dordrecht, Holland.
Tripp, D. A., Athay, R. G., and Peterson, V. L.: 1978, *Astrophys. J.* **220,** 314.
Tsubaki, T.: 1977, *Solar Phys.* **51,** 121.
Vernazza, J. E., Avrett, E. H., and Loeser, R.: 1973, *Astrophys. J.* **184,** 605.
Vernazza, J. E., Foukal, P. V., Huber, M. C. E., Noyes, R. W., Reeves, E. M., Schmidt, E. J., Timothy, J. G., and Withbroe, G. L.: 1975, *Astrophys. J. Lett* **199,** L123.
Vrabec, D.: 1974, in R. G. Athay (ed.), *Chromosphere Fine Structure,* D. Reidel, Dordrecht, Holland, p. 201.
White, O. R. and Athay, R. G.: 1979a, *Astrophys. J. Suppl.* **39,** 317.
White, O. R. and Athay, R. G.: 1979b, *Astrophys. J. Suppl.* **39,** 347.
Zirin, H. and Stein, A.: 1972, *Astrophys. J. Lett.* **178,** L85.

5. THE DYNAMICS OF ACTIVE REGIONS

G. E. Brueckner

5.1 INTRODUCTION

This chapter summarizes observations of the dynamics of active regions obtained by the Naval Research Laboratory's High Resolution Telescope and Spectrograph (HRTS) during two rocket flights in 1975 and 1978. The spectra obtained during these flights revealed a new view of the dynamic behavior of the chromosphere, transition zone, and corona of an active region, because of their high spatial resolution and simultaneously wide coverage of different temperature regimes. However, only a few examples of dynamic events in active regions can be described, because the spatial coverage of active regions obtained during these two sounding rocket flights was limited.

The general appearance of the spectra is discussed in Section 5.2. Chromospheric and transition-zone flows in sunspots and plages are discussed in Section 5.3. A clear distinction can be made between "flows" and "explosive events." Coronal flow patterns above an active region are given in Section 5.4. Possible consequences of the observed flow pattern on the formation of spectral lines are discussed in Section 5.5. The behavior of the He II 1640 Å line, which is different from that of all other lines, is described in Section 5.6. Finally, we try in the last section (5.7) to evaluate qualitatively the mass and energy balance of an active region.

5.2 THE SPECTRA

The ultraviolet region of the spectrum between 1175 and 1710 Å contains continua and lines covering a wide temperature regime (4×10^3 K to 1×10^7 K). The use of normal incidence optics allows the design of instruments that have simultaneously high spatial and spectral resolution, large spatial and spectral coverage, as well as high time resolution. The photographic recording covers a wide dynamic range, making it possible to observe on the same plate strong transition-zone emission lines and weak low-chromospheric lines, in order to establish a wavelength reference system for the transition-zone and coronal lines. Without such a reference system, all Doppler measurements are ambiguous because of systematic flows in the quiet sun (as will be described elsewhere: Brueckner, 1979).

The HRTS instrument has a spatial resolution of 1 arcsec along the 1000 arcsec–long slit and a spectral resolution of 0.06 Å over the entire spectral region from 1175 to 1710 Å. Exposure times varied from 0.1 s (for H Lyα) to 50 s (for weak coronal lines). Pointing stability during the first flight was better than the resolution of the telescope during a single exposure but insufficient over the duration of the rocket flight to obtain spectral time sequences. For the second flight, a fine-pointing system was added, which kept the spectrograph slit positioned on the solar surface within a fraction of the slit width for the duration of the rocket flight (4.5 minutes).

An Hα slit jaw television pointing system allowed the initial positioning of the spectrograph slit in real time. During the first and the second flight, the slit was pointed through active regions McMath 13766 and McMath 15139, respectively. In both cases, a sunspot and adjacent plage areas were covered. However, only a small fraction of the plage and sunspot areas could be observed. It is therefore not possible to generalize the results until more spatial coverage is available.

The format of the entire spectrum together with an on-band Hα slit jaw image is shown in Figure 5.1. An active region (McMath 13766) is located in the lower part of the spectrum. It was an old region with a single sunspot, and it did not exhibit any significant activity.

Slit jaw pictures in Hα and in the 1600 Å ultraviolet continuum (bandpass 40 Å) together with a small portion of the spectrum of McMath 15139 (covered during the second HRTS flight on 13 February 1978) are shown in Figure 5.2. This region had a rapidly developing, complex magnetic configuration. At 0150 UT, 15.5 hours prior to the rocket flight, a class 1B flare started in this region. X-ray emission of a small flare was observed from OSO 8 during the rocket flight, and the presence of Fe XXI emission (1354 Å, see Figure 5.2) on the spectra may be connected with this flare. However, there was no report of an Hα flare during the rocket flight.

5.3 FLOWS AND EXPLOSIVE EVENTS

Figure 5.3 shows a time sequence of active region spectra in C I (1656 Å) and C IV (1550 Å). In C IV, a continuum of downflow velocities from 0 to 150 km s^{-1} can be seen over the sunspot at *b,* where the maximum of the emission measure and velocity is observed over the "light bridge" (an area of the sunspot with strong enhancement of chromospheric and transition-zone lines). The velocities persist for the duration of the flight (260 s), and no significant change in intensity, velocity, or position is observed. The Doppler velocities are interpreted as flows confined in magnetic fields. A confinement of the streaming gas is required because there is no observation of significant changes in the pattern for 20 s, during which material would travel 3000 km. Although systematic downflows over the sunspot are the more likely explanation of the observed Doppler shifts, the possibility cannot be ruled out that some type of very fast magnetoacoustic wave causes systematic red-shifts.

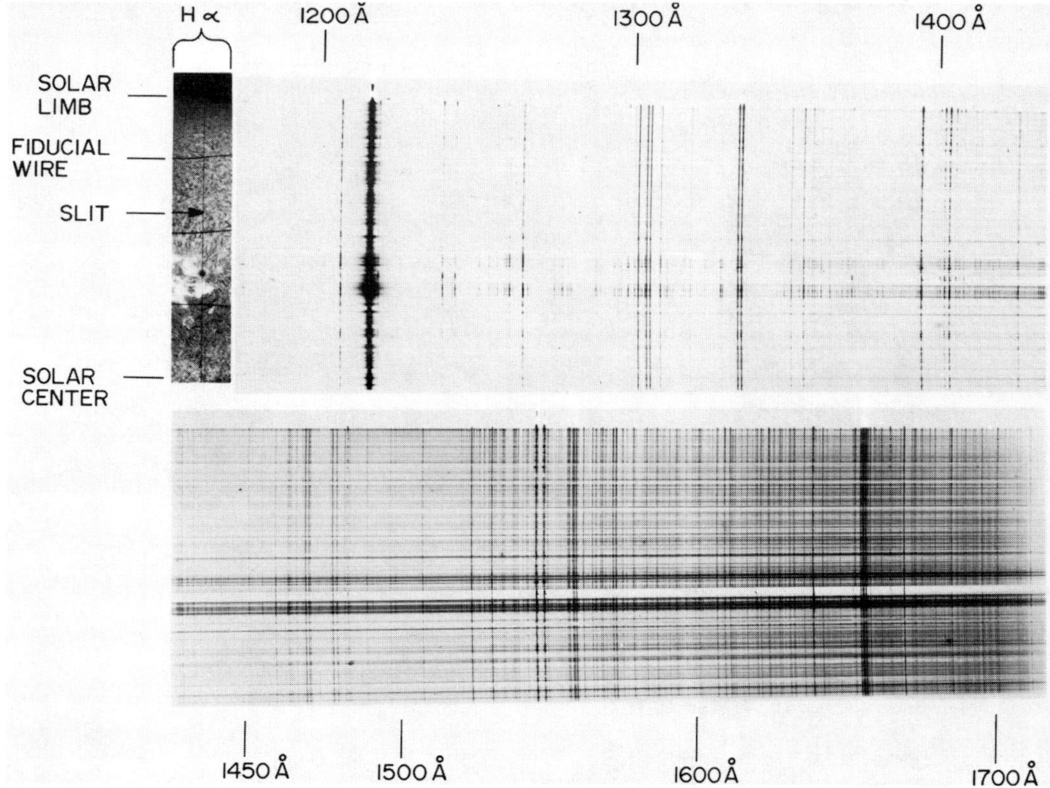

Figure 5.1. HRTS spectrum (20-s exposure) obtained during the first flight, 23 July 1975. Sun-center is at the bottom, limb is at the top. Hα(0.5 Å) slit jaw picture is to the left. (For convenience of display, the spectrum was cut into two sections.) The McMath 13766 active region is located in the lower half of the spectrum.

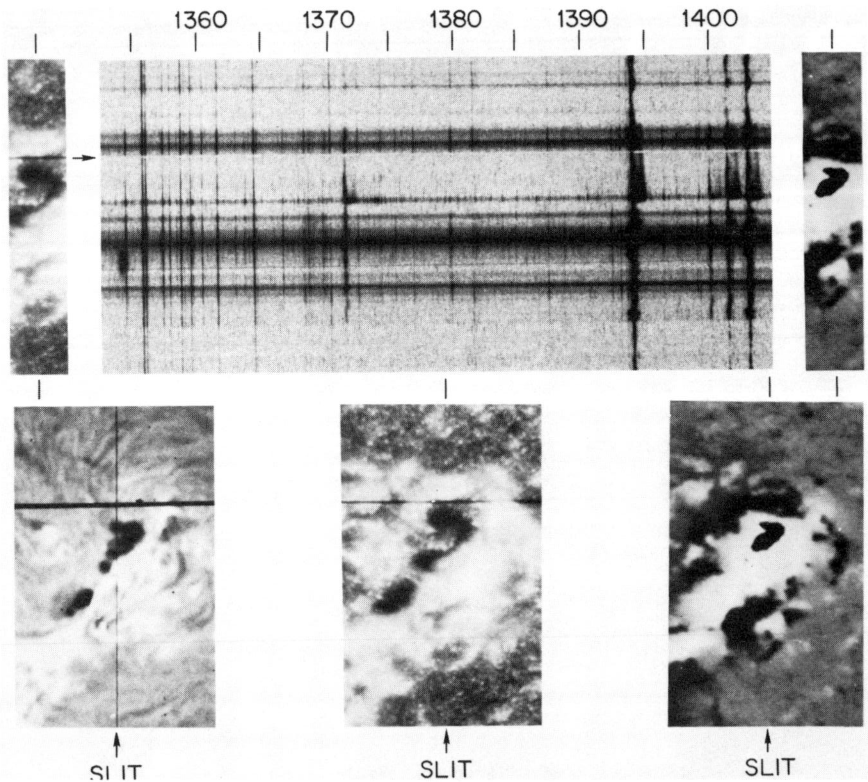

Figure 5.2. Top: *Spectrum of the active region McMath 15139 from 1355 Å to 1400 Å. The spectrum covers the Fe XXI flare line at 1354 Å.* Bottom: *The active region*—left: *Hα (bandpass 0.5 Å);* center: *ultraviolet continuum at 1600 Å (bandpass 40 Å);* and right: *magnetogram. The spatial scales of the spectrum and the images are matched. A fiducial wire (FDM), seen in the Hα and the UV continuum picture, can be recognized as a bright streak in the spectrum.*

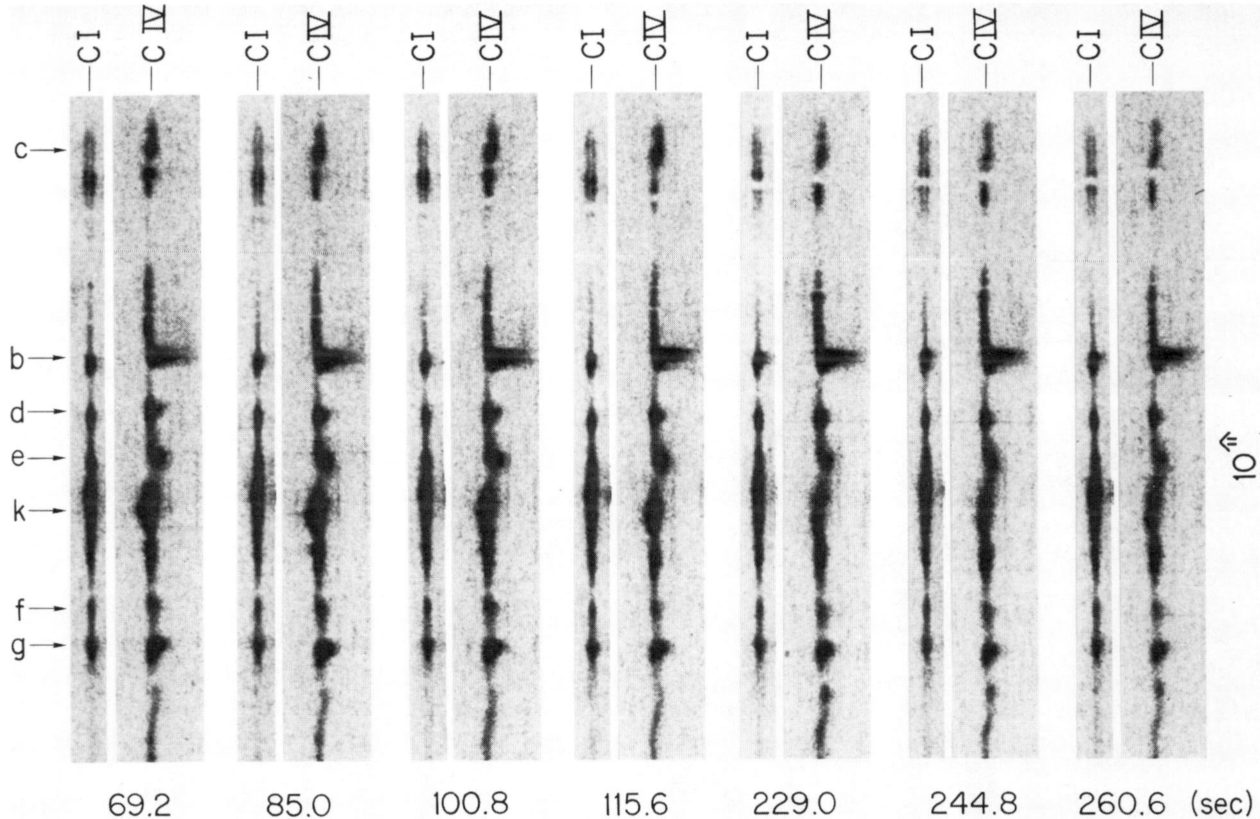

Figure 5.3. Time sequence of C I (1660 Å) and C IV (1550 Å) spectra of active region McMath 15139. T = 0 is equivalent to UT = 16 hrs. 46 min. 16 s, 13 February 1978.

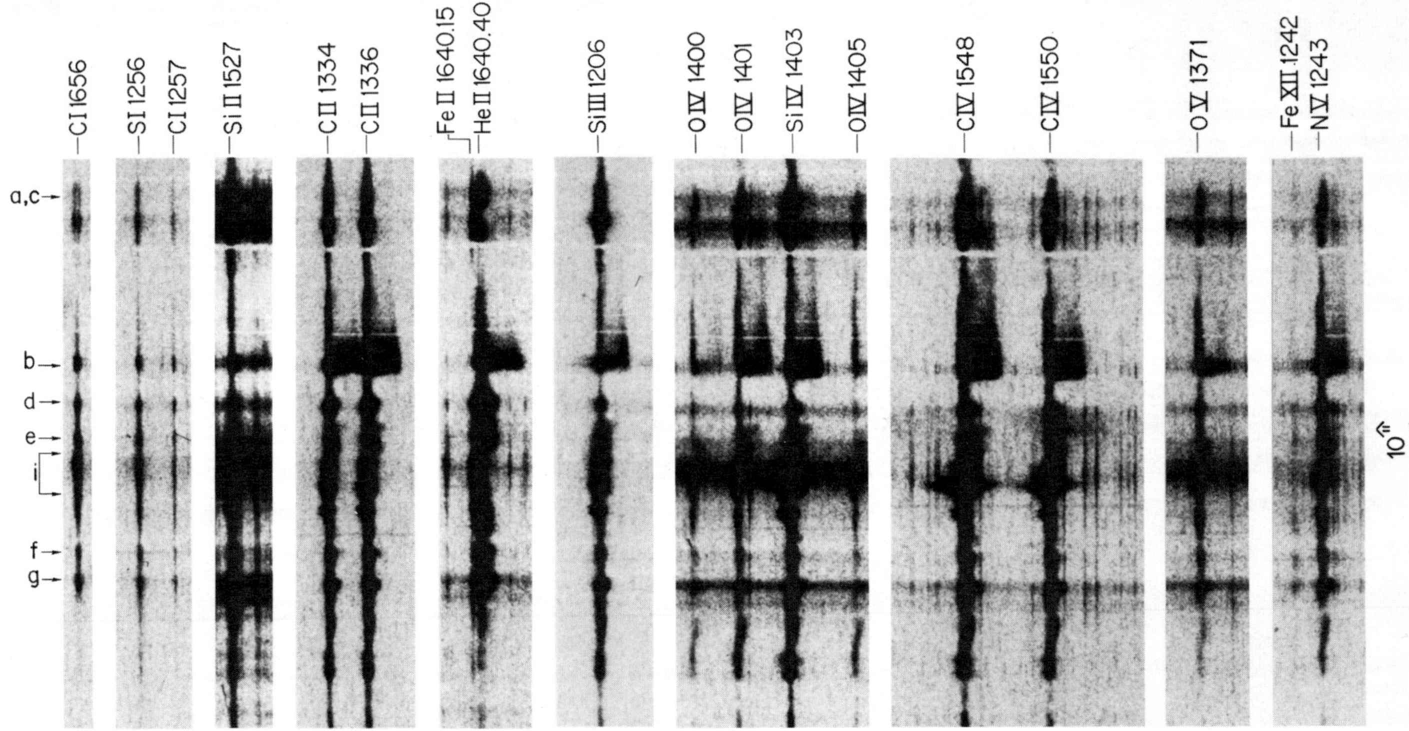

Figure 5.4. Active region McMath 15139 as seen in C I (1656 Å), S I (1256 Å), C I (1257 Å), Si II (1526 Å), C II (1334 Å), He II (1640 Å), Si III (1206 Å), Si IV, O IV (1400 Å), C IV (1548, 1550 Å), O V (1371 Å), N V (1243 Å), and Fe XII (1242 Å, bl. S I).

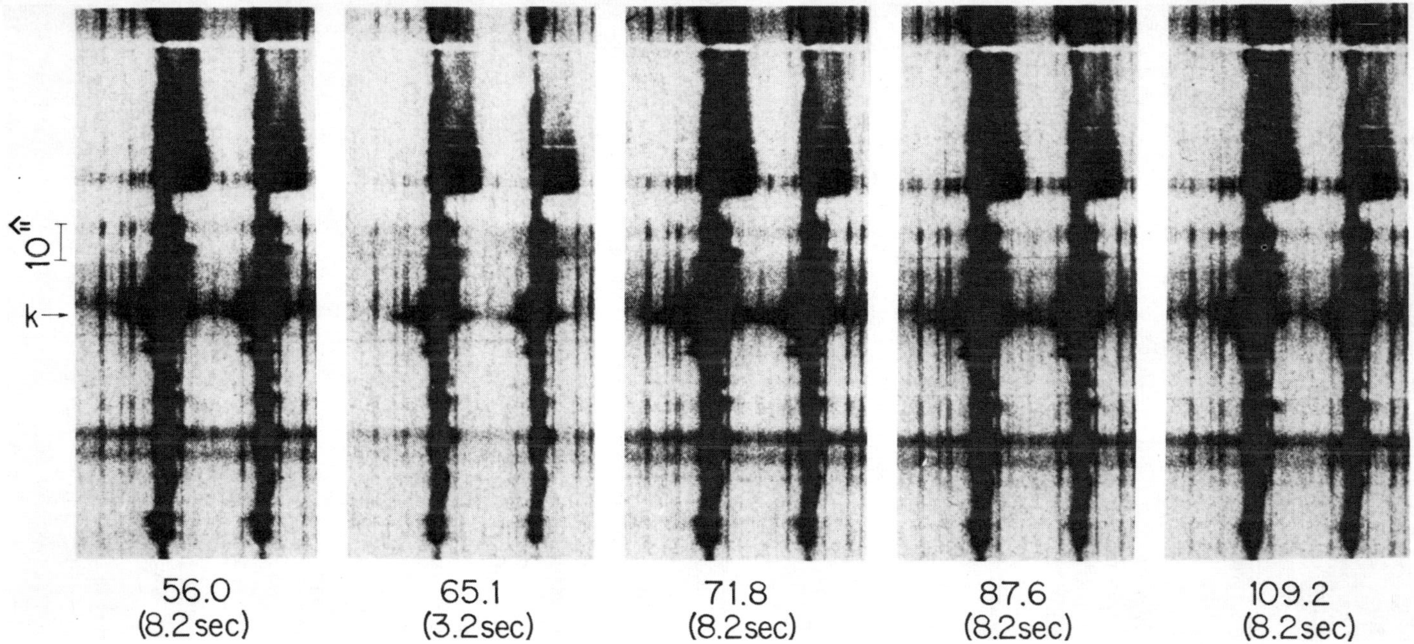

Figure 5.5. Short time sequence of C IV (1550 Å) spectra of an explosive event in McMath 15139. The time scale is the same as in Figure 5.3. (Exposure times are given beneath the images.)

At *c, d, e, f,* and *g* in Figures 5.3 and 5.4, downflows can be observed over the plage. Although there are observable changes in Doppler shifts and intensity with time in these areas, the general flow pattern lasts for the duration of the flight. The downflow velocities at C IV are generally 10–60 km s^{-1} over the plage, confined within areas 3 to 5 arcsec wide.

At 85 s in Figure 5.3, the C IV spectrum shows a broadening at position *k*. A short time sequence of this explosive event is shown in Figure 5.5. The event develops very rapidly within 10 s and lasts for 20 s. Supersonic velocities up to 200 km s^{-1} to the blue and 100 km s^{-1} to the red are observed at the peak of the event. Because of its short duration, we call this feature an "explosive event." Such events are common in the quiet sun; however, their velocities there are generally lower. Whether they play a significant role in the mass and energy balance of an active region cannot be decided at this point, because only a few events have been observed on the HRTS I and HRTS II spectra. Contrary to the flows, the explosive events have their largest emission measure in lines originating from 80,000 to 100,000 K. They are invisible in spin forbidden transition-zone lines, indicating a high density. They cannot be seen in chromospheric lines.

Flows over plages and sunspots show a differential velocity pattern as a function of electron temperature (T_e). As an example, the measured velocities for several ion species are plotted against T_e in Figure 5.6 for the downflow in the center of the HRTS I sunspot. It is obvious that velocity increases with decreasing temperature in these downflows. If we interpret the measured Doppler velocities over the center of the sunspot as free-falling material, it is possible to assign a height scale to the emission of different ions. (The scale in Figure 5.6 gives the distance in km from a point where the downflow starts, i.e., the top of a loop. The assumption $v = 0$ at the top of the loop is arbitrary. Even if the plasma is "pushed" over the top of the loop with a finite velocity, the results would not change drastically.) Also given in Figure 5.6 is the fractional ionization curve of C IV vs temperature (Jordan, 1969). Half of the C IV emission is emitted over a layer approximately 1200-km thick. From the absolute intensity of the 1548 C IV line as measured in the downfalling plasma, we obtain for ½ of the emission measure $\frac{1}{2} N_e^2 dh = 7.8 \times 10^{25}$ cm^{-5}. Assuming that this emission is generated in a 1200-km-thick layer, the density is $N_e = 8 \times 10^8$ cm^{-3}. The thermal energy of the plasma at 100,000 K is $E = 1.1 \times 10^{-2}$ erg cm^{-3}, while the radiation losses are $R = 5.1 \times 10^{-4}$ erg cm^{-3} s^{-1} ($T_e = 1 \times 10^5$ K, $N_e = 8 \times 10^8$ cm^{-3}). Half of the thermal energy is lost by radiation in the time interval

$$\tau = \tfrac{1}{2} E/R \sim 11 \text{ s}.$$

This is in good agreement with the 14-s fall time of the plasma through the 1200-km-thick layer.

It is therefore reasonable to assume that the downflow area in the center of the sunspot represents free-falling low-density plasma that is cooling by radiation. No additional energy sources or sinks are required. Thermal conduction from the corona does not play a role because of the thickness of

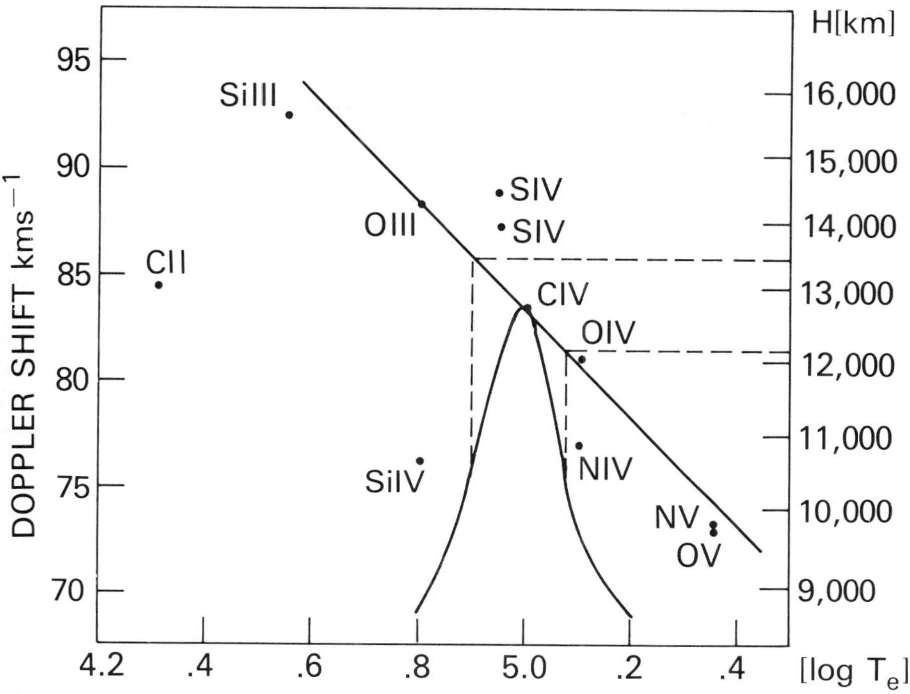

Figure 5.6. Downflow velocities in the center of the umbra of a sunspot in active region McMath 13766 (HRTS I) versus log T_e of the maximum ion abundance of each ion used for the velocity measurement. The scale gives distance in km from a fictitious starting point of the free-falling plasma.

the layer and the very low temperature gradient. Similar values can be derived for downflows over plages.

The emission measure has been derived using the assumption that the C IV line is formed by collisions. However, a revision of this relation for the case of a cooling, recombining plasma is needed.

5.4 VELOCITIES IN THE CORONA ABOVE PLAGES AND SUNSPOTS

The HRTS spectra allow the measurement of coronal Doppler velocities over plages and sunspots for the first time. The Fe XII line at 1349.40 Å has been used. Other coronal lines within the range of the spectrograph (Fe XII 1242.00 Å, Fe XI 1467.06 Å, and Si VIII 1445.75 Å) are less suited for Doppler velocity measurements, because they are blended with chromospheric lines. The shift of the Fe XII line has been determined at half maximum intensity. As a reference system, chromospheric lines of C I at 1315.918, 1328.834, 1354.288, 1357.134, 1357.659, and 1358.184 Å have been used; they were assumed to be at rest wavelength. Therefore, any systematic wavelength shifts in the lower chromosphere would change the observed Fe XII velocities. However, locally the C I lines show only very small Doppler shifts, not exceeding 5 km s^{-1}. The rest wavelength of Fe XII 1349.40 Å was determined by Sandlin *et al.* (1977) from Skylab spectra. The error of the Fe XII rest wavelength value, which was determined from many observations at the limb around the disk, should not exceed ±0.01 Å, or ±2.2 km s^{-1}.

From Figure 5.7, it is obvious that downdrafts are dominant over the plages adjacent to the sunspot. An average downdraft of 13 km s^{-1} exists in the plage areas, while the average Doppler shift over the sunspot is 2.5 km s^{-1} toward the blue. Locally larger excursions can be seen. Large velocities that stand out at only a single point cannot be considered real because of noise. However, in areas where three successive points show large velocities, they must be considered real. Note that the largest downdrafts are observed at the edges of the plage.

The average halfwidth of the Fe XII line over the plage is 0.190 Å, which is slightly larger than the expected thermal width of 0.175 Å for $T_e = 1.7 \times 10^6$ K. But this difference is negligible in view of the large Doppler shifts and can easily be explained by spatially unresolved motions in the line of sight. Therefore, the line broadening of coronal lines in active regions is caused by thermal effects only. This is contrary to the results of many other authors; they found large nonthermal line broadening in Skylab spectra above the limb. The discrepancy can easily be explained by the inferior spatial resolution of the Skylab spectra (60″× 2″) and the long line of sight at the limb.

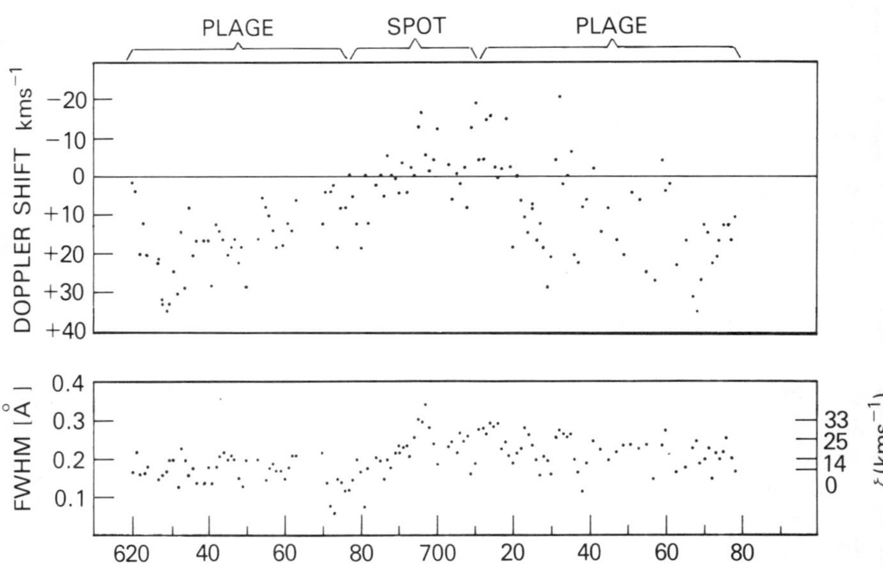

Figure 5.7. Doppler shifts at half maximum intensity (top) *and line width at half maximum intensity* (bottom) *of the Fe XII 1349.40 Å line observed over a plage region and sunspot (from HRTS I; see Figure 5.1).*

This result also means that no mechanical energy can be found in the coronal plasma above the plage that would manifest itself in line broadening.

However, over the sunspot, the coronal line shows an average halfwidth of 0.246 Å, which implies either a nonthermal broadening of 23 km s^{-1} or spatially unresolved Doppler flows.

5.5 INFLUENCE OF DYNAMIC EFFECTS ON THE FORMATION OF SPECTRAL LINES

There exists a conspicuous absence of large-scale blue shifts in active regions. In order to maintain a mass balance, it must be assumed that a large portion of the upward-moving gas is invisible in transition-zone and coronal lines. The explosive events cover only a small fraction (less than 1%) of the surface, but they could be the source of plasma being ejected upward. Because of their very short rise time (less than 10 s) and the high energies observed in their plasma turbulence (E_{kin} for carbon is 600 eV), they may be the location of very fast ionization processes and the production of nonthermal electrons. If chromospheric plasma is fully ionized instantaneously, the intermediate stages of ionization will not be visible, because their recombination time ($\approx$ s) is longer than the ionization time. The spectral signature of these events is the cooling portion of the plasma. While the plasma is being heated, it is invisible in transition-zone lines but afterwards becomes visible at coronal temperatures. The "explosive events" do show much less emission in the spin forbidden lines, which could be explained by high densities or the presence of high-energy, non-Maxwellian electrons that depopulate the metastable levels. They also show a different emission measure vs temperature distribution, which has no chromospheric emission below $T_e < 20{,}000$ K and a maximum around $T_e = 100{,}000$ K. They are invisible in forbidden coronal lines. The cooling portion of the plasma in the explosive events should radiate mostly in the temperature regime $100{,}000 < T_e < 200{,}000$ K, where the radiative losses have a maximum. The observed time scales and high energies indicate a plasma that is not in equilibrium. Therefore, all diagnostic methods available cannot be applied, and the amount of material moved upward may be *underestimated* if commonly accepted equilibrium ionization balance theories are used. If the observed Doppler shifts of the flows are interpreted as free-falling material, then they represent the other extreme of a nonequilibirum state of the plasma. There is no energy input into the falling material. Conduction from the corona can be estimated to be 2×10^3 (over sunspots) or 1.5×10^4 (over plages) erg cm^{-2} s^{-1}, which is small if compared with radiative losses of a free-falling plasma $F_R \sim 6.6 \times 10^4$ (at 1×10^5 K) (sunspot) and 6.2×10^4 (plage) erg cm^{-2} s^{-1}. However, both terms are negligible when compared with the kinetic energy flux of protons: $F_{kin} = 7.0 \times 10^6$ in sunspots or 1.1×10^5 erg cm^{-2} s^{-1} in plages, respectively. Therefore, two regimes in which the formation of spectral lines is dominated by completely different processes must be considered: (1) the area of free fall, where recombinations are dominant and (2) the "splashdown" layer, where the free-falling plasma is slowed rapidly in denser portions of the upper chromosphere. Here the spectrum will be formed by a combination of non-Maxwellian proton and electron collisions and recombinations. Figure 5.6 illustrates the slowdown process: the measured Doppler velocity at C II is smaller than that of Si III, indicating a slowdown at C II temperatures. The plasma falling down over the sunspot in McMath 15139 (Figure 5.4) has higher velocities and densities and therefore reaches lower altitudes before being slowed down than the plasma discussed above that was observed in McMath 13766 (Figure 5.1). The splashdown effect can actually be seen at the Si III level in Figure 5.4, where at the position of the light bridge (position *b*) Si III not only shows the strongest red shifts but also a pronounced blue shift. At 100 km s^{-1}, the proton kinetic energy is 50 eV; the thermal energy of the protons at 20,000 K where the splashdown occurs is 1.8 eV. The divergence of the kinetic energy in these layers will be very rapid because of the tremendous density difference between the falling plasma and the low-temperature, high-density chromosphere. This will aggravate the nonequilibrium conditions.

5.6 He 1640 EMISSION AND THE SEPARATION OF ELEMENTS IN DOWNFLOWS

Figure 5.8 shows that the intensity ratio

$$\frac{I_{\text{He II 1640 Å}}}{I_{\text{C IV 1550 Å}}}$$

varies considerably over different regions of the plage and sunspot (by about a factor of 10). Strong relative He II emission occurs in, or adjacent to, downflow areas, while the ratio is weak in areas of low Doppler shifts. A possible explanation of this effect may be the separation of light from heavy elements in the downflows, where they reach dense chromospheric layers and are slowed down.

Carbon atoms carry three times more kinetic energy than He atoms. Therefore, helium atoms will be slowed down earlier in the upper chromosphere than carbon atoms. Consequently, the relative abundance of helium will be larger at chromospheric layers with high temperatures, while carbon comes to rest at lower temperatures. Therefore, the stronger emission of helium is expected because of the enhanced abundance of the helium at higher temperatures in the downflow areas where helium comes to rest.

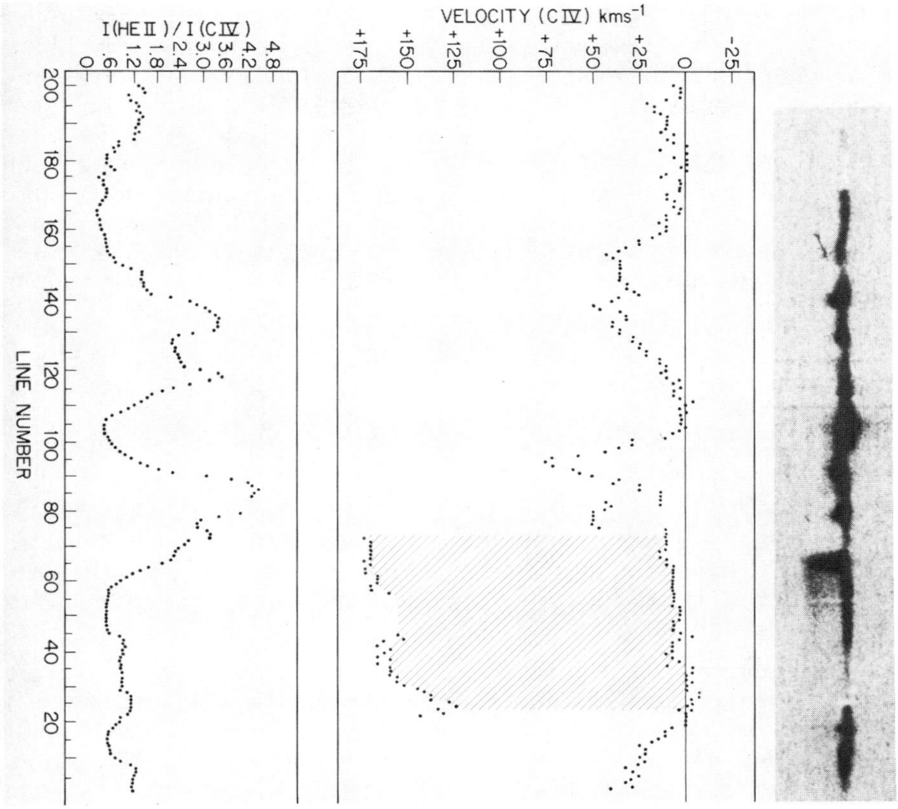

Figure 5.8. Top: *Spectrum of C IV, covering plage and sunspot of active region McMath 15139.* Center: *Doppler velocities of C IV 1550 Å, observed as line shifts at half maximum intensity. Above the sunspot, where a continuum of velocities is observed, maximum and minimum values are plotted.* Bottom: *Intensity ratio of He II 1640 Å over C IV 1550 Å, derived from integrated intensities.*

5.7 THE MASS AND ENERGY BALANCE OF ACTIVE REGIONS

If the observed transition zone downflows over active regions (20–60 km s^{-1}) and sunspots (up to 150 km s^{-1}) are free-falling material, the existence of loops with an altitude range of from 750 km to 40,000 km must be assumed. Large loops observed from Skylab have lifetimes of hours or longer. Our observations require that the material in these loops must be replenished in times of from 70 to 550 s.

We are confronted with this question: what physical process provides the mass and energy to move material up to the observed altitudes and can heat it to coronal temperatures?

The observations do not show permanent upflows. However, because of the limited spatial coverage of plages and sunspots, such upflows may have been missed. But the problem seems to be more fundamental. A very similar imbalance exists in the quiet sun (Brueckner, 1979). Much larger areas of the quiet sun than that of active regions have been observed during three rocket flights; therefore, statistics about motions in the quiet sun are much better than those for active regions. As in active regions, the Doppler motions in the quiet sun are predominantly downward. Only in explosive events is upward-moving material seen. The explosive events cover only a very small fraction (less than 1%) of the quiet sun's surface.

If the physical processes in active regions are qualitatively similar to those of the quiet sun, the explosive events may be considered as the source of material being propulsed upward and heated simultaneously. Because of the nonequilibrium conditions both in flows and explosive events, one may overestimate the amount of material falling down and underestimate the amount of material being moved upward. Therefore, a mass flow balance may exist between the explosive events and the downflows.

The spectral signature of the explosive events indicates that they are caused by MHD instabilities, rather than shock waves, because of their isotropic nature. The observed velocities are sufficient to eject material to heights above the solar surface, where the tops of loops are located.

If the syphon mechanism (Meyer and Schmidt, 1968) is responsible for the replenishment of loops, rather low upward velocities might be expected at one leg of the loop, where the pressure is higher. However, mass flow must be conserved. This would result in rather high densities of the upward-moving plasma, which would emit very strongly and therefore be visible in the observations as blue Doppler shifts.

Material may be moved upward in the explosive events. Where does it enter the loops? Because downflows are dominant in the observations, it may be possible that material is propulsed upward over a plage, captured by the strong magnetic fields, and returned along them to the surface. This requires that the upward-moving material cross magnetic field lines. Within the framework of conventional MHD theories, the rate at which material can cross field lines is too small to account for the downflows. However, the assumption of a simple loop structure that has enhanced density compared with its surroundings may be an oversimplification. It is possible that loops have a filamentary fine structure that cannot be observed but would enhance the rate at which material can cross magnetic field lines. D. Spicer (private communication) points out that a variety of plasma instabilities can increase the flow rate of material across magnetic field lines.

Finally, we should point out the analogy with many phenomena observed in flares. Very short X-ray bursts (0.1 s) indicate that very rapid ionization is possible. Flares may be driven by a process similar to (but more energetic than) the process generating the energy and mass motions in active regions. Skylab observations of strong transition zone line shifts in flares reported by Brueckner (1975) supports this analogy between active regions and flares.

REFERENCES

Brueckner, G. E.: 1975, in S. R. Kane (ed.), 'Solar Gamma-, X-, and EUV Rdiation', *IAU Symp.* **68,** D. Reidel, Dordrecht, Holland, p. 135.

Brueckner, G. E.: 1979, "Physics of the Chromosphere—Corona-Wind Complex and Mass-Loss in Stellar Atmospheres, A High Resolution View of the Solar Chromosphere and Corona," to be published in *Highlights of Astronomy* **5.**

Jordan, C.: 1969, *Monthly Notices Roy. Astron. Soc.* **142,** 501.

Meyer, F. and Schmidt, H. U.: 1968, *Z. Angew. Math. Mech.* **48,** 218.

Sandlin, G. D., Brueckner, G. E., and Tousey, R.: 1977, *Astrophys. J.* **214,** 898.

6. SPECTROSCOPIC DIAGNOSTICS OF THE ACTIVE REGION: TRANSITION ZONE AND CORONA

K. P. Dere and H. E. Mason

6.1 INTRODUCTION

Since the advent of solar space astronomy, spectroscopic observations in the X-ray and ultraviolet wavelengths have been made with steadily improving spatial and spectral resolution. Likewise, the techniques for interpreting the data have increased in number and sophistication. The basic physics underlying many of these methods has been presented in Zirin's *The Solar Atmosphere*, published in 1966. Numerous review papers have appeared that discuss the formation of the solar spectrum (recent examples include Jordan, 1979; Dupree, 1978; Feldman *et al.*, 1978a; and Doschek and Feldman, 1979). In this review, we will try to summarize most of the techniques currently used to analyze X-ray and ultraviolet spectroscopic data of solar features, with special reference to active regions. In particular, we will focus on the limitations and accuracy of the various methods.

6.2 TEMPERATURE AND EMISSION MEASURE DIAGNOSTICS

6.2.1 *Line Emission from a Low-Density Plasma*

The theoretical line emissivity $\epsilon(\lambda)$ (power per unit volume) is given by

$$\epsilon(\lambda_{ij}) = (hc/\lambda_{ij})A_{ji}N_j \quad , \tag{6.2.1}$$

where λ_{ij} is the wavelength of the transition between levels i and j, h is Planck's constant, c the speed of light, A_{ji} the spontaneous radiative transition probability, and N_j the number density of the upper emitting level j. This latter quantity is conveniently parameterized as

$$N_j(X^{+m}) = \frac{N_j(X^{+m})}{N(X^{+m})}\,\frac{N(X^{+m})}{N(X)}\,\frac{N(X)}{N(H)}\,\frac{N(H)}{N_e}\,N_e \quad , \tag{6.2.2}$$

where X^{+m} denotes the mth ionization stage of element X, $N_j(X^{+m})/N(X^{+m})$ is the population of level j relative to the total population of the ion X^{+m}, $N(X^{+m})/N(X)$ is the partial abundance of ionization stage X^{+m} relative to the total abundance of the element X, $N(X)/N(H)$ is the abundance of element X relative to the hydrogen abundance, and $N(H)/N_e$ is the hydrogen density relative to the electron density.

The solar values of the elemental abundances have been reviewed recently by a number of authors (Withbroe, 1971, 1976; Pagel, 1973; Trimble, 1975; Ross and Aller, 1976; Engvold, 1977; Lambert, 1978; Lambert and Luck, 1978). The compilations of photospheric abundances are generally deemed accurate to ±0.1–0.2 dex. The coronal abundances are equal to the photospheric abundances to within the accuracy of their measurement (±0.3 dex). At transition zone temperatures and above, H and He can be considered completely ionized, and thus the ratio $N(H)/N_e$ is about 0.89.

Densities in the solar transition zone and corona are insufficient to maintain a Boltzmann distribution of ionization stages and level populations (LTE). These must be calculated by requiring the detailed balance of the various collisional and radiative rates. In practice, collisional excitation and radiative decay proceed much more quickly than ionization and recombination, so that calculation of level populations and ionization equilibria can be performed separately. These latter calculations are discussed in Section 6.4.3.

The level populations N_i are governed by the rate equation

$$\frac{dN_i}{dt} = \sum_{j \neq i} N_j(N_e C_{ji} + F_{ji}) - N_i \sum_{j \neq i} (N_e C_{ij} + F_{ij}) + \sum_{j>i} N_j A_{ji} - \sum_{j<i} N_i A_{ij} \ , \tag{6.2.3}$$

where C_{ij} is the collisional excitation-rate coefficient from level i to level j, and F_{ij} is the photoexcitation rate from i to j (simulated emission is negligible). The populations are determined by assuming that a steady state ($dN_i/dt = 0$) has been reached. In exceptionally transient events, this may not be the case. Generally, the summations will extend over a limited number of levels.

In this section, we are concerned only with optically thin, allowed (electric dipole) transitions. For these transitions the radiative decay rates always greatly exceed the collisional decay rates. The excited levels are populated by electron collisions and depopulated by radiative decay. The populations of these excited levels are negligible compared to that of the ground level. The set of statistical equilibrium equations can be solved as a two-level system for each transition,

$$A_{ji} N_j = N_e C_{ij} N(X^{+m}) \tag{6.2.4}$$

The expression for the emissivity becomes

$$\epsilon(\lambda_{ij}) = (hc/\lambda_{ij})N_e C_{ij} N(X^{+m}) \quad . \tag{6.2.5}$$

More-complex systems, containing "metastable" levels, are discussed in Section 6.3.1.

The electron collisional excitation-rate coefficient is found by integrating the collision cross section over a Maxwellian velocity distribution. The electron excitation-rate coefficient from level i to level j is then

$$C_{ij} = 8.63 \times 10^{-6}\, T_e^{-1/2} (\Omega_{ij}/\omega_i) \exp(-\Delta E_{ij}/kT_e) \ \text{cm}^{-3}\,\text{s}^{-1} \quad , \tag{6.2.6}$$

where T_e is the temperature (in Kelvin), ω_i the statistical weight of the lower level i, ΔE_{ij} the energy difference between the levels, k Boltzmann's constant, and Ω_{ij} the collision strength, which has been assumed to be independent of the incident electron energy. Ω_{ij}, a dimensionless quantity symmetric in i and j, is related to the collision cross section σ_{ij} by

$$\sigma_{ij} = (\Omega_{ij}/\omega_i)\pi a_0^2/K_i^2 \quad , \tag{6.2.7}$$

where K_i^2 is the energy of the incident electron in Rydbergs and a_0 the Bohr radius. For allowed transitions, Ω is a slowly varying function of the incident electron energy, and the threshold value is often used in Eq. 6.2.6. Such an approximation usually agrees well with a full integration of the cross section over a Maxwellian velocity distribution. For example, Flower and Pineau des Forets (1973) estimated such errors as 20% for the XUV lines of Fe XIII. During the last several years, an increasing number of collision strength calculations have become available. These are discussed in Section 6.4.2. In the case of transitions where collision strength calculations are unavailable, Van Regemorter (1962) has determined an approximate formula based on the Bethe approximation

$$\Omega_{ij} = (8\pi/\sqrt{3})(I_H/\Delta E_{ij})g\omega_i f_{ij} \quad , \tag{6.2.8}$$

where I_H is the ionization energy of hydrogen, f_{ij} the atomic oscillator strength, and g an effective Gaunt factor. The Gaunt factor g is determined empirically as a function of the parameter $(K_i^2/\Delta E_{ij})^{1/2}$. For X-ray transitions, Van Regemorter finds g to be typically 0.2. Blaha (1969) has calculated g values in a simplified Coulomb-Born approximation for $\Delta n = 0$ transitions and finds values ranging between 0.3 and 1.0.

If the collision strengths of most of the strong lines are known, it is possible to determine the total radiative losses from a low-density optically thin plasma. Commonly used calculations are those by Cox and Tucker (1969) and Raymond *et al.* (1976). A recent calculation has been published by Summers and McWhirter (1979) that uses more accurate atomic parameters. The main uncertainty lies with the elemental abundances.

6.2.2 Emission Measure Analyses

A large number of the XUV lines emitted by active regions at $T > 10^5$ K are optically thin allowed lines. From Eqs. 6.2.2 and 6.2.4, the emissivity of such a line becomes

$$\epsilon(\lambda_{ij}) = \left[8.63 \times 10^{-6} \frac{hc}{\lambda_{ij}} \frac{\Omega_{ij}}{\omega_i} \frac{N(X)}{N(H)} \frac{N(H)}{N_e}\right] \cdot \left[\frac{N(X^{+m})}{N(X)} \frac{\exp(-\Delta E_{ij}/kT)}{T_e^{1/2}}\right] N_e^2 \tag{6.2.9}$$

or

$$\epsilon(\lambda) = \beta G(T) N_e^2 \quad , \tag{6.2.10}$$

where the dependency on density and temperature is explicitly shown. This formula is sufficiently accurate for most applications. The function $G(T)$, as defined by Pottasch (1964), is strongly peaked in temperature, with a width on the order of 0.3 dex. $G(T)$ is often referred to as the contribution function. The power in a line radiated from a volume of plasma is

$$P = \beta \int G(T) N_e^2 \, dV \quad . \tag{6.2.11}$$

If the volume contains isothermal plasma at a temperature T, the observed power becomes

$$P = \beta G(T) \int N_e^2 \, dV \quad . \tag{6.2.12}$$

If the temperature T is known, then the emission measure $\int N_e^2 \, dV$ can be derived from the observed power. For sharply peaked G functions, the temperature is often assumed to be that which maximizes G, which we denote here as T_m. In this manner, it is possible to derive "isothermal" emission measures as a function of temperature from a set of observed line intensities.

In the more-general case, we can make a change of variables as shown by Craig and Brown (1976):

$$P = \beta \iiint G(T) N_e^2 |\nabla T|^{-1} \, dS dT \quad , \tag{6.2.13}$$

or

$$P = \beta \int G(T) \xi(T) dT \quad , \tag{6.2.14}$$

where the differential emission measure ξ is defined as

$$\xi(T) = \iint_{S(T)} N_e^2 |\nabla T|^{-1} \, dS \quad . \tag{6.2.15}$$

The integration is performed over all surfaces S at temperature T in the observed volume. When $G(T)$ is independent of density, this relationship is valid. If G has a dependence on the density, the form of the equation remains if the density is a function only of temperature.

When the observed feature is a plane-parallel atmosphere, the line-of-sight differential emission measure can be integrated as

$$\xi_L(T) = N_e^2 \left(\frac{dT}{dh}\right)^{-1} \cos^{-1}\gamma \quad . \tag{6.2.16}$$

The angle between the radial direction and the line of sight is γ. A further assumption often made for a plane-parallel analysis of the transition zone and above is that the pressure $P = N_e T = P_0$ is a constant; then

$$\xi_L(T) = P_0^2 T^{-2} \left(\frac{dT}{dh}\right)^{-1} \cos^{-1}\gamma \quad . \tag{6.2.17}$$

The intensity of an optically thin spectral line emitted from a plane-parallel atmosphere is then determined by the pressure and the temperature gradient.

The equation relating spectral intensities to the differential emission measure,

$$I(\lambda) = \int e(\lambda,T)\, \xi(T)\, dT \quad , \tag{6.2.18}$$

is a Fredholm equation of the first kind and is known to be unstable. The straightforward method for solving this equation is by quadratures wherein only the inversion of a matrix is required. However, because of the unstable nature of the equation, as pointed out by Craig and Brown (1976), highly oscillatory unphysical solutions usually result. Solar researchers have proposed various ad hoc schemes for overcoming these difficulties.

In deriving the differential emission measure ξ, the general procedure is to measure a number of line intensities I_i $(i = 1,M)$, and from these to derive a set of N $(N \leq M)$ parameters that define ξ. A frequent choice is to set

$$\xi(T) = \xi_i \quad (T_i \leq T < T_{i+1}) \tag{6.2.19}$$

and then solve the linear least-squares problem. To avoid the obviously incorrect solutions containing negative values of ξ, it is possible to place constraints on the desired solution (for instance, Batstone *et al.*, 1970) and to search for a local minimum in χ^2. These constraints should be chosen with great care so that they do not unduly influence the derived values of ξ. Withbroe (1975) has employed a similar approach but requires an exact fit to the data through the variation of N ξ_i's, where in this case $N > M$. Such a procedure of solving for a number of unknowns greater than the number of observations is not legitimate. Another method is to represent the differential emission measure by some functional form of a set of free parameters that are determined by a nonlinear χ^2 optimization. The functional form can be conveniently chosen so that it is nonnegative but should be general enough so that a relatively unbiased solution is obtained. For a specific example, Dere (1978a) has shown that the differential emission measure can be derived for temperatures scanned by the T_m of the observed lines and will have errors on the same order as the errors in the line intensities. The elemental abundances can also be taken as additional free parameters and derived in the same fashion. An example of this procedure is found in Walker *et al.* (1974). Landini and Monsignori-Fossi (1979) have shown, following the suggestion of Brown (1974), that the currents induced in an X-ray ion chamber can be approximated as the Laplace transform of a function related to the differential emission measure and have used this to determine ξ in a flare.

A more elegant method has been proposed by Phillips (1962) and Twomey (1963). Phillips required that χ^2, as usually defined, plus the sum of the square of the second differences be a minimum. In other words, he required that the solutions obtained be smooth to a degree determined by an arbitrary smoothing parameter. Twomey showed how the method could be generalized for various conditions on the derived solution. The method has a drawback in that a smoothing parameter must be determined for each case and will depend on both the type of problem and the noise in the data. Kunasz *et al.* (1973) have recently discussed the application of this method to the case of the limb-darkening equation and have also suggested that the more sophisticated Backus-Gilbert method could be applied to those problems with worthwhile results. The value of this latter method is that it allows error bars to be placed on the solution. Lampton *et al.* (1976) have also provided a method for estimating the errors in derived parameters that might be of use. McKenzie *et al.* (1978) have employed a similar method for error estimation.

In a more intuitive sense, it is possible to estimate the amount of information one can expect to obtain on the shape of the differential emission measure. The kernel $G(\lambda, T)$ is a smoothly peaked function with a width on the order of 0.3 dex in T. Any variations in ξ that occur on a temperature scale less than 0.3 dex will be smoothed out. Hence, it should not be possible to detect any sharper variations from the analysis of spectral data. The theoretical static loop models offer a good example. These models all predict a singularity in ξ at $T = T_{max}$, where T_{max} is the maximum temperature at the top of the loop. As is necessary, the singularity is integrable, with $\xi(T) \propto (T_{max} - T)^{-1/2}$. When these curves for $\xi(T)$ are smoothed with a Gaussian of width 0.3 dex, these singularities become gentle bumps. Thus the maximum temperature in such loops is only capable of being determined to within ± 0.1 dex.

6.2.3 Determination of Electron Temperature

From Eqs. 6.2.5 and 6.2.6, the emissivity ratio for two optically thin allowed transitions excited from the same level (i) is

$$\frac{\epsilon(\lambda_{ij})}{\epsilon(\lambda_{ik})} = \frac{\Omega_{ik}}{\Omega_{ik}} \frac{\lambda_{ik}}{\lambda_{ij}} \exp\left[-(\Delta E_{ij} - \Delta E_{ik})/kT_e\right] . \tag{6.2.20}$$

This ratio is sensitive to changes in electron temperature if $(\Delta E_{ij} - \Delta E_{ik}) \sim kT_e$. This method was first applied to the Li-like ions (O VI, Ne VIII, Mg X) by Heroux and Cohen (1971). They studied the ratios of lines from 2s–2p transitions to those from 2p–3s, 2s–3p, 2p–3d.

When the observed intensity ratio $I(\lambda_{ij})/I(\lambda_{ik})$ is equated with $\epsilon(\lambda_{ij})/\epsilon(\lambda_{ik})$ of the form given in Eq. 6.2.23, there is an implicit assumption that the two lines are emitted with the same spatial distribution. A more exact formulation for $I(\lambda_{ij})/I(\lambda_{ik})$ is

$$\frac{I(\lambda_{ij})}{I(\lambda_{ik})} = \frac{\lambda_{ik}}{\lambda_{ij}} \frac{\int C_{ij} N(X^{+m})/N(X) N_e^2 \, dV}{\int C_{ik} N(X^{+m})/N(X) N_e^2 \, dV} . \tag{6.2.21a}$$

It is clearly possible for the two lines to be formed in different regions. In that case, the use of the line ratio as a temperature diagnostic is misleading. The magnitude of this effect is determined largely by the temperature range over which the ion has a significant abundance.

The lithium-like ions are abundant over a large temperature range, especially at temperatures above T_m. In particular, emission from the 2s–3p lines of O VI ($T_m = 3 \times 10^5$ K) has about equal strength from the corona as from the transition zone, as discussed by Heroux *et al.* (1972) and Gabriel and Jordan (1972). Flower and Nussbaumer (1975a) argue that if indirect excitation of the 3*l* levels via upper levels and the continuum is included, then it is possible to attribute all of the O VI 2*l*–3*l*′ emission to coronal sources. In contrast, only 20% of the O VI 2s–2p emission comes from the corona.

Flower and Nussbaumer (1975b) have suggested using the ratio of the transitions 3s–3p to 3p–3d in the sodium-like ions as a temperature diagnostic. Unfortunately, available solar data is not adequate for a meaningful analysis.

Lines of ions that exist over a narrow temperature range should all be formed in the same spatial element. Thus, ratios of these lines should provide a good indicator of the temperature of the region in which they are formed. However, their usefulness as a plasma diagnostic has been questioned by Feldman *et al.* (1978a). They point out that within the line of sight, there will generally be regions with temperature T_m corresponding to the ion of interest. As an example, they consider the line ratio I(172 Å)/I(630 Å) of O V, which is in the Be I sequence and is formed over a narrow temperature range. If a region with a temperature $T = 2T_m$ is to significantly affect the observed ratio, the emission measure at T must be an order of magnitude larger than the emission measure at T_m. On the other hand, these ratios should be an accurate means of determining T_m. This could serve as a needed check on ionization equilibrium calculations. In addition, such line ratios could also indicate transient ionization conditions that could be caused by diffusion and flows, as discussed in Section 6.4.4.

Ions in the boron I sequence have narrow contribution functions and should be useful for determining values of T_m. Flower and Nussbaumer

(1975c) used the O IV ratio I(790 Å)/I(554 Å) to determine that O IV was formed between 8 $\times$ 10^4 and 2.5 $\times$ 10^5 K. O IV is predicted to have T_m ~ 1.4 $\times$ 10^5 K. The line ratio I(991 Å)/I(686 Å) of N III was used by Nussbaumer and Storey (1979a) as a temperature diagnostic. They concluded that N III was formed at a temperature below T_m (7 $\times$ 10^4 K for N III). They mention that their results may be affected by Lyman continuum opacity at 686 Å, which Schmahl and Orrall (1979) have shown should be significant.

The satellite lines to the helium-like resonance line can be formed by either dielectronic recombination (see Section 6.4.3) or direct inner shell excitation. Intensity ratios of specific lines can be used to derive the electron temperature (T_e) and the temperature from the ionization ratio (T_z). A comparison of T_e and T_z should indicate whether the plasma is in equilibrium. The theory for this diagnostic technique was developed by Gabriel (1972) and Bhalla *et al.* (1975). This technique has application primarily to flare spectra; however, some active region spectra have also been analyzed. Both the high-resolution magnesium spectra analyzed by Bhalla *et al.* and the silicon spectra analyzed by Gabriel were consistent with an ionizing plasma. This result is difficult to believe for a nonflaring active region plasma. Some recent calculations of the atomic parameters for the dielectronic satellite lines of the Mg XI and Mg XII resonance lines have been carried out by Steenman-Clark *et al.* (1979) and Dubau *et al.* (1979), respectively.

6.2.4 Broadband X-Ray Diagnostics

A considerable amount of solar X-ray data has been obtained by broadband instruments such as ionization chambers, proportional counters, and photographic film. The power absorbed in the detector is given by

$$W = A \int \epsilon(\lambda) \int P(\lambda, T) N_e^2 \, dV \, d\lambda \quad , \tag{6.2.21b}$$

where A is the collecting area of the instrument, $\epsilon(\lambda)$ the photon efficiency, and P the incident flux per unit emission measure. The spectrum is produced collisionally and hence is proportional to the emission measure. When a theoretical spectrum P is integrated through the detector response, one obtains

$$W = \int F(T) N_e^2 \, dV \quad , \tag{6.2.22}$$

where

$$F(T) = A \int \epsilon(\lambda) P(\lambda, T) \, d\lambda \quad . \tag{6.2.23}$$

The spectrum P is the result of both line and continuum components. The X-ray continuum is produced by bremsstrahlung, radiative recombination, and two-photon emission from H- and He-like ions. Culhane (1969) has provided formulae for calculating thermal bremsstrahlung and radiative recombination. Since evaluation of these requires a certain amount of effort, Culhane and Acton (1970) have obtained a simplified approximation. Tabulations of the atomic data for calculating the line spectrum can be found in Tucker and Koren (1971a,b), Mewe (1972), Kato (1976), and Raymond and Smith (1977).

With Eq. 6.2.22, we are back to the usual Fredholm equation (6.2.18) for the differential emission measure but with some major differences. An observation consists of the output of a whole detector, rather than the intensity of some spectral line. To obtain more observations, it is necessary to add more detectors (or filters). Also, the convolved spectrum $F(T)$ is generally not a sharply peaked function in temperature like the line contribution functions $G(T)$. In some cases, F is a monotonically increasing function for all reasonable values of T. Consequently, we can expect that the amount of information that can be derived about the differential emission measure from broadband detectors will be considerably less than from the measurement of spectral line intensities. However, Kahler (1976) has used a somewhat different approach to the solution of Eq. 6.2.22 and has found that the response of an X-ray telescope is linearly proportional to the square of the pressure or energy density of a coronal plasma over a wide range of temperatures. Another practical alternative is the method of Landini and Monsignori-Fossi (1979) mentioned previously.

The ratio of the outputs from two detectors 1 and 2, when viewing an isothermal plasma at temperature T, is a function only of the temperature T, i.e.,

$$\frac{W_1}{W_2} = \frac{F_1(T)}{F_2(T)} = R_{12}(T) \quad , \qquad (6.2.24)$$

since the dependence on the emission measure cancels. The temperature can then be determined by taking the ratio of two detector outputs and comparing it with values calculated with a theoretical spectrum. With the temperature known, it is then possible to find the emission measure. Lately, there has been some question as to how meaningful these effective temperatures are (see Underwood and McKenzie, 1977 and 1978, and Gerassimenko and Nolte, 1978). Underwood and McKenzie contend that the temperatures derived are detector dependent and may or may not correspond to an emission measure weighted average of the temperature. Gerassimenko and Nolte show that the temperatures derived by Underwood and McKenzie from different detectors are all within expected statistical errors. However, experience with the SOLRAD X-ray ionization chambers shows that temperatures derived from the longer-wavelength detectors are systematically lower than those derived from shorter-wavelength detectors. This is attributable to the fact that longer-wavelength detectors are more sensitive to the thermal radiation produced at lower temperatures. Gerassimenko and Nolte further show that for narrow distributions of emission measure, the temperatures derived from detector ratios agree well with the temperature of maximum emission measure. However, they show this for distributions that are narrower than the detectable value of about 0.3

dex in T. As Underwood and McKenzie point out, this argument can be used to give information about a distribution only if it is known beforehand that it is sharply peaked, and this is not necessarily always the case. In conclusion, the ratio method can be used to indicate regions of generally hotter and cooler coronal material and can be calibrated against a theoretical thermal spectrum. The exact meaning of the derived temperature is not precisely defined but nevertheless probably useful as a plasma diagnostic.

6.2.5 Specific Active Region Diagnostics

The transition zone and corona over an active region are made up of magnetically confined plasmas. These loops require some, as yet undetermined, form of heat input. The simplicity and intensity of these active regions plasmas should offer prime observational material with which to study these heating mechanisms.

At present, considerable effort is being devoted to one-dimensional static energy balance (SEB) loop models. Most assume some empirical form for the heating function, which must balance conductive and radiative losses in a stationary plasma. Because of the highly nonlinear coefficient of thermal conductivity, the greatest part of the length of these loops is nearly isothermal, with sharp transition zones at the ends. Thus, the temperature at the top of the loop and the length should be easily observable. Rosner *et al.* (1978) and Craig *et al.* (1978) have shown how the length, maximum temperature, and pressure should be related for a uniformly heated loop. Craig *et al.* also show that it is possible to derive a limited amount of information on the shape of the heating function by measuring these three quantities. Rosner *et al.* have found general agreement to their formula for a number of loops with temperature between 2 and 3×10^6 K. A wider class of active region loops should be investigated.

The SEB loop models are also capable of more detailed predictions such as the temperature distribution along the loop. Once the pressure (assumed constant) is specified, the differential emission measure readily falls out of the calculations. In principle, it should be possible to test such predictions by measuring the intensity distribution of several well-chosen spectral lines. In practice, it is generally not possible to observe a single, isolated active region loop. The differential emission measure of a complete active region should then be the linear superposition of an ensemble of loops. This question has been discussed previously by Craig *et al.* (1978).

If the loops in an active region are adequately described by the SEB models, then it should be possible to find a weighting function that determines the distribution of loops in the active region. Such a test is clearly not as conclusive as a test with a single well-resolved loop. Typically, the static energy balance models are specified by five or more parameters. The following discussion is based on constant cross section geometries but can be generalized to the variable cross section geometries considered by some authors.

For transition zone and coronal lines, predicted intensities are sensitive to only two parameters, T_{max} the temperature at the top of the loop and the pressure p. Since the emission measure of the loops scale linearly with the pressure, the dependence can be absorbed by the unknown weighting function W if we assume that all loops with the same T_{max} have the same pressure. W will have units of area. Then the intensity emitted in a spectral line by a set of loops with the theoretical differential emission measure ξ_t is

$$I = \int W(T_{max}) \int \beta G(T)\xi_t(T, T_{max})\, dT\, dT_{max} \quad , \tag{6.2.25}$$

or

$$I = \int W(T_{max})\Gamma(T_{max})\, dT_{max} \quad , \tag{6.2.26}$$

where

$$\Gamma(T_{max}) = \beta \int G(T)\xi_t(T, T_{max})\, dT \quad . \tag{6.2.27}$$

For SEB loops of constant cross section, ξ_t is quite flat—aside from an integrable singularity at $T = T_{max}$, which actually contains little emission measure. When integrated, we get the approximate results

$$\Gamma(T_{max}) = \begin{cases} \beta < G\xi_t > \Delta T & T_{max} \geq T_m \\ 0 & T_{max} < T_m \end{cases} \quad . \tag{6.2.28}$$

In other words, Γ is essentially a step function, so that

$$I = \beta < G\xi_t > \Delta T \int_{T_m}^{\infty} W(T_{max}) dT_{max} \quad . \tag{6.2.29}$$

If a number of line intensities, I, with a decent range in T_m are measured, it should be a relatively simple matter to determine $W(T_{max})$, assuming static energy balance models are viable. Equation 6.2.26 is a return of Eq. 6.2.18 in a new guise. An attempt to invert this to solve for W will be especially imprecise because of the flat character of Γ. However, the question still to be answered is whether or not a suitable W exists.

Another method of comparing these loop models with active region loops is to directly compare the observed and predicted differential emission measures. From an assumed $W(T_{max})$, the differential emission measure is predicted. However, depending on the form of W, the predicted ξ can have singularities at temperature where $T = T_{max}$. These singularities cannot be observed by real spectral data. In order to make the observed and predicted forms of ξ comparable, the theoretical values should be smoothed over a temperature range of 0.1–0.3 dex. When this is done, the SEB models predict that $T\xi$ is practically flat up to T_{max}, where it goes to zero. This case has been discussed by Craig *et al.* (1978). Levine and Pye (1979) have compared the ξ derived from active region data with that predicted by various distributions of SEB loop models and find a wide variety of loop distributions consistent with the data.

6.3 ELECTRON DENSITY DIAGNOSTICS

6.3.1 Metastable Levels

Spectral emission lines can be grouped into three categories, based on the means by which their upper level is populated. This will determine the dependence of their emissivities on electron density. These categories are:

1. allowed lines that are collisionally excited from the ground level;
2. forbidden lines that originate from metastable levels; and
3. allowed lines that are excited from metastable levels.

As discussed in Section 6.2.1, the emissivity of an allowed line (Category 1) is proportional to N_e^2. Examples of allowed lines are the 2s ^2S–2p ^2P transitions in the lithium-like ions.

Forbidden lines have very small radiative transition probabilities compared to allowed lines. In the context of this review, a forbidden line is such that for a range of electron density values associated with the solar atmosphere, collisional deexcitation processes are comparable with or larger than the corresponding radiative decay rate. Thus, some intersystem transitions can also be included in this category. The population of the excited ("metastable") level from which the line originates and the intensity behavior of the forbidden line fall into three stages, depending on the electron density. For low electron density, radiative decay processes dominate, and the intensity of the line has the same behavior as an allowed line ($\propto N_e^2$). For intermediate electron densities, the collisional deexcitation processes and radiative processes are comparable. The population of the "metastable" level becomes significant, and the intensity of the line is proportional to N_e^β ($1 < \beta < 2$). For high electron densities, collisional processes dominate and the "metastable" level is in a pseudo-Boltzmann equilibrium. The line intensity then varies as N_e. These electron density ranges depend on the atomic parameters and differ for individual ions and transitions. Typical examples of forbidden lines are Fe XII 1242 Å ($3s^2 3p^3\ {}^4S_{3/2}$–${}^2P_{3/2}$) or O V 1218 Å ($2s^2\ {}^1S_0$–$2s2p\ {}^3P_1$).

The intensity for lines in Category 3 will depend on the population of the metastable level from which they are excited. When this population becomes significant, but not at its Boltzmann value, the line intensity will vary as N_e^β ($2 < \beta < 3$). Examples in this category are the $2s^2 2p^2\ {}^1D_2$–$2s2p^3\ {}^1P_1$ transition in carbon-like ions and the $3s3p\ {}^3P_2$–$3p^2\ {}^3P_2$ transition in magnesium-like ions. When the metastable levels reach a pseudo-Boltzmann equilibrium, the intensity of these lines in Category 3 varies as N_e^2.

For complex ions, it is necessary to solve a set of statistical equilibrium equations (Eq. 6.2.3) including all relevant collisional and radiative processes. The electron collisional excitation rate was defined in Section 6.2.1. For some forbidden transitions, indirect collisional excitation (cascading via excited configurations) can be more important than direct collisional excitation. For example, the indirect collisional excitation rate is almost an order of magnitude greater than direct electron excitation for the Fe XIV transition $3s^2 3p\ {}^2P_{1/2}$–${}^2P_{3/2}$. One must therefore consider carefully which levels to include in the solution of Eq. 6.2.3. Proton impact excitation is of importance for transitions between fine structure levels where $\Delta E/kT$ is small. For the fine structure transitions in the ground configurations of coronal ions, the proton excitation rates are comparable to the direct electron excitation rates. For more highly ionized systems such as Fe XXI, which is observed in solar flares, the proton excitation rates are even more important. A discussion of proton collisional calculations is given in Section 6.4.2.

For coronal and transition region ions, spontaneous radiative decay is a major depopulating mechanism. The accuracies of atomic structure calculations are outlined in Section 6.4.1. Excitations can also proceed by the absorption of a continuum photon. The continuum is strong only for wavelength greater than about 3000 Å; thus, photoexcitation should only be important for fine structure transitions, and then mainly at low densities. Assuming a blackbody continuum spectrum, the excitation rate is

$$F_{ij} = A_{ji}\, D(h)(\omega_j/\omega_i)[\exp(\Delta E_{ji}/k\, T_R)-1]^{-1} \quad , \tag{6.3.1}$$

where $D(h)$ (≤ 0.5) is a dilution factor for the background radiation field of temperature T_R. The stimulated emission rate is much smaller than the spontaneous emission rate for all wavelengths short of the infrared.

6.3.2 Electron Density Determination

A commonly used method for obtaining electron densities has been to divide the emission measure $N_e{}^2\Delta V$ by the volume ΔV derived from spatially resolved images. This technique assumes that the observed features are indeed fully resolved. For coronal loops, a resolution of 3″ appears to be sufficient. However, recent analysis of the transition zone indicates that the fine structure has yet to be resolved.

From the previous discussion, it is clear that if ratios of lines produced by different excitation mechanisms are taken, the ratios will be density-sensitive over a certain range of densities. The most straightforward method is to take the ratio of lines of the same ion, although this is not always possible. Using this line ratio technique, electron densities are derived that are independent of geometrical assumptions. Ideally, line ratios used for electron density determination should be independent of electron temperature. If this is not the case, then it becomes necessary to include in the analysis some model-dependent assumptions. This has been a major problem with studies of electron density diagnostics in the transition zone.

It was suggested by Gabriel and Jordan (1969) that the line intensity ratios from helium-like ions could be used as electron density diagnostics. The original calculations have been redone by Freeman *et al.* (1971) with corrected transition probabilities for the $^1S_0$–$^3S_1$ transitions. The transitions of interest are the resonance line (1s^2 ^{1}S–1s2p ^{1}P), forbidden line (1s^2 ^{1}S–1s2s ^{3}S), intercombination lines (1s^2 ^{1}S–1s2p ^{3}P), and the two-photon transition (1s^2 ^{1}S–1s2s ^{1}S). The level 1s2s ^{3}S should be metastable, and the intensity of the forbidden line can be reduced relative to the other lines at high densities. Specifically, the ratio of the intercombination to forbidden line intensity R has received the greatest interest. The electron density values deduced by Freeman *et al.* from active region spectra were between 10^{11} and 10^{13} cm^{-3}. These values are anomalously high. The atomic data was updated and intensity ratios recalculated by Gabriel and Jordan (1972). In these early publications, the populating mechanisms for the $n = 2$ levels were not all calculated explicitly, but they were accounted for by normalizing R to observations in the low-density limit. There is some uncertainty about what the low-density value of R actually is. For O VII, McKenzie *et al.* (1978) claim a significantly higher value (4.2 ± 0.2) than previous estimates (e.g., Acton *et al.*, 1972, $R = 3.6$). It was pointed out by Blumenthal *et al.* (1972) that the ratio R is also dependent on electron temperature because of cascades to the $n = 2$ levels following recombination or collisional excitation to higher levels. This was apparently confirmed by the observations by Acton *et al.* (1972), although Gabriel and Jordan (1973) maintain that the temperature dependence of R has been overestimated. The electron densities required to produce measurable deviations in R for O VII are near the upper limit of those expected in solar active regions. Acton and Catura (1976) claim the first conclusive observation of a decrease in R for O VII due to electron density effects, and McKenzie *et al.* (1979) have observed a factor of three decrease in R during a solar flare. Detailed investigations of the behavior of the He-like line intensities have recently been published by Mewe and Schrijver (1978a,b) and Peacock and Summers (1978). These authors also consider the effect of departures from ionization equilibrium on the intensity ratios.

Probably the most extensively studied ion in the solar UV spectrum is C III. Publications of atomic data and analyses of the spectral lines from C III are extensive. We do not reference them all here. For fuller reviews of earlier work see Gabriel and Jordan (1972), Dupree (1978), and Jordan (1979). The transitions of most interest are the resonance line at 977 Å ($2s^2\ ^1S_0$–$2s2p\ ^1P_1$), the triplet transitions at around 1176 Å ($2s2p\ ^3P$–$2p^2\ ^3P$), and the intersystem line at 1909 Å ($2s^2\ ^1S_0$–$2s2p\ ^3P_1$). In most solar features, the 977 Å and 1176 Å lines are predominantly excited by electron collisions from their respective lower levels. The levels $2s2p\ ^3P$ are metastable, making the ratio ϵ(1176 Å)/ϵ(977 Å) a possible electron density diagnostic for solar conditions. Unfortunately, this ratio is not very sensitive to the electron density, and errors in the electron excitation rates of only 25% can lead to orders of magnitude difference in the electron density deduced (Jordan, 1974). The excitation rates used in early analyses of C III were "distorted-wave" calculations by Eissner (1972). Dupree *et al.* (1976) carried out a semiempirical analysis of the ATM S055 data and suggested adjustments to several atomic rates to give results consistent with accepted solar models. Recent electron excitation calculations by Berrington *et al.* (1977) and Dufton *et al.* (1978) using a "close-coupling" code, give values consistent with those suggested by Dupree *et al.* Transition probabilities for C III through Fe XXIII have recently been recalculated by Nussbaumer and Storey (1978, 1979b,c).

It was pointed out by Loulergue and Nussbaumer (1974, 1976) that the line ratio ϵ(1176 Å)/ϵ(977 Å) is also dependent on temperature, which considerably complicates the interpretation of observed intensities. Recent studies by Doschek and Feldman (1977) of the widths of the 1176 Å line and the 1909 Å line indicate that they are formed in different spatial regions. Raymond and Dupree (1978) have considered the effect of mass flows on the C III spectral line intensity ratios, which are discussed further in Section 6.4.4. Cook and Nicolas (1979) have carried out an extensive study of the C III lines recorded by the ATM S082B instrument. They found that ratios involving the 1176 Å multiplet gave anomalously low values of electron density. This discrepancy is partially attributed to optical-depth effects (Jordan, 1979).

Another ion in the beryllium sequence that is useful as a density diagnostic for active regions is O V($T_m \sim 2.5 \times 10^5$ K). Atomic calculations for this ion were originally carried out by Malinovsky (1975) using the distorted-wave approximation. More recent results have been published by Berrington *et al.* (1977) and Dufton *et al.* (1978). The resonance transition at 629 Å ($2s^2\ ^1S$–$2s2p\ ^1P$) is used relative to the multiplet at 760 Å ($2s2p\ ^3P$–$2p^2\ ^3P$), and the region of density sensitivity is higher than for C III. At longer wavelengths and higher densities, the ratio of 1371 Å ($2s2p\ ^1P_1$–$2p^2\ ^1D_2$) to 1218 Å ($2s^2\ ^1S_0$–$2s2p\ ^3P_1$) is also useful. Dufton *et al.* (1978) and Dupree (1978) suggested that a major uncertainty in the O V calculation was the proton collision rates among the $2s2p\ ^3P_J$ levels. Landman and Brown (1979) recently calculated proton collision rates between the metastable ^3P_J levels for ions in the Be and Mg isoelectronic sequences. They found good agreement (±25%) with Malinovsky's O V results. These proton rates are in fact significantly smaller than the equivalent electron collision rates. A more serious effect on the O V $I(760)/I(630)$ ratio is the H Ly continuum absorption that could result in a factor three decrease in the expected ratio.

Ions in the boron isoelectronic sequence can also be used to determine electron density. The $2s2p^2\ ^4P$ levels in N III and O IV are metastable for electron densities appropriate to the transition region. Atomic calculations for O IV were published by Flower and Nussbaumer (1975b) and for N III by Nussbaumer and Storey (1979a). The advantage of using the ratios of lines between the $2s2p^2\ ^4P$ and $2s^22p\ ^2P$ levels as electron density indicators is that these ratios are insensitive to electron temperature, as discussed in Section 6.2.3. Unfortunately, the intensity ratios only change by a factor of three for three orders of magnitude change in electron density. The theoretical ratios are sensitive to changes in atomic parameters. In particular, Doschek and Feldman (1979) show that 20% changes in the radiative transition probabilities can lead to an order of magnitude error in the electron density deduced.

Extensive calculations have also now been carried out for ions with ground configurations $2s^22p^n$. Recent publications are for the boron isoelectronic sequence (Dankwort and Trefftz, 1976; Elwert and Raju, 1975; Flower and Nussbaumer, 1975c,d; Vernazza and Mason, 1978; Dere *et al.*, 1979), carbon isoelectronic sequence (Bhatia *et al.*, 1979a; Mason and Bhatia, 1978; Dere *et al.*, 1979), nitrogen isoelectronic sequence (Bhatia and Mason, 1979; Feldman *et al.*, 1978b), and the oxygen isoelectronic sequence (Bhatia *et al.*, 1979c).

The ion Si III ($T_m = 3.5 \times 10^4$ K) has a $3s^2$ ground configuration and is therefore similar to the beryllium-like ions discussed earlier. The levels 3s3p ^3P are metastable for the electron density range 5×10^9 to 10^{12} cm^{-3}. Nicolas (1977) and Nicolas *et al.* (1979) have carried out an extensive study of Si III lines recorded in the ATM S082B and HRTS data. The most useful line ratios are I(1301 Å)/I(1312 Å) and I(1301 Å)/I(1296 Å) (1301 Å:3s3p $^3P_1$–$3p^2$ $^3P_0$; 1312 Å:3s3p $^1P_1$–3s4s $^1S_0$; 1296 Å:3s3p $^3P_0$–$3p^2$ $^3P_1$). The ratio ϵ(1301 Å)/ϵ(1312 Å) varies by about a factor of 20 over three orders of magnitude change in electron density, but it is also dependent on electron temperature. The ratio ϵ(1301 Å)/ϵ(1296 Å) is less sensitive to electron density but has the advantage of being insensitive to electron temperature. The resonance line at 1206 Å was not used, since it is frequently optically thick. Nicolas *et al.* calculated the intensity ratios using a transition region model deduced from resonance line intensities assuming constant pressure. The pressure can then be deduced from the line ratios. The intensity ratios integrated through a theoretical solar atmosphere are compared with isothermal ratios calculated assuming the emission is at T_m, and the two methods yield electron densities within a factor two of each other.

The coronal iron ions (Fe IX–Fe XV) give rise to lines in the visible and UV spectral wavelength regions. Electron collision rates and radiative data have now been calculated for all of these ions—Fe IX (Flower, 1977b), Fe X, Fe XI and Fe XIV (Mason, 1975), Fe XII (Flower, 1977a), Fe XIII (Flower and Pineau des Foret, 1973), Fe XIV (Blaha, 1971), and Fe XV (Flower and Jordan, 1971). Extensive tables of excitation rates are also given by Davis *et al.* (1976). New atomic structure calculations have recently been published for Fe X and Fe XI (Bromage *et al.*, 1977) and Fe XII and Fe XIII (Bromage *et al.*, 1978).

The corresponding theoretical intensities and discussions of electron density-dependent line ratios have also been published for Fe IX (Feldman *et al.*, 1978c), Fe X (Nussbaumer, 1976), Fe XI and Fe XII (Kastner and Mason, 1978), Fe XIII (Flower and Nussbaumer, 1974), and Fe XV (Dere *et al.*, 1977; Bhatia and Kastner, 1979). All the useful electron density-sensitive ratios for lines in the wavelength region covered by the ATM S082A instrument (170–630 Å) were discussed in detail by Dere *et al.* (1979). For Fe X–Fe XIV, the metastable levels are in the ground configuration, and the UV lines reflect this density sensitivity. Fe IX and Fe XV have metastables in excited configurations. As discussed in Section 6.4.2, the atomic data that are likely to be most in error are connected with intersystem transitions. The populations of the upper levels in the ground configurations are likely to be most affected, and there is some evidence from analysis of the 1970 eclipse data that calculations for these levels are in error (Gabriel and Jordan, 1975).

A technique for determining electron density has been developed by Feldman *et al.* (1977) using spectra from the ATM S082B experiment. From Eq. 6.2.1, the power in a spectral line can be expressed as

$$P = \beta G(T_m) \langle N_e^2 \rangle \Delta V \quad , \tag{6.3.2}$$

which can be rewritten as

$$P = \gamma J(<N_e>, T_m) <N_e> \Delta V \quad , \tag{6.3.3}$$

where $J(<N_e>, T_m)$ is the value of $N_j(X^{+m})/N(X^{+m})$ at T_m and $<N_e>$, j being the emitting level. From Eqs. 6.2.1 and 6.2.2, the value of γ is,

$$\gamma = (hc/\lambda_{ij}) A_{ji} \frac{N(X^{+m})}{N(X)} \frac{N(X)}{N(H)} \frac{N(H)}{N_e} \quad , \tag{6.3.4}$$

where $N(X^{+m})/N(X)$ is also taken at T_m. Then, assuming $<N_e^2> = <N_e>^2$, we obtain

$$P = \gamma J(<N_e>, T_m) \frac{<N_e^2> \Delta V}{<N_e>} \quad . \tag{6.3.5}$$

For an allowed line, the quantity $\gamma J/<N_e>$ is a constant, and Eq. 6.3.5 simply restates that the power is proportional to the emission measure. However, for a number of forbidden and intersystem lines, the population of the emitting level can reach a pseudo-Boltzmann equilibrium. Then γJ is a constant, and $\gamma J/<N_e>$ is inversely proportional to the density. Consequently, such lines can be used as density diagnostics. The power P is measured and the emission measure $<N_e^2> \Delta V$ is obtained from an allowed line or several allowed lines that span the appropriate temperature range. The ratio $P/<N_e^2> \Delta V$ is then compared with calculated values of $\gamma J/<N_e>$. This ratio has the advantage of being strongly sensitive to the density. The upper bound on the electron density that can be measured by this method is determined by the sensitivity of the instrument.

This technique has been applied to lines of O IV, O V (Feldman *et al.*, 1977; Feldman and Doschek, 1978a), O III and C III (Doschek *et al.*, 1978), N IV (Doschek *et al.*, 1979), and S IV (Bhatia *et al.*, 1979b), which have been observed with the ATM slit spectrograph. Because of the limited spectral range of the instrument, it was often not possible to simultaneously observe an allowed line of the same ion in order to determine the emission measure. The emission measure of O IV was taken as the average of the C IV and N V emission measures, since these lines bracket O IV in temperature. In the case of C III and O III, the emission measure was found from Si IV. It was often necessary to normalize these curves to yield the observed ratios at their low-density limits found in the quiet sun, where the pressure is fairly well known. This normalization procedure corrects for errors in abundance, atomic data, calibration, etc. Part of the need for normalization also stems from the fact that the $G(T)$ curves for different ions are never identical. This introduces a temperature dependence in the ratio. To first order, this effect can be calculated by approximating the differential emission measure ξ by a power law over a small temperature interval. For N IV λ1486, ξ can be determined from C IV and N V, and Dere *et al.* (1980) have shown that the ratio $I(\text{N IV})/[I(\text{C IV}) + I(\text{N V})]$ is only weakly dependent on the slope of ξ in the low-density limit. Consistent results have been obtained for all of these ions except S IV. For this ion, $P/\langle N_e^2 \rangle \Delta V$ should not decrease until a fairly high density (5×10^{11} cm^{-3}) is reached. However, a decrease has been observed in regions of lower density as determined from other lines. The S IV EUV lines are due to $3s^2 3p\ ^2P - 3s3p^2\ ^4P$ transitions, and it has been suggested that the distorted-wave calculations are not accurate for this problem (Bhatia *et al.*, 1979b).

An advantage of this type of electron density diagnostic is its relative insensitivity to errors in atomic data. The electron density at which pseudo-Boltzmann equilibrium is reached is determined by A_{ji}/C_{ij}. An error of 25% in C_{ij} shifts the curves $P/\langle N_e^2 \rangle \Delta V$ by 25% along the N_e axis. A shift of 25% in A_{ji} shifts the curves by 25% along both axes. These uncertainties are not as serious as those discussed previously, such as for C III.

The process of Bowen fluorescence has been suggested by Raymond (1978b) as an electron density diagnostic. In this process, a Lyα photon of He II (303.771 Å) is absorbed by the transition $2p^2\ ^3P_2 - 2p3d\ ^3P_2$ (303.799 Å) of O III. The method rests on the measurement of a line produced mainly by fluorescence that is proportional to the column density of O III and the He II λ304 intensity and the measurement of another collisionally excited line proportional to the emission measure. The ratio of the two will be a function of density. However, high-spectral resolution data obtained with the NRL Skylab XUV spectroheliograph (Dere, 1978b) show that the feature at 374 Å used by Raymond is attributable to an unidentified coronal line and not O III. Furthermore, intensities of O III λ704 are highly subject to H Lyman continuum absorption. Unidentified lines in the solar spectrum at 303.63 Å and 395.56 Å, if attributable to O III, could be used to study the fluorescence mechanism.

6.3.3 Summary of Density Diagnostics

We have summarized in Figure 6.1 the electron density diagnostics discussed in this section. The full lines indicate methods that use line ratios from within the same ion. The dashed lines refer to the method developed by Doschek and Feldman (Section 6.3.2). The high-density limit for Fe IX and for the dashed lines that we have chosen is the density at which the intensity ratio of the forbidden to allowed line has fallen by an order of magnitude from its maximum value.

A summary of typical electron density values obtained recently from line intensity ratios is given in Tables 6.1 and 6.2. A detailed comparison of these results is difficult because of the diverse nature of active regions. The interpretation of these tables is also complicated by the fact that some spectra come from the disk and some from the limb. These tables are presented as a guide to typical electron densities. For detailed studies, the original papers should be consulted. The results of Dere (1979) are based on observations of active regions at the limb by the Skylab XUV spectroheliograph.

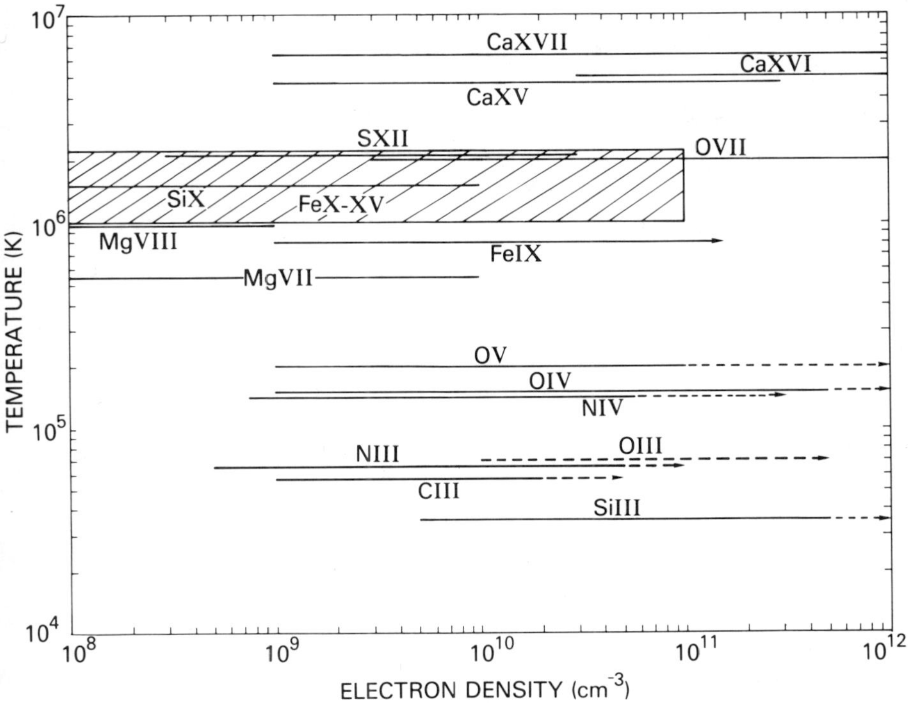

Figure 6.1. The range of sensitivity for a number of ions as a function of the characteristic temperature of emission T_m.

Nevertheless, a comparison of these values would be helpful. An appropriate way is to compare the electron pressures ($N_e T$) obtained from the derived densities and their corresponding values of T_m and to look for systematic trends. When this is done, it is found that, aside from some major exceptions, the measurements are consistent (to within a factor of two to three) with a constant pressure of about 6×10^{15} cm^{-3} K. Notable exceptions are the C III results of Dupree (1978) and Dufton *et al.* (1978), who use the $I(\lambda 1176)/I(\lambda 977)$ ratio, which is not especially sensitive at active region densities. Both Dupree and Dufton *et al.* rely on the measurements of Vernazza and Reeves (1978), whose ratios are equal (to within error) to the ratio calculated in the high-density limit by Cook and Nicolas (1979). The densities derived from the O V $I(\lambda 760)/I(\lambda 629)$ ratio are probably low because of the wavelength dependence of the H Ly continuum absorption. Feldman *et al.* (1978b) find densities from S X that are the same in both quiet and active regions. This result is not consistent with most other observations. The Si X densities of Vernazza and Mason (1978) are systematically low because of blending of the 347 Å line, which increases the intensity of the unresolved feature by about a factor of two.

Table 6.1
Typical Active Region Electron Densities: Line Ratios of a Single Ion

Ion	T_m(K)	N_e(cm^{-3}) Average	N_e(cm^{-3}) Intense	Source
Si III	3.5×10^4	$1\text{–}3 \times 10^{11}$	5×10^{11}	Nicolas *et al.* (1979)
C III	7×10^4	$0.4\text{–}2 \times 10^{11}$	3×10^{10}	Cook and Nicolas (1979)
		1.6×10^{10}	1.3×10^{10}	Dupree (1978)
		9×10^9		Dufton *et al.* (1978)
O V	2×10^5	$1\text{–}3 \times 10^{10}$	$1\text{–}3 \times 10^{10}$	Malinovsky (1975)
S X	1.3×10^6	1×10^9		Feldman *et al.* (1978b)
Si X	1.3×10^6	7×10^8	1×10^9	Vernazza and Mason (1978)
		$>3 \times 10^9$		Dere (1979)
Fe XII	1.5×10^6	$1\text{–}4 \times 10^9$		Gabriel and Jordan (1975)
		6×10^9		Dere (1979)
Fe XIII	1.6×10^6	3×10^9		Dere (1979)
Fe XIV	1.8×10^6	$1\text{–}3 \times 10^9$		Dere (1979)
O VII	2×10^6	7×10^9		Acton and Catura (1976)
S XII	2×10^6	3×10^9		Dere (1979)
Fe XV	2×10^6	6×10^9		Dere (1979)

Table 6.2
Typical Active Region Electron Densities: Ratio of Forbidden Line to Emission Measure

Ion	T_m(K)	N_e(cm^{-3})	Source
O III	6×10^4	1×10^{11}	Doschek *et al.* (1978)
C III	7×10^4	$\leq 3 \times 10^{11}$	Feldman and Doschek (1978b)
		$0.3\text{–}2 \times 10^{11}$	Cook and Nicolas (1979)
N IV	1.2×10^5	$1\text{–}2 \times 10^{11}$	Dere *et al.* (1980)
O IV	1.3×10^5	10^{11}	Feldman and Doschek (1978a)
		$\leq 1.5 \times 10^{11}$	Feldman and Doschek (1978b)
O V	2×10^5	$\leq 1.5 \times 10^{11}$	Feldman and Doschek (1978b)

6.4 ATOMIC PHYSICS

In this section, we review the methods commonly used for calculating atomic parameters. We have not attempted to review all available atomic data; reviews of this nature are cited in the text. Our aim is rather to indicate the accuracies and limitations of those calculations that have a direct relevance to the interpretation of solar spectra.

6.4.1 Atomic Structure

The accuracy of an atomic structure calculation is very dependent on which configurations have been included in the solution of the problem. As an element becomes more highly ionized, it approaches the hydrogenic limit for which the energy levels with the same principal quantum number are degenerate. Layzer (1959) pointed out that to obtain accurate wave functions for highly ionized systems, it is necessary to include all configurations in the same complex (same principal quantum numbers and parity). As an example, for Ca XV with ground configuration $2s^22p^2$, it is necessary to calculate the configuration interaction with $2p^4$. For the iron ions in the $n = 3$ complex, the number of configurations to be included is much larger (e.g., Fe XI—ground configuration $3s^23p^4$ interacts with $3p^6$, $3s^23p^23d^2$, $3s3p^43d$, $3p^43d^2$). In fact, to obtain really accurate wave functions, one should include an infinite set of configurations. This is clearly not feasible, and the skill in carrying out these calculations lies in choosing an optimum set of configurations. With a reasonable set of configurations, allowed transition probabilities are generally reliable to about $\pm 50\%$. Various techniques have been devised to improve still further the wave functions obtained with an optimum set of configurations. Some authors use a semi-empirical fitting procedure with observed energy levels (Bromage *et al.*, 1977, 1978). Others modify the potential for the outer orbitals, to simulate the effect of more configurations (Nussbaumer and Storey, 1978). In this way, radiative data can be calculated to an accuracy of 10%. A different method of obtaining very accurate atomic structure data is to use the techniques of collision theory to solve the bound state problem. Extensive calculations of this type, for low-charge ions, have been carried out by Seaton and coworkers at University College, London.

Atomic structure calculations for ions of interest in the solar atmosphere are usually carried out in the LS coupling scheme and often transformed into intermediate coupling. For highly ionized systems (i.e., coronal ions), it is important to allow for spin orbit interaction and other relativistic corrections. Such interactions are very sensitive to the accuracy of the LS coupling eigenvalues (energy positions). Small discrepancies in these energies can lead to significant differences in the radiative data. Some intersystem transitions (e.g., Fe XI, $3s3p^5\ {}^3P_1$–$3s^23p^4\ {}^1S_0$) are particularly vulnerable to cancellations in the summation to obtain the line strengths. These transition probabilities are usually very small, but large discrepancies (i.e., a factor of five) are obtained when different sets of configurations are considered (Mason, 1975). Electric quadrupole and magnetic dipole transition probabilities vary as λ^{-5} and λ^{-3}, respectively, so errors in these wavelengths are important.

A bibliography (Fuhr *et al.*, 1978) and a review (Wiese, 1979) of radiative transition probabilities have recently been published.

6.4.2 Electron and Proton Scattering

Work on electron-ion collisions has been reviewed by Seaton (1975). The main approximations used for ions are Coulomb-Born (CB), "distorted wave" (DW), and "close coupling" (CC). The CB and DW calculations are carried out neglecting coupling of the scattering-electron and target-state wave functions. The wave functions for the scattering electron are calculated in the field of a potential $V(r)$. For the CB approximation, $V(r) = -2z/r$ for all r, where z is the charge on the ion. In the DW approximation, an allowance is made for the screening of the nucleus by ionic electrons, and $V(r) = -2z/r$ asymptotically. In the close-coupling approximation, a set of integrodifferential equations is solved. To make the solution of the electron scattering problem feasible, it is again necessary to truncate the number of target terms included. Since the CC method is the most complex approximation, this truncation is much more severe for CC than for DW or CB. Pseudostates are often used in CC calculations to improve the target wave functions. Although the CC method is in principle more accurate ($\pm10\%$) than the DW method ($\pm25\%$), this depends on the target wave function used for particular calculations.

The accuracy of the DW approximation improves as an element becomes more highly ionized. However, when energy levels become very close, they are strongly coupled. Hayes and Seaton (1977) compared DW and CC calculations for the hydrogen sequence. They found large discrepancies near threshold for the 1s–2s transitions in low-charged elements ($z < 10$). For the 1s–2p transition at higher z values, the agreement was good ($<10\%$). The near degeneracy of energy levels was found to be a problem with the Li-like sequence. The DW calculations by Flower (1971) were not consistent with solar observations. Recent CC calculations by Hayes (1975) and Henry (1974) gave an increase of 50% in the N^{+4} 2s–3p collision strength.

A major complication in the electron scattering calculations is the resonance structure. To obtain the electron collision rate, the collision cross section is integrated over a Maxwellian velocity distribution. For the transition region ions, the most important contribution to the excitation rate coefficient comes from near threshold. The resonance structure ($3ln'l'$, $n' \geq 3$) for the 1s–2s and 1s–2p transitions in H-like ions was found to contribute 20% and 10%, respectively, to the excitation rate coefficients (Hayes and Seaton, 1978). However, the resonance structure for forbidden and intersystem transitions can be much more important. For example, Malinovsky (1975) estimated the resonance contributions to the O V excitation rates. The collision rates for $2s^2\ {}^1S_0$–$2s2p\ {}^3P_J$ and $2s2p\ {}^3P_J$–$2s2p\ {}^3P_{J'}$ were found to be enhanced by factors between 1.8 and 3.4. For such transitions, it is essential to estimate the resonance contributions to the electron excitation rates. These resonances are due to the capture of the incident electron by the ion X^{+m} to form a doubly excited state X^{+m-1}, followed by Auger break up to the continuum above the first excited level of X^{+m}. Many authors use an approximate method to allow for these resonances when calculating excitation rates (for example, Petrini, 1970). Quantum mechanical calculations of resonance structure are lengthy, but they are necessary if accuracy greater than 25% is required.

Cross sections for proton collisions are often calculated using a semiclassical approach (e.g., Seaton, 1964b; Bely and Faucher, 1970), based on the Coulomb excitation of nuclei (Alder *et al.*, 1956). Some authors solve the appropriate time-dependent Schrodinger equation explicitly (e.g., Masnou-Seeuws and McCarroll, 1972; Landman, 1975). Close-coupling calculations have been carried out for Fe XIII (Faucher, 1975). These results agree well with the calculations by Landman (Faucher and Landman, 1977). Quantum treatments of ions with np^2 ground configurations (Faucher, 1977) agree to within 30% with results obtained using Bely and Faucher's (1970) method for electric quadrupole transitions. Kastner (1977) and Kastner and Bhatia (1979) have recently developed some general approximations for proton excitation cross sections and rates based on available calculations.

6.4.3 Ionization Equilibrium

The ionization function $N(X^{+m})/N(X)$ in Eq. 6.2.2 is determined by the balance of ionization and recombination processes among the various species. $S(X^{+m})\,N(X^{+m}) = \alpha\,(X^{+m+1})\,N(X^{+m+1})$, where S is the ionization rate and α the recombination rate. Calculations of ionization equilibria have been carried out by numerous authors; the most commonly used publications are Jordan (1969, 1970), Allen and Dupree (1969), Landini and Monsignori-Fossi (1972), Summers (1972, 1974a), and Jacobs *et al.* (1977a,b, 1979). The difference between these sets of results are outlined below.

For solar conditions, almost all the population of a particular ion species is in the ground configuration. This enables the solution of excitation and deexcitation processes for lower levels to be treated separately from ionization and recombination processes.

The ionization rate generally includes direct collisional ionization, collisional excitation into autoionizing levels, and photoionization. Collisional ionization from the inner and outer shells of the ground configuration can be expressed as:

$$X_a^{+m} + \mathrm{e} \rightarrow X_b^{+m+1} + \mathrm{e} + \mathrm{e}' \quad . \tag{6.4.1}$$

Here, X_a^{+m} denotes an ion X^{+m} in a quantum state a, similarly for X_b^{+m+1}. Ionization processes were reviewed by Bely and Van Regemorter (1970). General formulas for ionization rates have been given by Seaton (1964a) (semiempirical formula—SEF), Lotz (1967), Burgess (1964b), and Burgess and Percival (1968) (exchange classical impact formulation—ECIP). SEF uses classical theory as a guide for extrapolation and is based on some theoretical and experimental measurements. It gives reasonably reliable results near threshold but does not have the correct high-energy behavior. Lotz made an empirical fit to the available experimental data, using classical theory to extrapolate for ions of higher charge. Lotz's values are in agreement with SEF near threshold, which is the astrophysically interesting regime. The ECIP formulation gives threshold values that are about half of the SEF or Lotz values, except for the hydrogen-like and helium-like results, where all formulations agree. A review by Burgess *et al.* (1977) compares the values obtained from these formulations with laboratory measurements. Crossed-beam experiments, which are capable of good accuracy (±10%), are presently limited to ions of charge one or two. Plasma spectroscopy experiments are used for more highly ionized species, but the accuracy is not as good (25%-50%). Burgess *et al.* conclude that on average the laboratory results favor the ECIP values. The more sophisticated Coulomb-Born calculations (e.g., Moores, 1972) seem to overestimate the laboratory cross sections near threshold, but give satisfactory agreement at high energies. A simple expression for the ionization cross sections of highly charged ions has been published by Sampson and Golden (1978) and Moores *et al.* (1979). This is based on calculations of the reduced cross section for hydrogenic ions, using the Coulomb-Born exchange method in the limit of $z \to \infty$. Cross sections obtained using this expression give excellent agreement with recent Coulomb-Born calculations (Moores, 1979, unpublished). Improved and more extensive laboratory measurements are needed to clarify which approximation is best. Burgess *et al.* discuss the use of the ECIP formulation to calculate the processes of both inner shell direct ionization and inner shell excitation followed by autoionization. These could significantly enhance the total ionization rate for particular ion species. Cowan and Mann (1979) have shown that for sodium-like ions, such as Fe XVI, ionization by collisional excitation into autoionizing levels proceeds at a rate about 50% greater than by direct collisional ionization.

The recombination rate includes both radiative and dielectronic processes. Burgess (1964a) showed that dielectronic recombination is the dominant process at high temperature. This process can be expressed as:

$$X_a^{+m+1} + e \leftrightarrows X_b^{+m} \rightarrow X_c^{+m} + \gamma \quad . \tag{6.4.2}$$

An electron attaches itself to the ion X_a^{+m+1}, forming a doubly excited state X_b^{+m}, which can either autoionize back to X_a^{+m+1} or undergo a stabilizing radiative transition to X_c^{+m}. This process is illustrated in Figure 6.2. The process of excitation followed by autoionization is illustrated by the dashed lines. A review of dielectronic recombination calculations was published by Seaton and Storey (1976). Most calculations of the ionization balance in the solar corona use the general formula for the dielectronic recombination rate given by Burgess (1965), which has an estimated accuracy of about 30%. The process usually considered is such that the initial state X_a^{+m+1} and the final state X_c^{+m} are ground states; however, Jacobs *et al.* (1977a,b) considered the possibility that these are both arbitrary states. In particular, they calculate the reduction in the dielectronic recombination rate due to the doubly excited state autoionizing into the continuum associated with an excited state of the recombining ion. If the primary excitation is via a $\Delta n = 1$ transition, there can be significant autoionization to continua associated with the $\Delta n = 0$ as well as the $\Delta n = 1$ transitions. This possibility is included in the early formulation by Burgess (1964a), but it could not be taken into account in the simple general approximation (Burgess, 1965) that is widely used. Raymond (1978a) considered another process in which the stabilizing transition of the recombining electron goes via $\Delta n = 0$ transitions. That is, in the case where $n' = n$, the state $X_b^{+m}(b = nl + 1, n''l'')$ can decay to $X_d^{+m}[d = nl + 1, n'''l''']$. This process results in an increase to the dielectronic recombination rate of less than 30%.

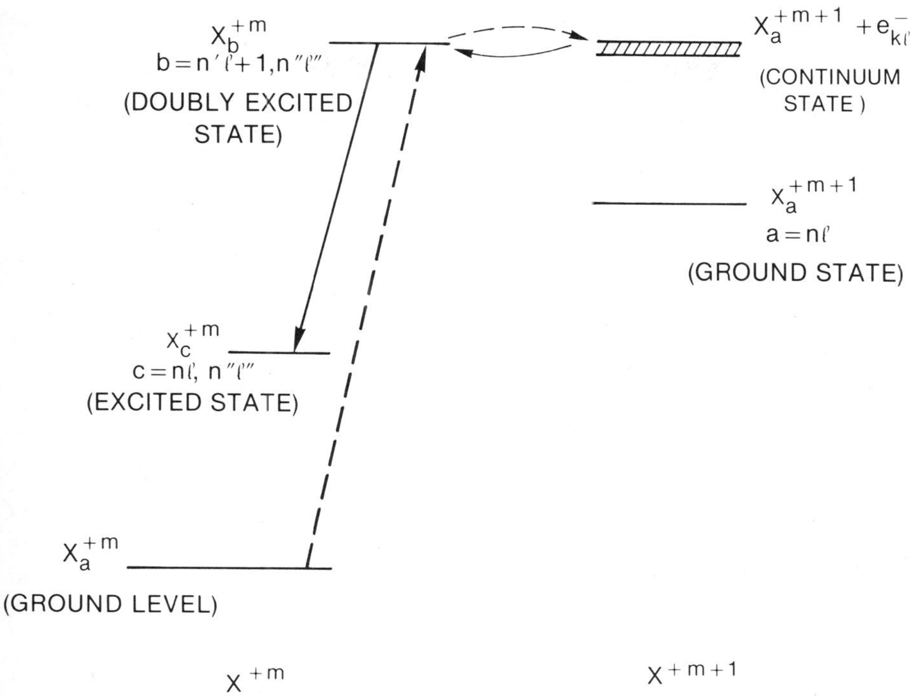

*Figure 6.2. The process of dielectronic recombination (*solid arrow*) and its inverse, autoionization (*dashed arrow*).*

The process of radiative recombination is

$$X_a^{+m+1} + e \rightarrow X_b^{+m} + \gamma \tag{6.4.3}$$

In this process, a free electron in the field of a positive ion makes a radiative transition to a bound level, with the photon carrying away the excess energy. The inverse process is photoionization. Since this is essentially a two-body process, it can be evaluated with better precision than, for example, collisional ionization. Aldrovandi and Pequignot (1973) and Burgess and Seaton (1964) have provided approximate expressions for calculating the radiative recombination rate.

Recombination occurs mainly to highly excited levels. Burgess's general formula is valid in the low-density limit. As the electron density increases ($>10^8$ cm^{-3}) the recombination coefficient will be reduced through rapid ionization of these upper levels by collisions. Burgess and Summers (1969) and subsequent papers by Summers (1972, 1974a,b) carried out a full treatment of this problem. They solved the set of equilibrium equations for the populations of individual states by reducing it to a manageable size using matrix condensation techniques. In most ionization equilibrium calculations, the density effects are estimated by excluding the recombination contributions for levels higher than a critical value n_t. The level n_t is defined such that collisional reionization is as likely as a stabilizing radiative transition. Jordan (1969) used an expression for the collision limit given by Griem (1964, p. 160). Jacobs *et al.* (1977a,b, 1979) use the Inglis-Teller (1939) limit (Griem, 1964, p. 125), which considers the magnitude of the quasistatic linear Stark shifts relative to the level separation. In the case of Fe XV, the Inglis-Teller limit is about a factor of three higher than the collision limit. Jordan (1970) did not exclude levels above n_t and therefore has overestimated the dielectronic recombination rate.

Burgess and Summers (1969) considered another process that introduces an electron density dependence in the recombination rate. This is collisional redistribution of electrons among the angular momentum states of doubly excited levels. Jacobs *et al.* (1976) discuss the enhancement in the dielectronic recombination rate due to the *l*-redistribution effect of quasistatic ion electric microfields.

To summarize, the differences between some of the ionization equilibrium calculations are listed below. Jordan (1969, 1970) uses SEF, Jacobs *et al.* (1977a,b, 1979) and Summers (1972) use Lotz, but Summers (1974a,b) uses ECIP ionization rates. Burgess *et al.* (1977) investigated what difference a factor of two in the ionization rate can have on the ionization equilibrium calculations. Using neon as an example, they concluded that the greatest effect would be on highly charged ions. As an example, at 10^6 K, temperatures corresponding to the same population ratio are lower by about 25% if 2 × ECIP rates are used rather than the actual ECIP values. Jordan and Jacobs *et al.* both use approximate methods to allow for the electron density effects on the dielectronic recombination rate. Jacobs *et al.* consider an additional process involving dielectronic recombination, which reduces the rate. Jacobs *et al.*'s values for the temperature of maximum equilibrium abundance are somewhat lower than Jordan's. Summers (1972, 1974) gives a full treatment of the electron density dependence for dielectronic recombination but does not take account of the additional processes discussed by Jacobs *et al.* His dielectronic rates are therefore probably overestimates. These rates, combined with smaller ionization rates, lead to temperatures for maximum equilibrium abundance that are significantly higher than Jordan's and Jacobs *et al.*'s values.

Feldman and Doschek (1977b) and Cheng *et al.* (1979a,b) have carried out extensive studies of line widths, using the ATM S082B data in an attempt to determine the temperatures of maximum abundance for different ion species. Their results, which are discussed in Section 6.5.1, favor Jordan's as opposed to Summers' calculations.

Nussbaumer and Storey (1975) discuss two further processes that can influence ionization equilibrium calculations. These are photoexcitation and photoionization by the solar radiation field and the effect of significant populations in metastable levels. From their study of carbon ions, they found the former process to be the most important, particularly in active regions. The latter process is dependent on electron density and could be significant for high-density features (i.e., flares). A recent publication by Vernazza and Raymond (1979) investigates the effect of metastable levels on the ionization equilibrium for other ion species.

6.4.4 Effects of Diffusion and Flows on Ionization Balance

Analyses of solar ultraviolet spectra, whether in quiet or active regions, generally show small emission measures at transition zone (10^5 K) temperatures. For a one-dimensional atmosphere, this requires steep temperature gradients. Chapman (1958) calculated the thermal diffusion factor for ionized gases and showed that such a temperature gradient would produce increased concentrations of highly ionized ions in the higher-temperature regions. The effects of diffusion have been studied for homogeneous one-dimensional models derived from ultraviolet line intensities by Delache (1967) and Tworkowski (1975, 1979). These authors showed that heavier ions will diffuse into higher-temperature regions and spectral lines would be emitted at electron temperatures higher than those determined from an assumption of ionization equilibrium. The calculations assumed a net outward flow velocity (the origin of the solar wind), which enhanced this effect, but diffusion was the dominant mechanism. Mariska (1980) has searched for variations in solar abundances that could be caused by such processes and found no significant variation.

Raymond and Dupree (1978) have considered the effect of mass flows through a temperature gradient. They note that velocities on the order of 10 km s^{-1} are often observed in the transition zone, where steep temperature gradients on the order of 10^3 K km^{-1} are deduced. The time scales for temperature changes are then on the order of the typical ionization and recombination times, and thus one might expect transient ionization to be present. In particular, they have performed calculations for the flow of a C III emitting plasma through typical plane parallel solar atmospheres. They found that, for example, the line intensity ratio $I(\lambda 1176)/I(\lambda 977)$ is strongly affected by downflows but relatively unaffected by upflows.

The severity of these effects is directly related to the steepness of the temperature gradient in the transition zone. This is generally derived on the assumption that the quiet sun is homogeneous and one dimensional. However, such assumptions have generally been inconsistent with limb brightening curves of transition zone lines (Withbroe, 1970; Brueckner and Nicolas, 1973; Feldman *et al.*, 1979). Considerable network transition zone emission originates in unresolved spicular structures that follow the magnetic field. Temperature gradients perpendicular to the magnetic field would then be large, but gradients parallel to the field, those responsible for diffusive and flow effects, would be considerably less than those calculated from homogeneous models. The relevance of plasma flows through temperature gradients can be questioned on a number of grounds. The temperature gradients derived from spectral data are all based on the assumption of ionization equilibrium. If these plasmas are in transient ionization, the derived gradients must then be suspect. Further, the effect of a plasma flow would be to wipe out any such temperature gradient due to the associated enthalpy and kinetic energy fluxes. Finally, the temperature gradient deduced from Doppler-shifted lines is moving with the same velocity as the plasma; in other words, there is no flow through a temperature gradient in that case. Only over sunspots have two-component flows been observed, and the most plausible explanation is the existence of multiple structures along the line of sight.

6.5 LINE PROFILES

6.5.1 Optically Thin Profiles

The profiles of optically thin transition zone and coronal lines are determined mainly by Doppler effects produced by both random and ordered velocity distributions.

The profiles can generally be reproduced by one or more Gaussian components of width

$$\Delta\lambda^2 = \Delta\lambda_I^2 + (4 \ln 2)(\lambda/c)^2(2kT_i/M_i + \xi_0^2) \quad , \tag{6.5.1}$$

where $\Delta\lambda$ is the full width at half maximum (FWHM), $\Delta\lambda_I$ is the instrumental width, ξ_0 is the most probable velocity for random mass motion, T_i is the ion temperature, and M_i is the ion mass. In equilibrium, $T_i = T_e$, and it should generally also be true that $T_i = T_m$, where T_m is the temperature corresponding to $G(T)_{max}$ (Section 6.2.3). The values of T_m are somewhat uncertain, varying by about 0.1 dex between different authors (Section 6.4.3).

Figure 6.3 shows a profile of N IV λ1486 obtained over a sunspot by the HRTS instrument. The line profile is made up of two components, both with a FWHM of 0.2 Å and both Doppler-shifted. Such profiles demonstrate the need for high spectral resolution, so that physically different plasmas are not simply blended together to yield a single line intensity.

Feldman and Doschek (1977b) have searched for turbulent free emission from transition zone ions using the NRL S082B data. The narrowest lines they found originated from quiescent prominences and certain active regions. Their analysis favors Jordan's (1969) ionization equilibrium calculations, since they do not find examples where $T_m > T_{ion}$. This is not the case for values of T_m from Summers (1974a,b). In general, optically thin transition zone and coronal lines show excess line widths indicative of nonthermal velocities between 10 and 40 km s^{-1}. These tend to increase with height above the limb. The interpretation of these nonthermal velocities is not clear. They may result from the existence of a number of velocity fields within the 1 arcmin field-of-view or from turbulence. The nonthermal broadening of quiet sun lines has also been ascribed to the propagation of acoustic or MHD waves that might be responsible for heating the solar atmosphere (Boland *et al.*, 1975; Mariska, Feldman, and Doschek, 1978; and Cheng, Doschek, and Feldman, 1979a).

Elzner (1975) calculated the effect on the O VI λ1032 line profile of outward-propagating acoustic shock waves passing through the solar atmosphere. The resultant profiles had an asymmetry that depended on the pulse time, with the center of gravity of the line red shifted for a pulse time less than 110 s and blue shifted for a pulse time more than 110 s. Byerley *et al.* (1978) considered the outward passage of isothermal and adiabatic sound

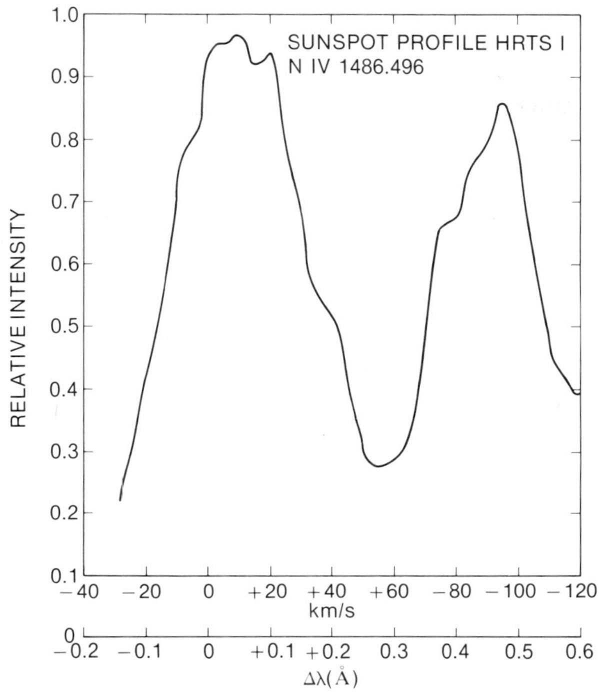

Figure 6.3. The profile of N IV λ1486 over a sunspot. Both components are considerably wider than predicted from thermal Doppler broadening. This profile also emphasizes the need for instrumentation that can at least separate such components.

waves again through an O VI emitting region. In both cases, the λ1032 line profile was asymmetric and blue shifted. Bruner and McWhirter (1979) have searched for such asymmetries in the C IV λ1548 line but have found the observed profiles to be symmetric and Gaussian in shape. They conclude that in equilibrium conditions, acoustic waves alone cannot supply enough energy to balance the radiative and conductive losses from the corona. The data are, however, consistent with heating by Alfvén waves.

Stark broadening should be unimportant for multiply ionized species but has been used by Feldman and Doschek (1977a) to determine active region densities from the hydrogen Balmer series formed at around 8000 K.

Zeeman shifts are on the order of $\Delta\lambda \sim 10^{-9}\,\lambda^2 B$ (Griem, 1964, p. 299) with λ in angstroms and B in kilogauss. In order for these shifts to be comparable to the thermal Doppler width of a line, it is necessary that $B \sim 200\ T^{1/2}\lambda^{-1}$. For EUV (1500 Å) transition zone (10^5 K) lines, detectable Zeeman splittings require 4×10^4 gauss, and for XUV (500 Å) coronal (1×10^6 K) lines, the required B fields are 4×10^5 gauss. These field strengths are considerably higher than have been observed even in sunspots. The limits of detectability can be brought down by making use of the different polarizations of the Zeeman components, as was done with the Babcock magnetograph. This technique will be attempted on the Solar Maximum Mission (launched 14 February 1980).

6.5.2 Optical Depth

Aside from several of the strongest lines, transition zone and coronal lines have a small optical depth ($\tau < 1$). However, knowledge of the optical depth in a line can provide information on $\int N_e dl$, the electron density integrated along the line of sight. The opacity in these lines arises from two sources: absorption by H I and He I for wavelengths short of 912 Å and 504 Å, respectively, and resonance scattering. Evidence for Lyman continuum absorption in the quiet sun has been published by Withbroe (1977) and Kanno (1979) and was attributed to spicules, dark mottles, or fibrilles. However, Schmahl and Orrall (1979) find that absorption shortward of 912 Å is manifest everywhere on the sun. This has serious consequences for the interpretation of the EUV spectrum from the transition zone.

In the case of resonance scattering, the optical depth $\tau = \int K dl$ is given by

$$\tau = (\pi e^2/mc^2)\, f_{ij}\, \lambda_{ij}^2\, \phi(\lambda) \int N_i(X^{+m}) dl \quad , \tag{6.5.2}$$

where

$$\phi = \exp(-(\lambda - \lambda_{ij})^2/\Delta\lambda^2)/(\sqrt{\pi}\Delta\lambda) \quad , \tag{6.5.3}$$

and $\Delta\lambda$ is the Doppler line width.

For the small optical depths expected, the equation of transfer simplifies, since the scattering mainly serves to remove photons from the line of sight, and the specific intensity is

$$I(\lambda) = (1/4\pi) \int_0^{L} \epsilon(\lambda)\phi(\lambda) \exp(-Kl)\, dl \quad . \tag{6.5.4}$$

For constant emissivity over the length L,

$$I(\lambda) = (1 - \exp(-\tau_0\phi))\tau_0^{-1}\, \Delta\lambda\epsilon(\lambda) L/(4\sqrt{\pi}) \quad , \tag{6.5.5}$$

where τ_o is the optical depth at line center. This expression is independent of the partial ion abundance. Thus, once the intensity and optical depth are known, the quantity $N_e L$ can be determined.

The ratios of the doublets from the same ion, such as the Na I and Li I isoelectronic sequences, are a function only of τ_o:

$$\frac{I_0(^2S_{1/2}-{}^2P_{3/2})}{I_0(^2S_{1/2}-{}^2P_{1/2})} = \frac{1 - \exp(-2\tau_o)}{1 - \exp(-\tau_o)} \quad , \tag{6.5.6}$$

where I_0 is the intensity at line center and τ_o is the optical depth at line center of the $^2S_{1/2}$–$^2P_{1/2}$ transition. This ratio provides a good diagnostic for $0.1 < \tau_o < 2$, where the ratio varies from a value of about 2 to a value of about 1.

Feldman and Doschek (1978b) have published the peak intensities of lines emitted from two active regions near the limb, which can be used to determine optical depths. The relevant ions and temperatures of formation are Al III (4×10^4 K) and Si IV (6×10^4 K) in the Na I sequence and C IV (1.0×10^5 K) and N V (1.6×10^5 K) in the Li I sequence. The observations extend from 12 arcsec inside the limb to 20 arcsec outside the limb. Typically, little opacity is found (from Eq. 6.5.6) at the extreme limits of the observations inside and outside the limb, and the greatest opacity is found directly at the limb. A maximum value of $\tau_0 = 0.8$ is found for Al III and C IV, with N V showing little opacity. With the high densities reported in these active regions by Feldman and Doschek, the path lengths determined from the opacity range in extent from tenths to several arcsec, and the filling factor of the 2″ × 60″ slit is considerably less than unity.

Self reversed profiles of the resonance line of C IV 1548 Å have been observed from OSO-8 (Lites *et al.*, 1980). The intensity and degree of self-reversal indicate a density of 10^{10} cm^{-3} in an active region at the solar limb. If the slit were not completely filled, the necessary density would be shifted toward the higher values found by Feldman and Doschek (1978b).

Acton and Catura (1976) have obtained observations of the ratio of the O VII forbidden plus intercombination lines to the resonance line. This ratio increases near the limb, and they attribute this increase to resonance scattering. Acton and Brown (1978) have shown that otherwise only non-equilibrium situations affect this ratio significantly.

Even as the optical depth in a line becomes significant, the probability that photons in that line will finally emerge may still be close to unity, since collisional processes that can "destroy" the photon are still slow. The chief means for destroying a photon is by collisional deexcitation of the excited ion before spontaneous reemission occurs. The probability P for reemission is then

$$P = \frac{A_{ul}}{A_{ul} + N_e C_{ul}} \quad . \tag{6.5.6}$$

This can be estimated in the Van Regemorter approximation ($\bar{g} = 0.8$) for a two-level ion as

$$P = (1 + 1.65 \times 10^{-23} \lambda_{Å} N_e / T^{1/2})^{-1} \tag{6.5.7}$$

and is very close to unity for typical transition zone and coronal conditions. Thus, for a symmetric object, the effect of opacity will be to reduce the intensity at line center while leaving the total emergent intensity unchanged. In this case, the lines are considered to be "effectively thin."

6.6 REMARKS

The structure of the solar transition region and corona has generally been determined from the analysis of XUV and EUV line intensities under the assumptions that the observations completely resolved the solar fine structure and that ionization equilibrium was maintained. However, the structure of the transition zone at the limb is apparently related to inhomogeneous spicular structures. Feldman *et al.* (1979) have shown that these structures cover only 1% of the solar surface and probably consist of elongated threads of plasma that follow the magnetic field. Such a configuration eliminates problems such as diffusion and runaway electrons (Spicer, 1979) but leads to the requirement of local heating at least in regions of Si III formation (Nicolas *et al.*, 1979). It is highly probable that the active region is made up of similar unresolved structures.

Recent observations of the sun in C IV (Bartoe *et al.*, 1979) with high temporal and spatial resolution have revealed the highly transient nature of the transition zone. This behavior brings into question the relevance of steady-state models of the transition zone. Transient ionization may also be present especially in the explosive events observed. The development of unambiguous diagnostics for ionizing and recombining transition zone plasmas would be worthwhile for studying these evolving plasmas.

Acknowledgements

H. E. Mason acknowledges the support of the Science Research Council.

We thank G. A. Doschek for his careful reading of the manuscript and helpful comments and Mrs. Lori Sizemore and Miss Joyce Brown for their patient and careful typing of the manuscript.

REFERENCES

Acton, L. W. and Brown, W. A.: 1978, *Astrophys. J.* **225,** 1065.

Acton, L. W. and Catura, R. C.: 1976, *Phil. Trans. Roy. Soc. London A* **281,** 383.

Acton, L. W., Catura, R. C., Meyerott, A. J., and Wolfson, C. J.: 1972, *Solar Phys.* **26,** 183.

Alder, K., Bohr, A., Huus, T., Mottelson, B., and Winther, A.: 1956, *Rev. Mod. Phys.* **28,** 432.

Aldrovandi, S. M. V. and Pequignot, D.: 1973, *Astron. Astrophys.* **25,** 137.

Allen, J. W. and Dupree, A. K.: 1969, *Astrophys. J.* **155,** 27.

Bartoe, J.-D. F., Brueckner, G. E., and Van Hoosier, M. E.: 1979, *Bull. Amer. Astron. Soc.* **11,** 397.

Batstone, R. M., Evans, K., Parkinson, J. H., and Pounds, K. A.: 1970, *Solar Phys.* **13,** 389.

Bely, O. and Faucher, P.: 1970, *Astron. Astrophys.* **6,** 88.

Bely, O. and Van Regemorter, H.: 1970, *Ann. Rev. Astron. Astrophys.* **8,** 329.

Berrington, K. A., Burke, P. G., Dufton, P. L., and Kingston, A. E.: 1977, *J. Phys. B* **10,** 1465.

Bhalla, C. P., Gabriel, A. H., and Presnyakov, L. P.: 1975, *Monthly Notices Roy. Astron. Soc.* **172,** 359.

Bhatia, A. K., Doschek, G. A., and Feldman, U.: 1979a, *Astron. Astrophys.* **76,** 359.

Bhatia, A. K., Doschek, G. A., and Feldman, U.: 1979b, *Astron. Astrophys.* (in press).

Bhatia, A. K., Feldman, U., and Doschek, G. A.: 1979c, *Astron. Astrophys.* **80,** 22.

Bhatia, A. K. and Kastner, S. O.: 1979, *Solar Phys.* (submitted).

Bhatia, A. K. and Mason, H. E.: 1979, *Monthly Notices Roy. Astron. Soc.* (in press).

Blaha, M.: 1969, *Astrophys. J.* **157,** 473.

Blaha, M.: 1971, *Solar Phys.* **17,** 99.

Blumenthal, G. R., Drake, G. W. F., and Tucker, W. H.: 1972, *Astrophys. J.* **172,** 205.
Boland, B. C., Dyer, E. P., Firth, J. G., Gabriel, A. H., Jones, B. B., Jordan, C., McWhirter, R. W. P., Monk, P., and Turner, R. F.: 1975, *Monthly Notices Roy. Astron. Soc.* **171,** 697.
Bromage, G. E., Cowan, R. D., and Fawcett, B. C.: 1977, *Phys. Scripta* **15,** 177.
Bromage, G. E., Cowan, R. D., and Fawcett, B. C.: 1978, *Monthly Notices Roy. Astron. Soc.* **183,** 19.
Brown, J. C.: 1974, in G. Newkirk, Jr. (ed.), 'Coronal Disturbances', *IAU Symp.* **57,** 395.
Brueckner, G. E. and Nicolas, K.: 1973, *Solar Phys.* **29,** 301.
Bruner, E. C. and McWhirter, R. W. P.: 1979, *Astrophys. J.* **231,** 557.
Burgess, A.: 1964a, *Astrophys. J.* **139,** 776.
Burgess, A.: 1964b, *Proceedings of the Symposium on Atomic Collision Processes in Plasmas*, Culham, AERE Rept. 4818,63.
Burgess, A.: 1965, *Astrophys. J.* **141,** 1588.
Burgess, A. and Percival, I. C.: 1968, *Adv. Atomic Mol. Phys.* **4,** 109.
Burgess, A. and Seaton, M. J.: 1964, *Monthly Notices Roy. Astron. Soc.* **127,** 355.
Burgess, A. and Summers, H. P.: 1969, *Astrophys. J.* **157,** 1007.
Burgess, A., Summers, H. P., Cochrane, D. M., and McWhirter, R. W. P.: 1977, *Monthly Notices Roy. Astron. Soc.* **179,** 275.
Byerley, A., McWhirter, R. W. P., and Wilson, R.: 1978, *J. Phys. B* **11,** 613.
Chapman, S.: 1958, *Proc. Phys. Soc. London* **72,** 353.
Cheng, C.-C., Doschek, G. A., and Feldman, U.: 1979a, *Astrophys. J.* **227,** 1037.
Cheng, C.-C., Feldman, U., and Doschek, G. A.: 1979b, *Astrophys. J.* **233,** 736.
Cook, J. W. and Nicolas, K. R.: 1979, *Astrophys. J.* **229,** 1163.
Cowan, R. D. and Mann, J. B.: 1979, *Astrophys. J.* **232,** 940.
Cox, D. P. and Tucker, W. H.: 1969, *Astrophys. J.* **157,** 1157.
Craig, I. J. D. and Brown, J. C.: 1976, *Astron. Astrophys.* **49,** 239.
Craig, I. J. D., McClymont, A. N., and Underwood, J. H.: 1978, *Astron. Astrophys.* **70,** 1.
Culhane, J. L.: 1969, *Monthly Notices Roy. Astron. Soc.* **144,** 375.
Culhane, J. L. and Acton, L. W.: 1970, *Monthly Notices Roy. Astron. Soc.* **151,** 141.
Dankwort, W. and Trefftz, E.: 1976, *Astron. Astrophys.* **47,** 365.
Davis, J., Kepple, P. C., and Blaha, M.: 1976, *J. Quant. Spectrosc. Rad. Transfer* **16,** 1043.
Delache, P.: 1967, *Ann. Astrophys.* **30,** 827.
Dere, K. P.: 1978a, *Astron. Astrophys.* **70,** 439.
Dere, K. P.: 1978b, *Astrophys. J.* **221,** 1062.
Dere, K. P.: 1979 (in preparation).
Dere, K. P., Bartoe, J.-D. F., Brueckner, G. E., and Van Hoosier, M. E.: 1980 (to be submitted).
Dere, K. P., Horan, D. M., and Kreplin, R. W.: 1977, *Astrophys. J.* **217,** 976.
Dere, K. P., Mason, H. E., Widing, K. G., and Bhatia, A. K.: 1979, *Astrophys. J. Suppl.* **40,** 341.
Doschek, G. A. and Feldman, U.: 1977, *Astron. Astrophys. Letters* **58,** L13.
Doschek, G. A. and Feldman, U.: 1979, NRL Report No. 8307.
Doschek, G. A., Feldman, U., Bhatia, A. K., and Mason, H. E.: 1978, *Astrophys. J.* **226,** 1129.
Doschek, G. A., Feldman, U., and Mason, H. E.: 1979, *Astron. Astrophys.* **78,** 342.
Dubau, J., Loulergue, M., and Steenman-Clark, L.: 1979, *Monthly Notices Roy. Astron. Soc.* (in press).
Dufton, P. L., Berrington, K. A., Burke, P. G., and Kingston, A. E.: 1978, *Astron. Astrophys.* **62,** 111.
Dupree, A. K.: 1978, *Adv. Atom. Molec. Phys.* **14,** 393.
Dupree, A. K., Foukal, P. V., and Jordan, C.: 1976, *Astrophys. J.* **209,** 621.
Eissner, W.: 1972, *Physics of Electronic and Atomic Collisions VII*, North Holland, Amsterdam, p. 406.
Elwert, G. and Raju, P. K.: 1975, *Astrophys. Space Sci.* **38,** 369.
Elzner, L. R.: 1975, *Solar Phys.* **45,** 93.
Engvold, O.: 1977, *Physica Scripta* **16,** 48.

Faucher, P.: 1975, *J. Phys. B* **8,** 1886.

Faucher, P.: 1977, *Astron. Astrophys.* **54,** 589.

Faucher, P. and Landman, D. A.: 1977, *Astron. Astrophys.* **54,** 159.

Feldman, U. and Doschek, G. A.: 1977a, *Astrophys. J.* **212,** 913.

Feldman, U. and Doschek, G. A.: 1977b, *Astrophys. J. Letters* **216,** L119.

Feldman, U. and Doschek, G. A.: 1978a, *Astron. Astrophys.* **65,** 215.

Feldman, U. and Doschek, G. A.: 1978b, *Astrophys. J. Suppl.* **37,** 443.

Feldman, U., Doschek, G. A., and Behring, W. E.: 1978a, *Space Sci. Rev.* **22,** 191.

Feldman, U., Doschek, G. A., and Mariska, J. T.: 1979, *Astrophys. J.* **229,** 369.

Feldman, U., Doschek, G. A., Mariska, J. T., Bhatia, A. K., and Mason, H. E.: 1978b, *Astrophys. J.* **226,** 674.

Feldman, U., Doschek, G. A., and Rosenberg, F. D.: 1977, *Astrophys. J.* **215,** 652.

Feldman, U., Doschek, G. A., and Widing, K. G.: 1978c, *Astrophys. J.* **219,** 304.

Flower, D. R.: 1971, *J. Phys. B* **4,** 697.

Flower, D. R.: 1977a, *Astron. Astrophys.* **54,** 163.

Flower, D. R.: 1977b, *Astron. Astrophys.* **56,** 451.

Flower, D. R. and Jordan, C.: 1971, *Astron. Astrophys.* **14,** 473.

Flower, D. R. and Nussbaumer, H.: 1974, *Astron. Astrophys.* **31,** 353.

Flower, D. R. and Nussbaumer, H.: 1975a, *Astron. Astrophys.* **39,** 295.

Flower, D. R. and Nussbaumer, H.: 1975b, *Astron. Astrophys.* **42,** 265.

Flower, D. R. and Nussbaumer, H.: 1975c, *Astron. Astrophys.* **45,** 145.

Flower, D. R. and Nussbaumer, H.: 1975d, *Astron. Astrophys.* **45,** 349.

Flower, D. R. and Pineau des Forets, G.: 1973, *Astron. Astrophys.* **24,** 181.

Freeman, F. E., Gabriel, A. H., Jones, B. B., and Jordan, C.: 1971, *Phil. Trans. Roy. Astron. Soc. Lond. A* **270,** 127.

Fuhr, J. R., Miller, B. J., and Martin, G. A.: 1978, NBS Special Pub. **505.**

Gabriel, A. H.: 1972, *Monthly Notices Roy. Astron. Soc.* **160,** 99.

Gabriel, A. H. and Jordan, C.: 1969, *Monthly Notices Roy. Astron. Soc.* **145,** 241.

Gabriel, A. H. and Jordan, C.: 1972, in McDaniel, E. W. and McDowell, M. R. C. (eds.), *Case Studies in Atomic Collision Physics II*, North Holland, Amsterdam, p. 210.

Gabriel, A. H. and Jordan, C.: 1973, *Astrophys. J.* **186,** 327.

Gabriel, A. H. and Jordan, C.: 1975, *Monthly Notices Roy. Astron. Soc.* **173,** 397.

Gerassimenko, M. and Nolte, J. T.: 1978, *Solar Phys.* **60,** 299.

Griem, H. R.: 1964, *Plasma Spectroscopy*, McGraw-Hill, New York.

Hayes, M. A.: 1975, *J. Phys. B* **8,** L8.

Hayes, M. A. and Seaton, M. J.: 1977, *J. Phys. B* **10,** L573.

Hayes, M. A. and Seaton, M. J.: 1978, *J. Phys. B* **11,** L79.

Henry, R. J. W.: 1974, *J. Phys. B* **7,** L439.

Heroux, L. and Cohen, M.: 1971, *Phil. Trans. Roy. Soc. Lond. A* **270,** 99.

Heroux, L., Cohen, M., and Malinovsky, M.: 1972, *Solar Phys.* **23,** 369.

Inglis, D. R. and Teller, E.: 1939, *Astrophys. J.* **90,** 439.

Jacobs, V. L., Davis, J., and Kepple, P. C.: 1976, *Phys. Rev. Letters* **37,** 1390.

Jacobs, V. L., Davis, J., Kepple, P. C., and Blaha, M.: 1977a, *Astrophys. J.* **211,** 605.

Jacobs, V. L., Davis, J., Kepple, P. C., and Blaha, M.: 1977b, *Astrophys J.* **215,** 690.

Jacobs, V. L., Davis, J., Rogerson, J. E., and Blaha, M.: 1979, *Astrophys. J.* **230,** 627.

Jordan, C.: 1969, *Monthly Notices Roy. Astron. Soc.* **142,** 501.

Jordan, C.: 1970, *Monthly Notices Roy. Astron. Soc.* **148,** 17.

Jordan, C.: 1974, *Astron. Astrophys.* **34,** 69.

Jordan, C.: 1979, in Hanle, W. and Kleinpoppen, H. (eds.), *Progress in Atomic Spectroscopy*, Plenum, New York.

Kahler, S.: 1976, *Solar Phys.* **48,** 255.

Kanno, M.: 1979, *Publ. Astron. Soc. Japan* **31,** 115.

Kastner, S. O.: 1977, *Astron. Astrophys.* **54,** 255.

Kastner, S. O. and Bhatia, A. K.: 1979, *Astron. Astrophys.* **71,** 211.
Kastner, S. O. and Mason, H. E.: 1978, *Astron. Astrophys.* **67,** 119.
Kato, T.: 1976, *Astrophys. J. Suppl.* **30,** 397.
Kunasz, C. V., Jefferies, J. T., and White, O. R.: 1973, *Astron. Astrophys.* **28,** 15.
Lambert, D. L.: 1978, *Monthly Notices Roy. Astron. Soc.* **182,** 249.
Lambert, D. L. and Luck, R. E.: 1978, *Monthly Notices Roy. Astron. Soc.* **183,** 79.
Lampton, M., Margon, B., and Bowyer, S.: 1976, *Astrophys. J.* **208,** 177.
Landini, M. and Monsignori-Fossi, B. C.: 1972, *Astron. Astrophys. Suppl.* **7,** 291.
Landini, M. and Monsignori-Fossi, B. C.: 1979, *Astron. Astrophys.* **72,** 171.
Landman, D. A.: 1975, *Astron. Astrophys.* **43,** 285.
Landman, D. A. and Brown, T.: 1979, *Astrophys. J.* **232,** 636.
Layzer, D.: 1959, *Ann. Phys.* **8,** 271.
Levine, R. L. and Pye, J. P.: 1979, *Solar Phys.* (to appear).
Lites, B. W., Hansen, E. R., and Shine, R. A.: 1980, *Astrophys. J.* **236,** 280.
Lotz, W.: 1967, *Astrophys. J. Suppl.* **14,** 207.
Loulergue, M. and Nussbaumer, H.: 1974, *Astron. Astrophys.* **34,** 225.
Loulergue, M. and Nussbaumer, H.: 1976, *Astron. Astrophys.* **51,** 163.
Malinovsky, M.: 1975, *Astron. Astrophys.* **43,** 101.
Mariska, J. T.: 1980, *Astrophys. J.* **235,** 268.
Mariska, J. T., Feldman, U., and Doschek, G. A.: 1978, *Astrophys. J.* **226,** 698.
Masnou-Seeuws, F. and McCarroll, R.: 1972, *Astron. Astrophys.* **17,** 441.
Mason, H. E.: 1975, *Monthly Notices Roy. Astron. Soc.* **170,** 651.
Mason, H. E. and Bhatia, A. K.: 1978, *Monthly Notices Roy. Astron. Soc.* **184,** 423.
McKenzie, D. L., Broussard, R. M., Landecker, P. B., Rugge, H. R., Young, R. M., Doschek, G. A., and Feldman, U.: 1979 (in preparation).
McKenzie, D. L., Rugge, H. R., Underwood, J. H., and Young, R. M.: 1978, *Astrophys. J.* **221,** 342.
Mewe, R.: 1972, *Solar Phys.* **22,** 459.
Mewe, R. and Schrijver, J.: 1978a, *Astron. Astrophys.* **65,** 99.
Mewe, R. and Schrijver, J.: 1978b, *Astron. Astrophys.* **65,** 115.
Moores, D. L.: 1972, *J. Phys. B* **5,** 286.
Moores, D. L.: 1979 (unpublished calculations).
Moores, D. L., Golden, L. B., and Sampson, D. H.: 1979, *J. Phys. B* (in press).
Nicolas, K. R.: 1977, Ph.D. Thesis, University of Maryland.
Nicolas, K. R., Bartoe, J.-D. F., Brueckner, G. E., and Van Hoosier, M. E.: 1979, *Astrophys. J.* **233,** 741.
Nussbaumer, H.: 1976, *Astron. Astrophys.* **48,** 93.
Nussbaumer, H. and Storey, P. J.: 1975, *Astron. Astrophys.* **44,** 321.
Nussbaumer, H. and Storey, P. J.: 1978, *Astron. Astrophys.* **64,** 139.
Nussbaumer, H. and Storey, P. J.: 1979a, *Astron. Astrophys.* **71,** L5.
Nussbaumer, H. and Storey, P. J.: 1979b, *Astron. Astrophys.* **74,** 244.
Nussbaumer, H. and Storey, P. J.: 1979c, *Astron. Astrophys.* (submitted).
Pagel, B. E. J.: 1973, *Space Sci. Rev.* **15,** 1.
Peacock, N. J. and Summers, H. P.: 1978, *J. Phys. B* **11,** 3757.
Petrini, D.: 1970, *Astron. Astrophys.* **9,** 392.
Phillips, D. L.: 1962, *J.A.C.M.* **9,** 63.
Pottasch, S. R.: 1964, *Space Sci. Rev.* **3,** 816.
Raymond, J. C.: 1978a, *Astrophys. J.* **222,** 1114.
Raymond, J. C.: 1978b, *Astrophys. J.* **224,** 259.
Raymond, J. C., Cox, D. P., and Smith, B. W.: 1976, *Astrophys. J.* **204,** 290.
Raymond, J. C. and Dupree, A. K.: 1978, *Astrophys. J.* **222,** 379.
Raymond, J. C. and Smith, B. W.: 1977, *Astrophys. J. Suppl.* **35,** 419.
Rosner, R., Tucker, W. H., and Vaiana, G. S.: 1978, *Astrophys. J.* **220,** 419.
Ross, J. E. and Aller, L. H.: 1976, *Science* **191,** 1223.

Sampson, D. H. and Golden, L. B.: 1978, *J. Phys. B* **11,** 5.
Schmahl, E. J. and Orrall, F. Q.: 1979, *Astrophys. J. Letters* **231,** L41.
Seaton, M. J.: 1964a, *Planetary Space Sci.* **12,** 55.
Seaton, M. J.: 1964b, *Monthly Notices Roy. Astron. Soc.* **127,** 191.
Seaton, M. J.: 1975, *Adv. Atom. Mol. Phys.* **11,** 83.
Seaton, M. J. and Storey, P. J.: 1976, in Burke, P. G. and Moiseiwitsch, B. L. (eds.), *Atomic Processes and Applications*, North Holland, New York.
Spicer, D. S.: 1979, *Solar Phys.* **62,** 269.
Steenman-Clark, L., Bely-Dubau, F., and Faucher, P.: 1979, *Monthly Notices Roy. Astron. Soc.* (in press).
Summers, H. P.: 1972, *Monthly Notices Roy. Astron. Soc.* **158,** 255.
Summers, H. P.: 1974a, *Monthly Notices Roy. Astron. Soc.* **169,** 663.
Summers, H. P.: 1974b, Appleton Laboratory Internal Memo 367.
Summers, H. P. and McWhirter, R. W. P.: 1979, *J. Phys. B* **12,** 2387.
Trimble, V.: 1975, *Rev. Mod. Phys.* **47,** 877.
Tucker, W. H. and Koren, M.: 1971a, *Astrophys. J.* **168,** 283.
Tucker, W. H. and Koren, M.: 1971b, *Astrophys. J.* **170,** 621 (erratum).
Twomey, S.: 1963, *J.A.C.M.* **10,** 97.
Tworkowski, A. S.: 1975, *Astrophys. Letters* **17,** 27.
Tworkowski, A. S.: 1979, *Monthly Notices Roy. Astron. Soc.* (submitted).
Underwood, J. H. and McKenzie, D. L.: 1977, *Solar Phys.* **53,** 417.
Underwood, J. H. and McKenzie, D. L.: 1978, *Solar Phys.* **60,** 311.
Van Regemorter, H.: 1962, *Astrophys. J.* **136,** 906.
Vernazza, J. E. and Mason, H. E.: 1978, *Astrophys. J.* **226,** 720.
Vernazza, J. E. and Raymond, J. C.: 1979, *Astrophys. J. Letters* **228,** L89.
Vernazza, J. E. and Reeves, E. M.: 1978, *Astrophys. J. Suppl.* **37,** 485.
Walker, A. B. C., Rugge, H. R., and Weiss, K.: 1974, *Astrophys. J.* **188,** 423.
Wiese, W. L.: 1979, in Hanle, W. and Kleinpoppen, H. (eds.), *Progress in Atomic Spectroscopy*, Plenum, New York.
Withbroe, G. L.: 1970, *Solar Phys.* **11,** 208.
Withbroe, G. L.: 1971, in Gebbie, K. B. (ed.), *The Menzel Symposium on Solar Physics, Atomic Spectra, and Gaseous Nebulae*, NBS Special Publication No. 353, Washington, D. C.
Withbroe, G. L.: 1975, *Solar Phys.* **45,** 301.
Withbroe, G. L.: 1976, Center for Astrophysics Preprint 524.
Withbroe, G. L.: 1977, in Hansen, E. and Schaffner, S. (eds.), *Proceedings of the November 7-10, 1977 OSO-8 Workshop*, Laboratory for Atmospheric and Space Physics, University of Colorado, Boulder, Colorado, p. 2.
Zirin, H.: 1966, *The Solar Atmosphere*, Blaisdell, Waltham, Mass.

7. ACTIVE REGION STRUCTURES IN THE TRANSITION REGION AND CORONA

David F. Webb

7.1 INTRODUCTION

The characteristic emissions of the transition region and corona arise at ultraviolet and X-ray wavelengths and are best observed from above the earth's atmosphere. Since the initiation of spaceborne experimentation, studies of the solar atmosphere have benefited from greater spatial, spectral, and temporal resolution and increasingly sophisticated analyses and data modeling. The data relating to the photosphere, chromosphere, flows, and loop models are presented in other chapters of this book; the previous chapter emphasized spectroscopic observations of the active region of the transition region and corona. This chapter is an overview of the observational aspects of the active region's upper atmosphere emphasizing imaging of the plasma loops that comprise this atmosphere.

Images of the lower corona reveal that the emission is confined to closed loops and arcades (Figure 7.1) and connects regions of opposite magnetic polarity. In the corona, the plasma $\beta \ll 1$, and one expects the magnetic field to be force-free. Thermal conduction perpendicular to the field is inhibited. Because of these factors, material reaching coronal heights will be constrained along the field lines. Therefore, it is not surprising that loop structures are observed in the corona. We should differentiate, however, between "flux loops" and "plasma loops." A flux loop is an isolated strand of magnetic field surrounded by a relatively field-free medium (Spruit, 1977). The simplest flux loop model is a potential field bounded by a current sheet. A plasma loop is the observed emitting structure that traces one or more flux loops in an unknown manner. Hereafter in this chapter the term "plasma loop" will be assumed when loops are discussed. It is now apparent that the classic question of the heating of the corona is equivalent to the question of the origin, heating, and evolution of these individual plasma loops.

Some important questions that were asked of the Skylab-era data during the Workshop were: What is the basic structure of an active region? How closely do plasma loops resemble flux loops? How is the material organized within a loop? Is the plasma in some loops hotter than others, and if so why? What can we say about the thermal gradients in a loop; what are at the footpoints of the loops? How do active region loops evolve, and what physical processes determine each phase of this evolution?

These questions remain largely unanswered, but important progress was made towards their solution during the Workshop. The next section describes the search for an answer to the first question in terms of morphological and thermal classifications of active region loops, including umbral structures. In Section 7.3, I discuss what is known observationally about the thermal structure of loops. The temporal evolution of active region loop structures is reviewed in Section 7.4 with an emphasis on ephemeral regions and the emergence of active regions. Section 7.5 summarizes this material and concludes with a brief discussion of scheduled spaceborne observations of the transition region and corona of active regions.

7.2 WHAT IS THE BASIC STRUCTURE OF AN ACTIVE REGION?

A major observational thrust during the Workshop was to identify and classify individual loops and arcades in images characteristic of the transition region and corona. Three main conclusions arise from these studies: (1) The number of *complete* loops that can be identified in an image increases with the temperature to which a heliogram is sensitive until coronal temperatures ($T_e > 10^6$ K) are reached. This fact is related to the result that most of the emission from an active region comes from "hot" material ($T_e > 10^6$ K) (see Section 7.3). (2) Hot loops and loops observed in lines characteristic of transition region temperatures generally are not cospatial. Indeed, their physical characteristics are quite different. Because of this physical distinction, hereafter I will use the term "hot loops" to mean those loops with $T_e > 10^6$ K and the term "cool loops" to mean those loops with $T_e < 10^6$ K. It has been suggested that "typical" active region loops might have cool cores (e.g., Van Hoven *et al.*, 1977; Hood and Priest, 1979) and that some observations support this view (e.g., Foukal, 1975). However, there seems to be no clear recent observational evidence that hot nonumbral loops have cool cores. But this statement is based on a limited set of Skylab data restricted to a few spectral lines widely separated in temperature. This subject is discussed in detail in Section 7.3.3. (3) The brightest hot loops form low-lying arcades, with the ends of each loop lying in bright plage of oppositely directed field. The brightest loops often have one foot in a pore or spot penumbra. The atmosphere over a large umbra does not contain hot material but can contain an excess of material at transition region temperatures that appears to be hydrodynamically unstable.

Sheeley (1980), Cheng (1980), Cheng *et al.* (1980), Dere (1980), and Mariska *et al.* (1980) have analyzed Skylab NRL EUV spectroheliograms and spectra of several active regions, arriving at the conclusions summarized above. Figure 7.1 shows typical loop structures associated with a simple bipolar region in lines characteristic of transition region and coronal temperatures. The Ne VII image ($T_e = 5 \times 10^5$ K)* shows loop-like structures, most of

*Hereafter, all indicated values for line-emission electron temperatures represent the temperature of the maximum of the contribution function for that transition. It is important to note that the width of the contribution function can be broad (Chapter 8), meaning that the emission in a given image can arise from material emitting over a broad range of temperatures.

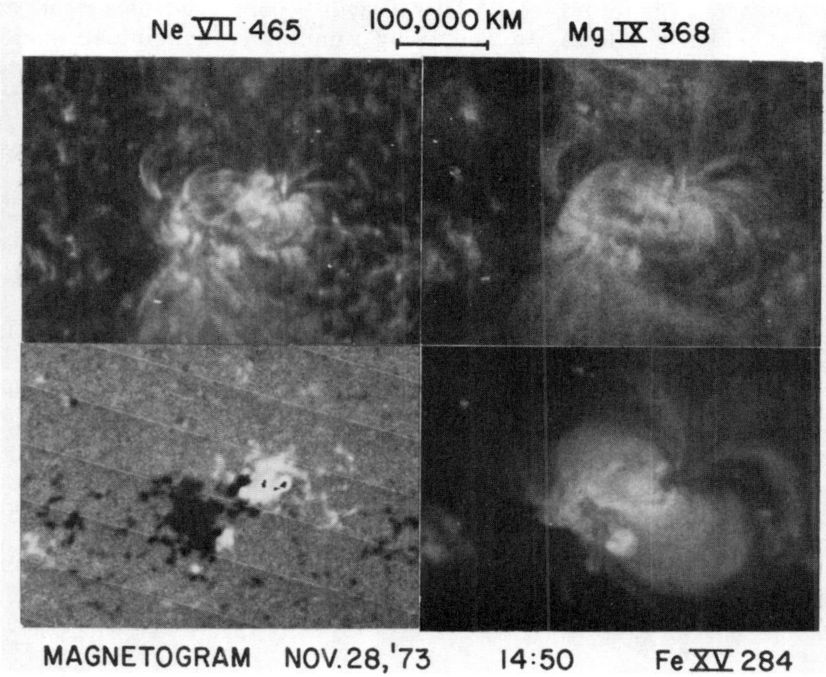

Fig. 7.1. Simultaneous Ne VII 465 Å, Mg IX 368 Å and Fe XV 284 Å NRL S082-A spectroheliograms from Skylab at 1450 UT, and a Kitt Peak ground-based magnetogram at 1542 UT on 28 November 1973 of McMath region 12628 (from Sheeley, 1980).

which are incomplete, with the emission concentrated in the legs. This low-lying emission gives the region a "spiky" appearance, with the legs clustered about the strong field regions. Mg IX ($T_e = 1 \times 10^6$ K) shows that the corresponding features are longer, but more diffuse and that different sets of complete loops join the opposite poles of the region. Fe XV ($T_e = 2 \times 10^6$ K) emission is diffuse but concentrated in complete loop systems that range from very compact arcades to large, fainter loop systems. Mariska *et al.* examined a region at the limb and discovered that the brightest hot arcades were low-lying and no higher than the Ne VII structures (10^4–10^5 km), whereas the fainter, more diffuse complete loops in Mg IX and Fe XV could reach much greater heights. This coronal morphology has, of course, been observed previously in X-rays (e.g., Figures 7.5 and 7.6).

The Ne VII spikes and loops generally are not cospatial with the hot loops. For instance, in Figure 7.1 the Fe XV emission tends to be faint or absent where the Ne VII emission is brightest. This is particularly obvious in the lower left of the region, where there is a large fan of emissio. Ne VII and Mg IX but no emission in Fe XV.

In a typical static loop model (Chapter 8), the large pressure scale height ($\sim 5000\ T_{max}$ cm) and high thermal conductivity ($T^{5/2}\ dT/ds$) of the corona suggest that the coronal portions of such loops are isothermal (i.e., $\nabla T \times$

$T^{5/2}$ is constant). Therefore, cool emission will be observed only near the feet of such idealized loops. Indeed, using a uniformly heated loop model in which the Ne VII "spikes" are the legs of coronal loops, Dere (1980) found that the predicted Ne VII emission would be restricted to the foot of a loop whose temperature is near its maximum over most of its length (Figure 7.2). However, this model underestimates the loop lengths observed in both Fe XV and Ne VII.

Cheng (1980) and Cheng *et al.* (1980) have emphasized that often one can observe bright patches in Ne VII (and Mg IX) at the bases of hot coronal arcade loops (Figure 7.3, *arrows*). These patches lie within bright chromospheric plage (Hα and He II). Cheng *et al.* found that there were two classes of Ne VII structures: the arcade patches, which lay in areas of strong field near the neutral line; and longer loop segments, which ended in the periphery of the active region (Figure 7.3). The complete Ne VII loops had aspect ratios r (loop radius/loop length) of 0.006, whereas the hot loops had $r \sim 0.1$–0.2 (see also Golub *et al.*, 1980). In summary, the complete cool loops were longer, thinner, less stable, and less numerous than the compact hot loops. They were physically different structures, located in different parts of the active region.

Using several sources of EUV data, Schmahl and Orrall (1979) have found evidence for absorption caused by significant amounts of nonprominence neutral hydrogen at coronal heights over the solar disk. The absorption occurred over $4.8 < \log T < 5.8$ and was greatest over active regions. The amount of attenuation implies neutral H column densities $> 3 \times 10^{17}$ cm^{-2}. Evidence that at least some of this very cool material exists within complete

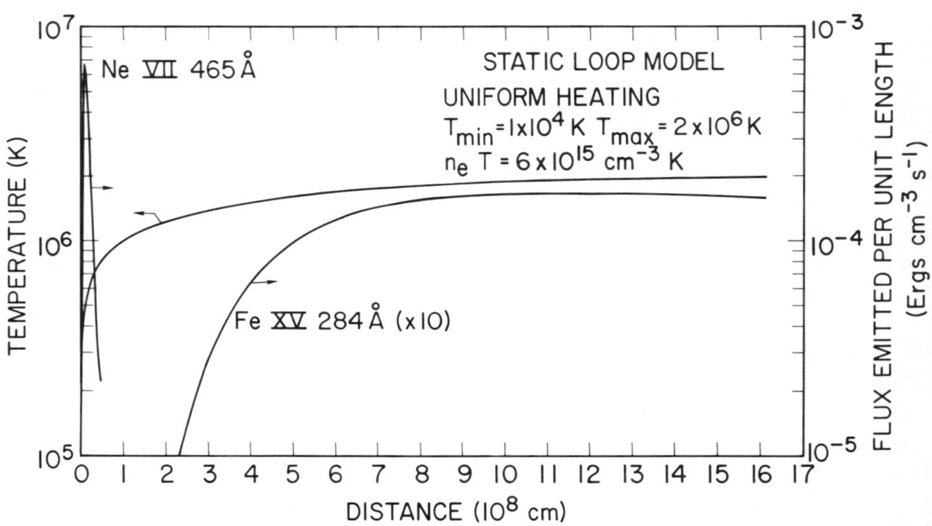

Fig. 7.2. Calculated temperature as a function of distance from the base of a uniformly heated loop with the parameters shown. Also shown are the calculated emissivities in the lines of Ne VII and Fe XV (from Dere, 1980).

quiescent loop systems comes from Hα coronagraph images at the limb (Orrall, private communication) and from high-resolution (1 arcsec) Lyα filtergrams from a recent sounding rocket payload (Bonnet *et al.*, 1980). Foukal (1978) and Cheng *et al.* (1980) also found dark loop-like structures in Skylab EUV images, which they attributed to HI (λ < 912 Å) and HeI (λ < 504 Å) continuum absorption. The existence at coronal heights of large amounts of such cool, dense material could explain some of the ubiquitous downflows over active regions observed with the OSO-8 and HRTS rocket data (Chapters 4 and 5).

Images in broadband X-rays and white light of the hot component of active regions reveal a hierarchy of loops ranging from bright, low-lying, compact arcades to long, faint loops connecting regions far from the active region (Van Speybroeck *et al.*, 1970; Krieger *et al.*, 1971; Rosner *et al.*, 1978). It is well known that these loops connect regions of opposite-polarity photospheric magnetic field and that to first order their geometry can be approxi-

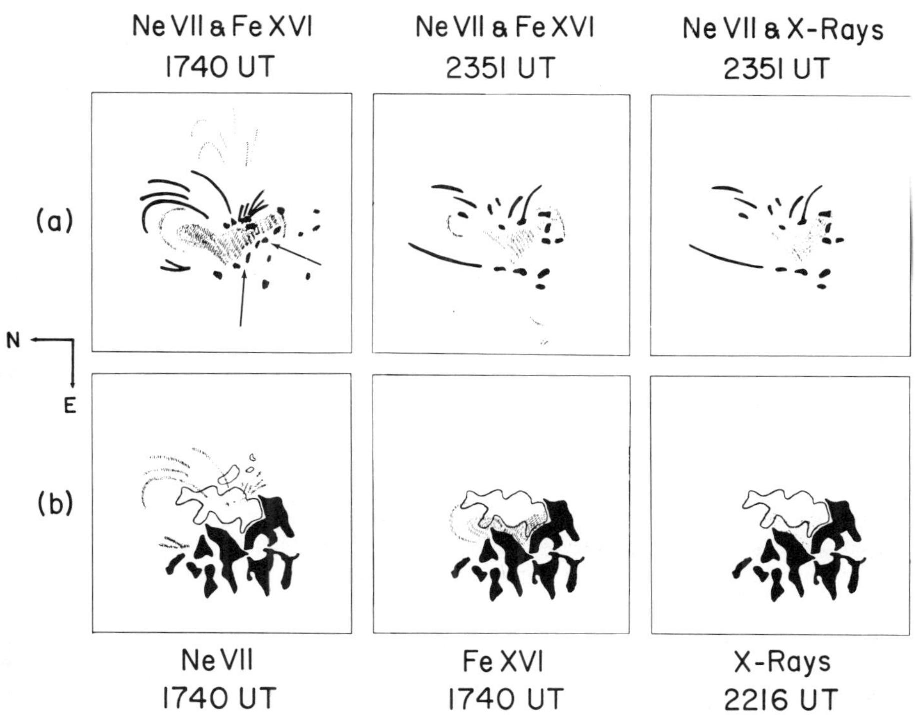

Fig. 7.3. Data from Cheng et al. *(1980) for McMath region 12379 on 14 June 1973. (a) A schematic of the spatial relationship of structures observed in Ne VII, and Fe XVI 335 Å NRL S082-A spectroheliograms and S056 soft X-ray images (8–16 Å) from Skylab. The Ne VII structures are solid black and the coronal loops are shaded. The arrows in the first column indicate a row of Ne VII elements lying near the base of the Fe XVI loops. (b) A schematic of the locations of the Ne VII, Fe XVI, and X-ray structures (all shaded) relative to areas of strong photospheric magnetic fields (Kitt Peak magnetogram at 1423 UT).*

mated by a potential (current-free) field (e.g., Poletto *et al.*, 1975). However, before the Workshop a detailed classification of active region coronal loops and a study of their footpoints had never been done. Webb and Zirin (1980) compared chromospheric filtergrams, magnetograms, and Skylab X-ray images of a typical active region with a dominant p spot and f plage and classified its loops into five categories (Figure 7.4): (1) The majority of the loops were low-lying and connected the bright plage to the spot *penumbra.* (2) A few short, bright loops connected the penumbra with areas of emergent flux or strong fields. (3) Emerging flux regions (EFRs) were associated with compact X-ray arcades. (4) Stable, but sometimes sheared arcades overlay filaments and their neutral lines. (5) Large, faint loops interconnected other regions, emanating from weak plage on the periphery of the active region. The feet of the brighter loops were in areas of strong and/or changing magnetic fields that appeared as bright patches in the chromosphere (Hα) and transition region (C III, O IV, O VI). Nicholas *et al.* (1979) have used HRTS data to measure the energy balance and pressure in an active region; they infer that they are observing the transition zone in Si III ($T = 3.5 \times 10^4$ K) at the feet of loops that are isobaric from the photosphere into the corona.

The results of these studies indicate that the bases of at least the brighter coronal loops are imbedded in strong fields and contain cool material emitting over a range 10^4 K $< T \lesssim 6 \times 10^5$ K. These structures vary (possibly in response to rapid field changes) on time scales short with respect to the associated hot loop. Note in Figure 7.3 how the "feet" (Ne VII patches) of the coronal arcade change between 1740 and 2351 UT, whereas the brightnesses of the arcade loops themselves are unchanged. Cheng *et al.* suggest that their observations indicate a sharp temperature gradient at the feet of the arcade loops. But two points need to be made. First, such gradients must be time-varying, since the base emission appears to vary much more rapidly than the loop itself. Second, because we expect such gradients to be quite steep (e.g., Dere, 1980), our present resolution ($\gtrsim$1 arcsec) is not sufficient for constraining loop models.

Our discussion of active region structures would be incomplete without mention of recent contributions at radio frequencies. Recent developments of synthesis techniques and the construction of large radio interferometers have greatly increased the sensitivity and spatial resolution available for solar studies at radio frequencies. The slowly varying component of the microwave sun peaks over active regions with a spectral maximum at about 10 cm. Observations have shown that the brightest centimetric components are associated with sunspot umbral and penumbral structures (see section 7.3), transverse fields over neutral lines or filaments (Kundu *et al.*, 1977; Kundu *et al.*, 1980a,b), and emerging flux regions (Kundu and Velusamy, 1980). The fact that such features are also found to be associated with X-ray loops and that thermal bremsstrahlung can produce emission at both microwave and X-ray wavelengths suggests that at least some of the microwave emission in active regions arises in coronal loops. In fact, Kundu and Velusamy claim to have imaged a complete coronal loop using the VLA at 6 cm. The emission from the top of the loop had a brightness temperature

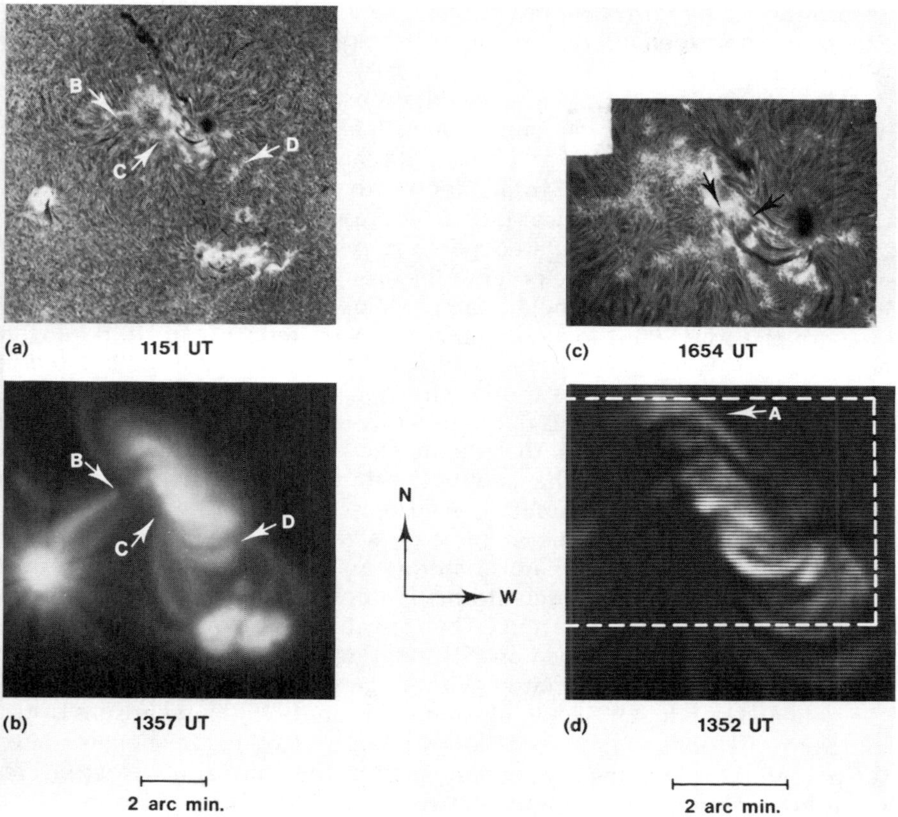

Fig. 7.4. An active region complex on 6 July 1973 in Hα (top) and soft X-rays (bottom) illustrating classes of hot loops and their footpoints (from Webb and Zirin, 1980). (Left side) A large-area view of three interconnected regions: McM 12427, east of center; McM 12417, north; and McM 12414, southwest. (a) is an Hα filtergram at line center. (b) is an AS&E S054 X-ray image through Skylab Filter 3 (passband 2–32, 44–54 Å). Co-aligned arrows (B, C, and D) point to the McM 12417 terminations of the interconnection loops. (Right side) An enlarged view of McM 12417: (c) a full field Hα image (the arrows denote two new f pores); and (d) a Skylab Filter 1 (2–17 Å) computer-enhanced X-ray image. The dashed outline marks the position of the Hα box on the X-ray field. Arrow A shows a bright arcade arch overlying a filament. Hα images are from Big Bear Solar Observatory.

(T_b) of $\sim 10^6$ K and was attributed to thermal bremsstrahlung. The footpoint emission had a higher T_b and was attributed to thermal gyroresonance absorption. One foot was in bright plage near an EFR, while the other foot ended in a spot penumbra, similar to the bright X-ray loops observed by Webb and Zirin.

An exciting use of the microwave data is as a probe of the coronal magnetic field strengths in these loops. Kundu *et al.* (1980a) inferred from microwave maps and Skylab EUV and X-ray data that the bright radio emission in an active region was coming from a loop with one foot in a spot penumbra. Deducing that the emission was due to gyroresonance absorption from optically thick layers, they calculated a field strength of 300 G in the loop and a magnetic scale height on the order of the observed width of the hot coronal loops (about 10^9 cm). In a similar manner, Marsh *et al.* (1979) deduced a field strength of about 300 G in a coronal arch associated with an Hα flare with $N_e = 1.2 \times 10^{10}$ cm^{-3}. This value is typical of field strengths calculated in flare plasmas assuming the source emission is gyrosynchrotron (Svestka, 1976). The agreement of the field strengths in both quiescent and flaring active region loops suggests that the magnetic field strengths in loops are relatively constant and unchanged by the flare process. This fact could be used to set limits on the production of currents in such loops.

Another use of interferometric radio data is to use meter and decameter wavelength observations of noise storms as a probe of high loops over active region complexes. Such storms are common and persistent, sometimes lasting for several rotations. Their spectrum is a continuum, with fine structure bursts such as Type I and III superimposed on it. Such bursts have been inferred to be associated with large-scale coronal loop systems (Gergely and Kundu, 1975; Gergely and Erickson, 1975). Comparisons of the locations of storms with coronal loops imaged at other wavelengths have the potential for locating the source region of the nonthermal electrons responsible for the storm emission relative to the thermal coronal loops.

Several such studies have been made recently using interferometric radio data from Culgoora and Clark Lake Radio Observatories and Skylab soft X-ray images. Webb and Kundu (1978) analyzed an event in which a storm at 40–70 MHz was associated with a filament eruption and an X-ray coronal transient. They found that the filament eruption was the source of the nonthermal particles that propagated through closed coronal structures to high heights, producing the storm radiation. Stewart and Vorpahl (1977) compared Culgoora radioheliograph and Skylab S056 X-ray data and found that the source positions of Type-I, -III (U and J), and -V bursts lay radially above the X-ray active region and interconnecting loops. They modeled the regions as occurring in current-carrying flux tubes at heights of $\lesssim 1.5\ R_\odot$ (Type IIIs) and $\lesssim 0.6\ R_\odot$ (Type Is). Gergely *et al.* (1980) used Clark Lake radio data and Skylab AS&E X-ray images and arrived at a similar conclusion. In addition they found that the occurrence of storms is not associated with any particular evolutionary phase of a loop system and that not every large-scale system is associated with a storm.

During the Workshop, Jackson (1979) compared Culgoora radio data and Skylab AS&E X-ray images for a period in late November 1973. Between 25

November, 2300 UT, and 26 November, 0555 UT, recurrent storm activity occurred in McMath 12628, a rapidly evolving region with bright, internal X-ray loop systems and numerous interconnecting loops (Figure 7.5). On Figure 7.6 are plotted the extrapolated centroids of the right-hand (*R*) and left-hand (*L*) polarized Type-I storms and the positions of the Type-III bursts (*3*) for this period on a negative of the soft X-ray image (*a*) taken on 26 November, 0542 UT (Figure 7.5). The radio spectra (c) are shown for reference. A

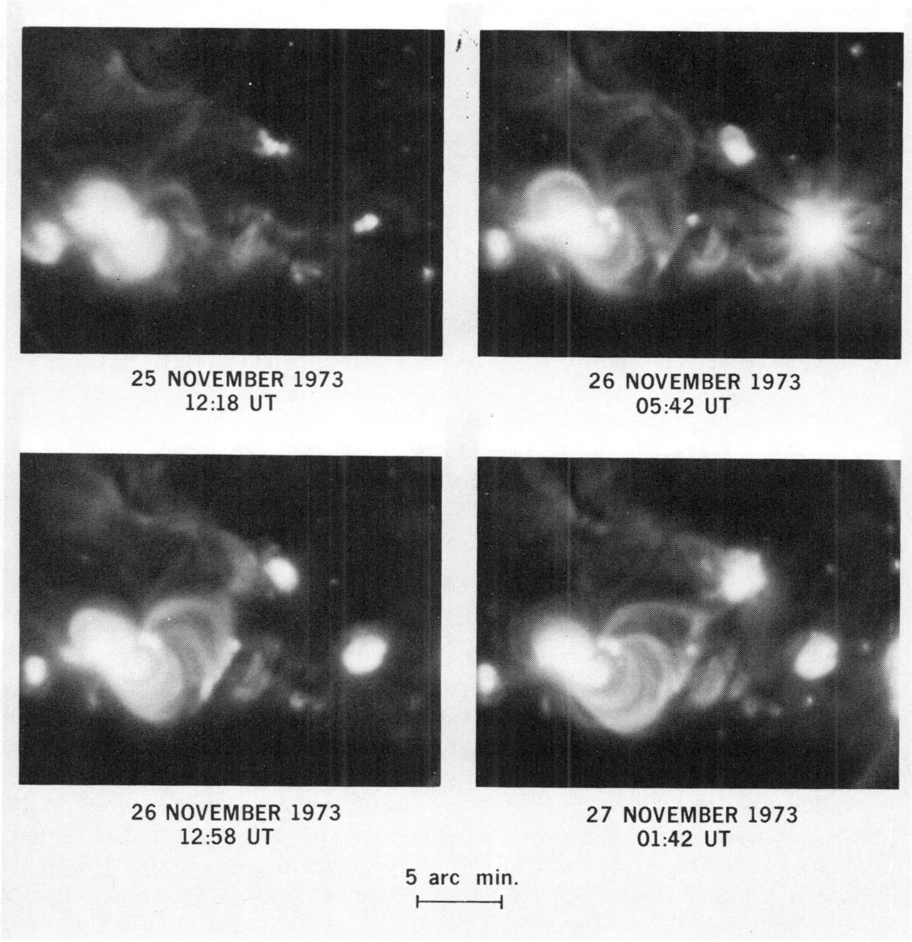

Fig. 7.5. Consecutive AS&E X-ray images through Filter 3 showing McMath 12628 and its extensive coronal interconnections from 25–27 November 1973 from Skylab. The image on 26 November, 0542 UT is the same as in Figure 7.6a.

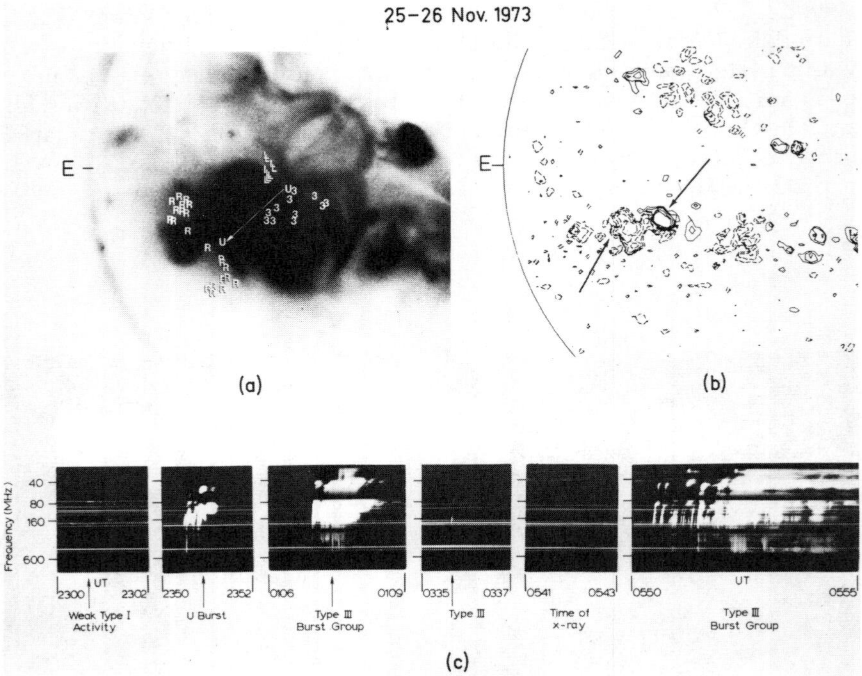

Fig. 7.6. Comparison of coronal X-ray active region structure in McMath region 12628 and meter-wave radio burst activity on 25–26 November 1973: (a) *projected 160-MHz Culgoora heliograph source positions superposed on a negative AS&E X-ray image on 26 November, 0542* UT, (b) *Mt. Wilson photospheric magnetogram on 25 November, 1830* UT *(the arrows denote the locations of preceding and following sunspot groups—see Figure 7.7b), and* (c) *Culgoora radio spectra of burst activity from 25 November, 2300* UT, *to 26 November, 0555* UT *(from Jackson, 1979).*

magnetogram (*b*) for 25 November, 1830 UT, is shown, on which are indicated the locations of the large preceding spot group and a small following polarity spot (see Figure 7.7b). To the north and west of the *p* spot group numerous flaring and surging activity in satellite spots occurred as shown by the intense, compact X-ray activity on Figure 7.5.

Radiation from the radio bursts probably escapes from higher in the corona than the region from which the particles originate (e.g., Bougeret and Steinberg, 1977). In plotting the source positions on Figure 7.6, the 160 MHz radiation was assumed to come from a height of 0.35 $R_\odot$ above the surface. The positions were then projected *radially* down to a height of 0.1 $R_\odot$, assumed to be the average height of the X-ray structures. The onset and end positions of the U-burst are shown (*U and arrow*). U-bursts may be produced by exciter electrons confined to flux loops. This U-burst appears to connect the active area to the northwest of the *p* spots and *f* polarity near the *f* spot. Underlying it in X-rays were several intense loop-like structures crossing the main neutral line.

Jackson concluded that high-level loops overlying the X-ray loops are the sites of the Type-I storms. In this case, these loops may have been associated with a coronal transient (the bright X-ray loops to the northeast of the active region overlay a disappearing filament). The Type-III bursts were 1%–10% left-hand polarized, implying field lines directed outwards, and were clustered about the *p* spot group. They represent emission from electrons that were able to escape along open field lines, which might be strongly diverging (Dulk *et al.*, 1979). Some open fields are apparently associated with active regions (see Levine, 1977, and references therein); it is interesting that the open fields in McMath 12628, as possibly delineated by the Type IIIs, were associated with the strongest unipolar surface fields in the region. It remains for future work to consider the question whether such open active region fields preferentially occur over sunspots.

Finally, concerning basic active region structure we need to address the question: What is known about the structure of the atmosphere over a sunspot? My comments will be brief because sunspots were not a major topic of the Workshop. However, sunspots are clearly important to our understanding of active regions, since they contain most of the magnetic flux of the active region. (But very little of the plasma over an active region appears to be associated with sunspots.)

In a series of papers (Foukal *et al.*, 1974; Foukal, 1975, 1976, and 1978), Foukal has given ample evidence from Harvard EUV results that the atmosphere over sunspots contains large amounts of material emitting at transition region temperatures. However, some confusion has arisen among Skylab researchers about whether there is hot material over spots, as well as about the spatial coincidence of hot and cool loops. This confusion may in a large part be due to the interpretation of lines such as Mg X (625 Å and $T_e \sim 1.5 \times 10^6$ K) as *purely coronal* lines. Although the atomic physics calculations of lines like Mg X indicate that its temperature sensitivity has a relatively sharp low-*T* cutoff and a high-*T* tail, a large amount of material emitting in the range $6 \times 10^5 \text{ K} < T < 1 \times 10^6$ K, such as over spots, might be visible in an Mg X image. In fact, Webb and Levine (1979) and Webb and Zirin (1980) have presented evidence that many structures in Mg X images are not spatially coincident with those in X-ray images, especially structures associated with spots. Therefore, there must be very little material at coronal temperatures in structures overlying spots [Foukal, 1976 (Figure 1); Pallavicini *et al.*, 1979; Webb and Zirin, 1980], unlike the situation with other active region loops.

Foukal (1976) has suggested that the strong magnetic field associated with a sunspot inhibits heating of the corona. Van Hoven *et al.* (1977) developed a steady-state model of sunspot loops with positive radial pressure gradients which demonstrates that stable loops can exist over spots. However, the effect on such models of the large observed downflows over spots is not known (see Foukal, 1978, and other chapters in this book). Also, if as was mentioned above some of the field lines rooted in the sunspot are open and strongly diverging, then new models of the sunspot atmosphere must be developed. There is some evidence that the core emission observed over a spot (which was originally called a sunspot plume) is entrained on strongly diverging field lines [Figure 7.1 and Foukal, 1976 (Figure 1)].

The Skylab EUV and X-ray observations of the umbral atmosphere have also created controversy regarding the source of the peak microwave emission in active regions. Until very recently, the prevailing opinion of radio observers was that bright cores of microwave emission were coincident with umbrae. However, the Skylab observations of large columns of cool material and an absence of X-ray emission over sunspots would appear to cause problems with the two popular models for the bright microwave emission (Pallavicini *et al.*, 1979; Pallavicini, 1980). The thermal gyroresonance absorption model requires high field strengths ($\sim 10^3$ G), such as exist over umbrae, but is incompatible with the observed cool EUV plumes. A nonthermal process, such as gyrosynchrotron emission from nonthermal electrons, seems incompatible because it requires continuous electron acceleration in the active region.

As noted earlier, the recent high-resolution microwave observations have revealed that some umbrae may have bright emission over them and others, even in the same active region, may not (Kundu *et al.*, 1980a,b; Kundu and Velusamy, 1980). The brightest microwave emission often arises in the penumbra. The situation is further complicated by Pallavicini's (1980) recent reanalysis of Skylab EUV observations indicating that cool columns of material typically cover only a fraction of the umbral area. So it is possible that the cool transition region material is related to the large spot flows and is confined only to a limited number of plumes or loops over the spot at any given time. It is clear that correlated, high-resolution EUV, X-ray, microwave, and magnetic field data are necessary to resolve these conflicting observations of the umbral atmosphere.

7.3 THERMAL STRUCTURE OF LOOPS

7.3.1 Scaling Laws and Empirical Data

Recent studies of the physics of active region structures have involved the development of static loop models in which energy balance is assumed between thermal conduction on the one hand and radiative loss and energy input (heating) on the other. Experimentally, researchers have tried to compare measurements supposedly pertaining to individual loops with different static loop models to gain insights into the loop heating function and the form of the radiative-loss function (since thermal conduction is considered only as a means of redistributing energy along a loop).

Because the heating function is poorly understood, some results of static loop models have been formulated to be independent of the heating term. This has led to loop scaling laws, which were considered useful in relating measured loop plasma parameters with the models. Rosner *et al.* (1978) first developed a scaling law of the form

$$T_{max} = C(pL)^{1/3} , \tag{7.3.1}$$

where p is the loop's pressure, L is its length, T_{max} is the maximum temperature (which in these models occurs at the loop apex), and C is a function of the assumed heating function and depends on the boundary condition for the conductive flux F_c at the footpoint and the flux tube divergence Γ. The exponent is relatively insensitive to changes in these latter parameters and in the form of the radiative-loss function used. This scaling law has been used extensively to compare theoretical and observed loop parameters, with general agreement.

However, one finds a large amount of scatter on such comparison plots (e.g., Rosner *et al.*; Emslie and Machado, 1979), and no error bars indicating the rather large measurement uncertainties that can exist for the X-ray data. From our experience with the Skylab S-054 X-ray data, we have found that such a plot of, say, T *vs* (pL) using observed data shows a large scatter, which is consistent with a large range in the exponent of Eq. 7.3.1 (M. Gerassimenko, private communication). The point is that the inaccuracy of the present empirical data combined with the insensitivity of the scaling law and the rather *ad hoc* assumptions of static loop models make such comparisons of limited value (Roberts and Frankenthal, 1980).

7.3.2 Emission Measure Analyses

Another useful method for comparison of empirical data with models is the differential or volume emission measure analysis. The amount of material (number of electrons) in a given volume and over a specified temperature range can be found by using the relation

$$Q(T)\,dT = N_e^2\,dV \ , \qquad (7.3.2)$$

where $Q(T)$ is the differential emission measure. Empirically, $Q(T)$ is found by measuring spectral intensities in lines or with broadband filters whose temperature dependencies are presumed known. In practice such analyses are most successful using a combination of the narrower lines typical of the EUV and the broader soft X-ray lines integrated over an entire active region (Withbroe, 1975; Levine and Pye, 1980). Such an analysis for individual plasma loops has never been done because of the difficulty of imaging cospatial material over a wide range of temperatures (Section 7.2), and the lack of sufficiently high combined spatial resolution.

All such emission measure analyses of individual active regions have shown that a large majority of the material in an active region above the chromosphere is emitting at $T_e > 10^6$ K; i.e., most of the extended material is "hot" and therefore emits predominantly in the soft X-ray region of the spectrum. The typical quiescent active region is characterized by material at $T_e \approx 2\text{–}3 \times 10^6$ K with very little material emitting at $T_e > 4\text{–}5 \times 10^6$ K (Davis *et al.*, 1975; Wolfson *et al.*, 1977; Withbroe, 1975; Pye *et al.*, 1978; Levine and Pye, 1980). In one of the first analyses of this type, Withbroe (1975) used OSO-4 and OSO-6 XUV emission lines to deduce that the mean coronal temperature for several active regions was 2.1×10^6 K, with one active region having a majority of material between 2 and 3.5×10^6 K. Davis *et al.* combined Skylab S-054 X-ray images with rocket-borne X-ray spectra of McMath region 12336, an old, spotless active region, and found $2.0 < T_e < 2.8 \times 10^6$ K and $0.5 < N_e < 4.5 \times 10^9$ cm^{-3}. Gerassimenko *et al.* (1976) then used the same X-ray data to compare the estimated flux density at 9.1 cm, if the observed X-ray flux were due to thermal bremsstrahlung, with the observed 9.1 cm, radio flux density. They concluded that most, if not all, of the observed emission was caused by thermal bremsstrahlung, and that because of the agreement in flux densities coupled with the different dependencies with temperature at the two wavelength extremes (T^2 for X-rays and $T^{-1/2}$ at 9.1 cm), most of the material was at $T_e > 10^6$ K.

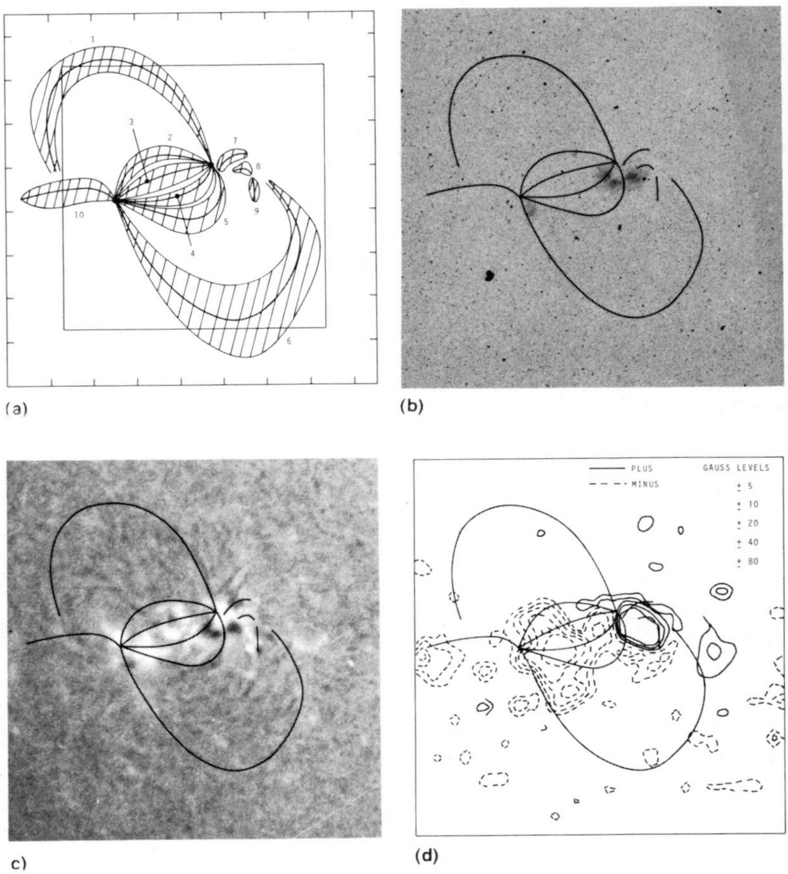

Fig. 7.7. A description of McMath 12628 from Pye et al. *(1978). A schematic* (top left) *of ten hot loops as seen in the AS&E X-ray image on 26 November, 0542* UT *(Figure 7.5). The central axes of the loops are superposed in the other parts of the figure. The sunspot group in white light from the AS&E experiment* (top right)*; the Hα chromosphere* (bottom left) *at 0541* UT *from Carnarvon, Australia; the photospheric magnetic field* (bottom right) *on 25 November, 1830* UT*, from Mt. Wilson. Note that the loops terminate in bright plage regions, not in the sunspot umbrae.*

As an opposing example of a young, rapidly evolving region, McMath 12628 (Figure 7.5) was studied in two complementary emission measure analyses. Pye *et al.* (1978) compared Skylab S-054 X-ray images and rocket-borne X-ray line spectra, and Levine and Pye (1980) compared the same rocket spectra with Skylab Harvard EUV images. The Pye *et al.* data were sensitive to material at $T_e > 10^6$ K. Figure 7.7, from their paper, is a schematic representation of the locations of the main hot loops in region 12628 (*top left*) relative to the photosphere (sunspots—*top right*), chromosphere (*bottom left*), and the photospheric magnetic field (*bottom right*). Figure 7.8 shows their derived volume emission measure, showing that most of the hot material was at $T_e < 3.5 \times 10^6$ K and a small amount at temperatures up to about 5×10^6 K. They could exclude the presence of hotter material because of the absence of ions such as Si XIV and S XV. The Levine and Pye data were sensitive to material at $10^5 \leq T_e \leq 5 \times 10^6$ K. They concluded that most of the plasma was in the range 8×10^5 K to 2×10^6 K (Figure 7.9), thus fixing the lower temperature bound of the material. Further, they were able to determine that any material above 5×10^6 K accounted for <1% of the total emission measure, again in agreement with Pye *et al.* Together these two papers provide probably the most detailed analysis to date of the integrated characteristics of an active region atmosphere.

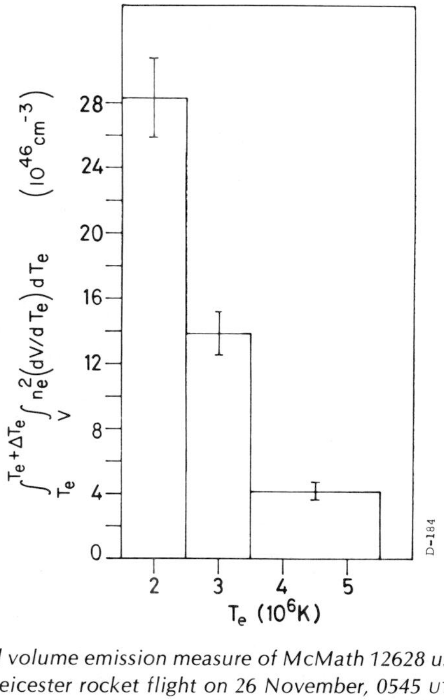

Fig. 7.8. The integrated volume emission measure of McMath 12628 using AS&E X-ray images and spectral data from a Leicester rocket flight on 26 November, 0545 UT. The spectra of the active region material were convolved with the rocket spectrometer field function; the error bars indicate the uncertainty due to Poisson statistics in the input line strengths (from Pye et al., 1978).

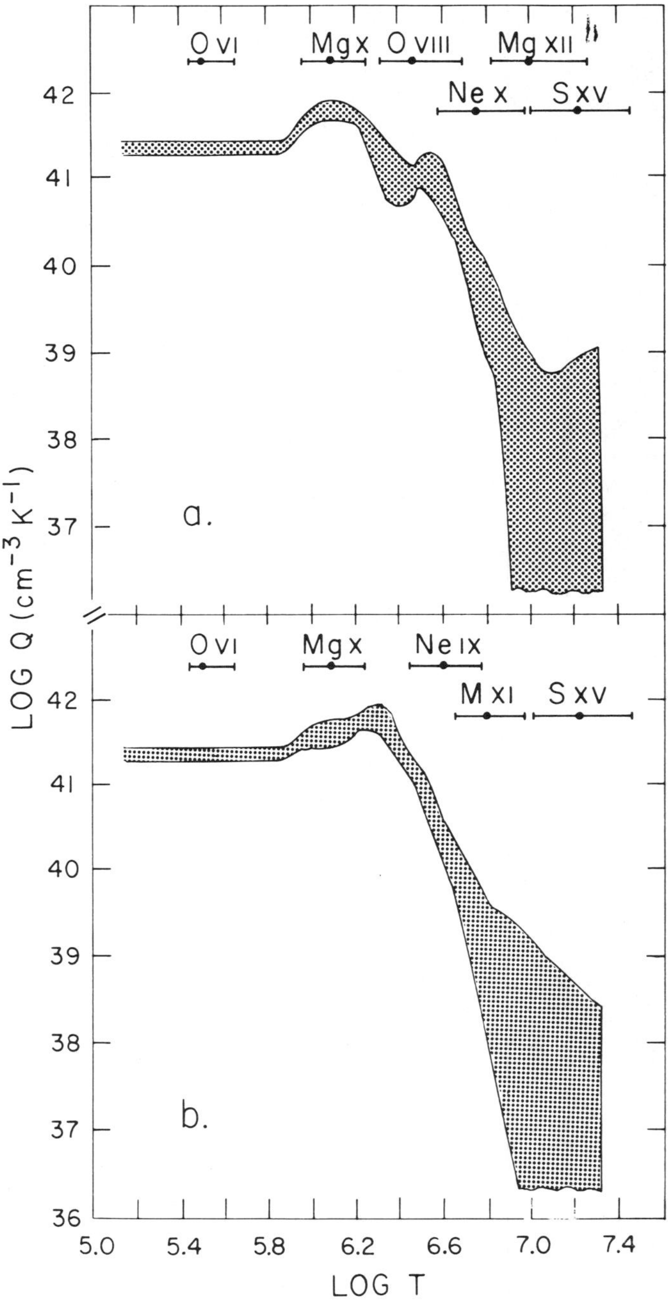

Fig. 7.9. The differential emission measure of McMath 12628 using Harvard S055 EUV spectroheliograms from Skylab on 26 November, 0527 UT, and the same X-ray rocket spectra as in Figure 7.8: (a) using hydrogen-like X-ray lines and (b) using helium-like X-ray lines. The lines used are shown, along with the temperature of the peak of each ionization function and its FWHM. The vertical band represents the range of curves obtained when the input intensities were randomly varied within the experimental uncertainties (from Levine and Pye, 1980).

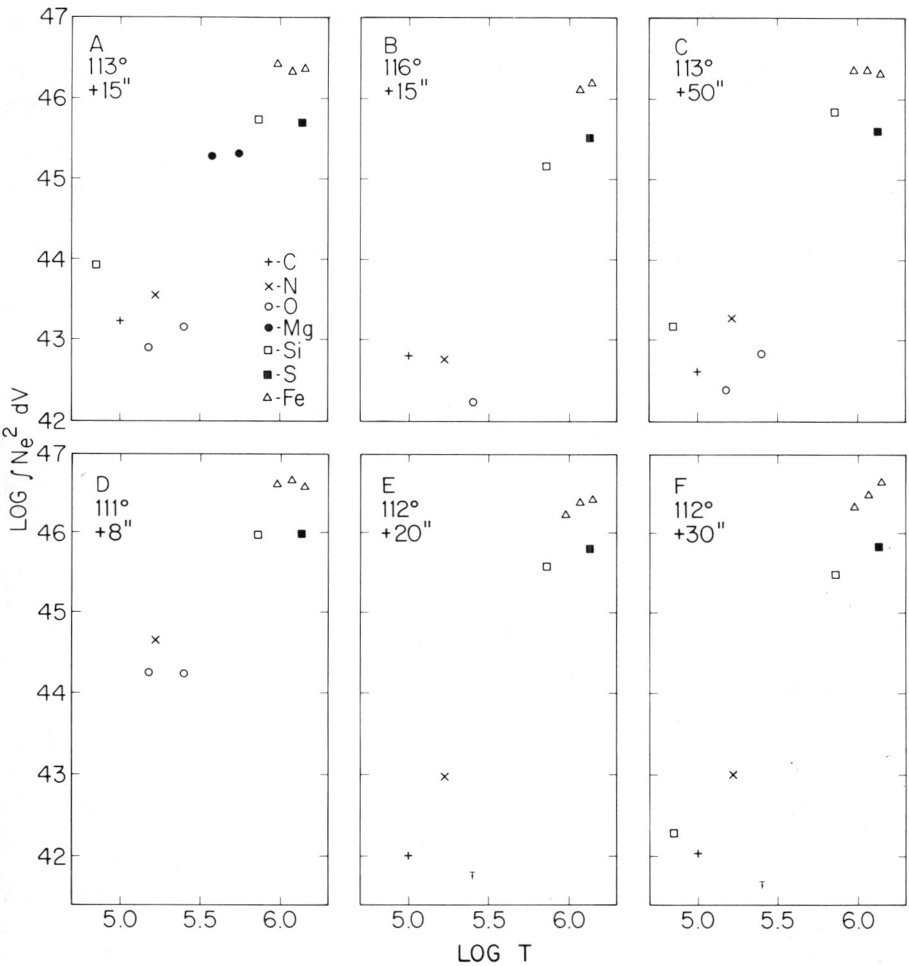

Fig. 7.10. The volume emission measure as a function of temperature and height of McMath 12690 at the east limb on 5–6 January 1974 using the NRL S082-B spectrograph on Skylab. In the upper left corner of each plot are given the position designation, the position angle of the slit, and the height above the limb in arcsec. The values for O V at positions E *and* F *are upper limits (from Mariska* et al., *1980).*

Using spectra from the Skylab NRL slit spectrograph, Mariska *et al.* (1980) have extended this kind of analysis to the volume emission measure of an active region as a function of height. Figure 7.10, from their paper, shows the volume emission measure *vs* T_e as a function of position and ion for McMath region 12690 observed at the limb. They find that the emission above a height of 6×10^3 km, which is concentrated in loops, is dominated by material at $T_e > 7 \times 10^5$ K. Also, the emission measure of the hot material is nearly invariant as a function of position.

7.3.3 Thermal Structure of Individual Loops

Empirical emission measure analyses can be used as a probe of the conductive flux F_c into the chromosphere and of the geometry of plasma loops (Chapter 8; Levine and Pye). However, this type of analysis is inherently limited in what one can learn about individual loop structures (Levine and Pye). We therefore must turn to results from imagery of high spatial resolution to learn more about the thermal structure of loops.

Skylab provided the best available data set for this kind of information. Unfortunately the results to date are limited and inconclusive in many respects. From soft X-ray data, it is apparent that observations of the hot component of loops are consistent with their having constant *longitudinal* gas pressure (e.g., Rosner *et al.*, 1978). This is expected in the corona: the pressure scale height is large compared to the observed heights of active region loops, and the thermal conductivity is high, tending to smooth out any temperature variations along the loop (assuming that plasma and flux loops coincide). Therefore, with our present 2–5 arcsec (1.5–3.7 $\times 10^3$ km) resolution, it is difficult or impossible to detect any longitudinal pressure or temperature gradient along a quiescent active region loop (e.g., Figure 7.2; Van Hoven *et al.*, 1977; Cheng, 1980). The hot loops do appear to narrow and fade toward their bases, but this is a subtle effect. Evidence was presented in Section 7.2 that compact, bright elements emitting in chromospheric or transition zone lines are occasionally seen at the bases of hot loops. The size of these features is typically on the order of the instrumental resolution limit, indicating that their true size is less than 1–2 Mm. Therefore, we can only put a weak lower limit of about 2×10^{-2} K/cm on the gradient at the loop base.

So the typical quiescent active region corona is characterized by individual plasma loops that are, to first order, isobaric and isothermal and whose temperatures cluster in a narrow range of 2–3 $\times 10^6$ K. Pye *et al.* (1978) derived temperatures of nine loops in a rapidly evolving active region (Figures 7.5 and 7.7) and found their average $T_e = 2.9 \times 10^6$ K with a total range of only $\pm 0.3 \times 10^6$ K. Cheng (1980) studied a simple, small active region and found that four loops, or loop ensembles, had $T_e = 2.0$–2.2×10^6 K. Little and Krieger (1977) found that the integrated temperature of newly born X-ray active regions clustered closely around 2.45×10^6 K. These new active regions probably initially contain only a few flux loops.

But the brightness of different quiescent loops at a given time, and of the same loop over a period of time, can vary by factors of at least two or three, much greater than the observed temperature variations. Davis *et al.* (1975), Little and Krieger (1977), Pye *et al.* (1978), and Cheng (1980) found that these brightness variations are due to variations in the density between loops or within the same loop over a period of time. Important questions for future work are: why are the temperatures of hot loops relatively fixed, and what causes the density variations between loops and in a given loop with time? Models for quiescent active region loops that attempt to explain these observations have been developed but will not be discussed here (see Krieger, 1977, for a review).

Finally, I briefly discuss the evidence for *radial* thermal gradients across a loop (i.e., transverse to the loop axis). I have already mentioned the controversy relating to the internal constitution of a loop. Does the loop temperature vary with radius and if so how, or does the diffuse structure we see contain intertwined subloops? Theoretical studies have shown that hot loops with cooler cores can be MHD stable (e.g., Van Hoven *et al.*, 1977; Hood and Priest, 1979). But just what do the *observations* tell us about the core structure of loops?

Foukal (1975) has argued that coronal loops may have *positive* radial temperature gradients (a hot sheath and a cooler core) at a constant density, yielding a positive pressure gradient. But Cheng (1980) did not detect any loop temperature gradients in the Skylab NRL data, and Pye *et al.* observed no pressure gradients, implying that any low-pressure core must be smaller than half a loop radius. Levine and Withbroe (1977) have produced empirical models indicating that any cool cores must be very thin (i.e., $\lesssim 20\%$ of the radius).

In Section 7.2 we noted that the NRL researchers had found no evidence for radial thermal gradients in their Skylab EUV data and in fact observed that hot loops were physically distinct from cool loops. How can we reconcile these apparently conflicting Skylab results? It could be that these studies have involved different *classes* of loops with quite dissimilar heating functions (Priest, 1978). Most of the loops studied by Foukal, and Levine and Withbroe, were sunspot, or post-flare, loops whose characteristics are quite different from "quiescent" loops. For instance, such loops could be characterized by thermal nonequilibrium leading to cool plasma circulation in the loop (Foukal, 1978; Priest, 1978; Hood and Priest, 1979; Roberts and Frankenthal, 1980). This could explain the relatively rapid "flows" of cool material observed by Sheeley (Figure 7.11) and Levine and Withbroe (1977), and the observation that complete cool loops have shorter lifetimes than hot loops (Section 7.4). At any given time, most of the loops in an active region will be in relative thermal and hydrostatic equilibrium ("quiescent"), explaining the observation that most of the material in individual loops or integrated over the active region is at coronal temperatures with a small or zero radial gradient and with no detectable cool core. In this view, any significant amount of cool material would be associated only with sunspot or "transient" loops.

The main point of this section is that our present observations of loop gradients, especially of radial gradients, are insufficient to understand the physics of loops. One must be very careful to ensure that data characteristics of different temperature regimes are cospatial and cotemporal, and that these observations delineate between sunspot and transient loops and other more stable active region loops. These points have not been adequately addressed in many past studies.

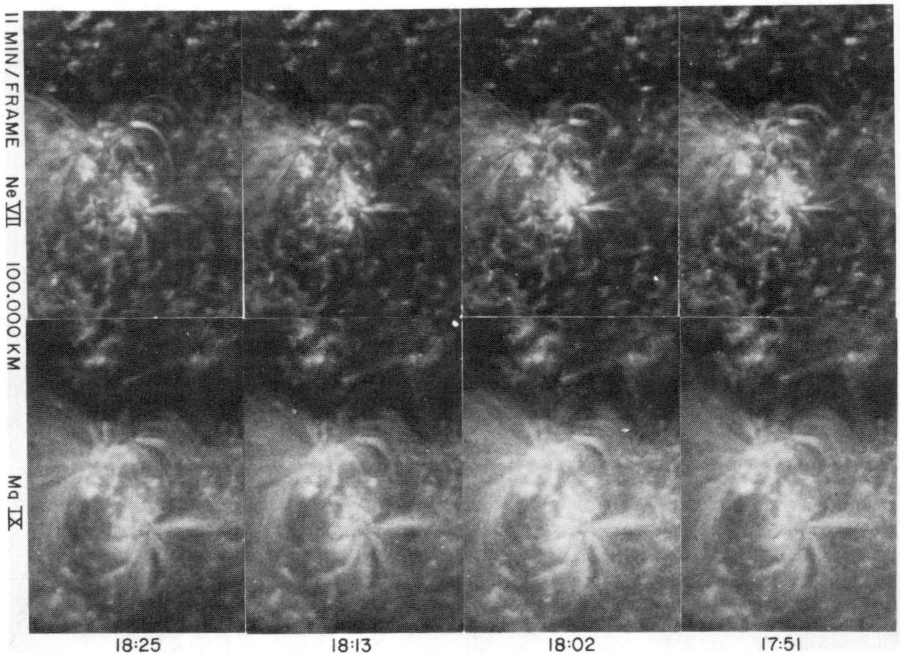

Fig. 7.11. Simultaneous evolution of the Ne VII and Mg IX emission from the region in Figure 7.1 between 1751 and 1825 UT on 28 November 1973 (from Sheeley, 1980).

7.4 TEMPORAL BEHAVIOR OF LOOPS

7.4.1 The Evolution of Loops in a Mature Active Region

I have touched upon the observed differences in the temporal behavior of loops at temperatures characteristic of the transition region and of the corona. In this section, I will describe what we know about the evolution of these loops and about the emergence of active regions over the short term and the long term.

Sheeley (1980) has summarized the observations of the temporal behavior of active region loops as seen with Skylab NRL EUV data and is in basic agreement with the studies of Cheng, Dere, and Mariska. Sheeley compared cospatial and cotemporal EUV images in the lines of Ne VII ($T_e = 5 \times 10^5$ K), Mg IX ($T_e = 1 \times 10^6$ K) and Fe XV ($T_e = 2.0 \times 10^6$ K) with the time resolution varying from 1.5 minutes to 3 hours. He found that the Ne VII emission was confined to spikes or legs of loops and occasionally complete loops, and changed on time scales of minutes, with complete structures appearing and disappearing over 30 minutes. Figure 7.11 from Sheeley (1980), shows the temporal evolution of an active region in Ne VII and Mg IX (Figure 7.1). The most notable feature is the long Ne VII loop at 1825 UT. This loop was progressively "filled" with emission moving up from one footpoint, while the brightness of its complete Mg IX counterpart remained constant throughout this interval.

On the other hand, the diffuse Fe XV emission formed complete loop systems whose overall patterns persisted for days, as is also observed in soft X-rays. Individual Fe XV loops evolved on a time scale of about six hours. The Mg IX emission was intermediate between that in Ne VII and Fe XV. In Ne VII, changes occurred by the upward or downward proper motion of emission, which could be interpreted either as heat flows or material motions (Figure 7.11). The Fe XV loops, however, seemed to brighten and fade as a whole, with no observed motions.

Levine and Withbroe (1977), Gerassimenko *et al.* (1978), and Nolte *et al.* (1979) studied the short-term evolution of individual active region loops. They found evidence for steady-state heating of hot loops. Gerassimenko *et al.* studied three active region loops with 90-s time resolution and found that their brightnesses were constant over periods comparable to their radiative lifetime. Nolte *et al.* reached the same conclusion for small-scale loops (X-ray bright points). In addition, both small-scale (Nolte *et al.*) and large-scale (Levine and Withbroe) loop emission could suddenly decay radiatively, which they attributed to a sudden disruption of the heating mechanism. Hood and Priest (1979) and Roberts and Frankenthal (1980), however, have argued against this interpretation.

Activity at the base of even the long-lived hot loops can occur on time scales as short as minutes. The Ne VII "spiky" structures, which evolve on the order of minutes, may be the legs of hotter loops. Webb and Zirin (1980) noted that bright chromospheric patches typically appeared at the feet of bright coronal loops. These features could change rapidly, whereas their associated X-ray loops remained unchanged for periods of hours to days. However, there was a correlation between the bright loops and persistent footpoint activity. Whether this base activity in the chromosphere and transition region is symptomatic of the heating process of the loop remains unknown.

7.4.2 The Emergence of Active Regions

A. A Review. This section is a brief review of observations relating to the emergence of flux and the early development of active regions, emphasizing results gathered just prior to and during the Workshop. This review is based on material presented primarily by Weart (1970), Glackin (1973), Harvey and Martin (1973), Harvey *et al.* (1975), Martin and Harvey (1979), and Golub (1980). Following the review, recent results pertaining to the thermal and temporal behavior of newly born active regions will be discussed with emphasis on soft X-ray results.

The sizes of active regions on the sun appear to form a continuous distribution. Golub (1980) has emphasized from studies of Skylab data that the distribution of the number of features emerging per unit time as a function of either size or lifetime is a monotonically decreasing function. In other words, there are many more smaller regions on the sun at any given time than larger ones, and these smaller regions tend to have shorter lifetimes. A somewhat arbitrary statistical distinction has been made between the more numerous shorter-lived region and the long-lived regions. Regions living less than about one day are called bright points or ephemeral regions (ER), and those living more than four days are called active regions. Two important recent studies, which will be discussed later, involve whether one can distinguish these two classes by their plasma parameters as they emerge, and what the distribution in space and time of the bright points can tell us about magnetic flux emergence on the sun and about basic dynamo models.

Historically the term "emerging flux region" (EFR, from Weart, 1970) has been applied to those regions that develop into long-lived active regions. Such regions usually consist of two bright plages connected by an arch filament system; they brighten and expand rapidly in area (hours), often accompanied by flaring and surging. The bright points, of course, remain small after emergence and within a short time fade into the background. They are best observed in soft X-rays at high contrast against the solar disk, where they are called X-ray bright points (XBP, from Krieger *et al.*, 1971 and Golub *et al.*, 1974). XBP are correlated with bipolar structures that appear on magnetograms as ER (Harvey and Martin, 1973). Bright points are also visible in the chromosphere and transition region (Harvey and Martin, 1973; Golub *et al.*, 1974; Bohlin *et al.*, 1975; Habbal and Withbroe, 1980), where they form a part of the bright emission network. There is a recent report of the marginal detection of XBP in microwaves at $T_b = 8 \times 10^4$ K (Marsh *et al.*, 1980).

The significance of bright points is that they represent emerging magnetic flux, and their numbers imply that at times most of the flux emerging on the sun is associated with these short-lived features rather than with active regions. For instance, during Skylab about 80% of the emerging flux was contributed by regions living less than two days. Comparison of XBP with Kitt Peak magnetograms has shown that the lifetime of an XBP is linearly correlated with its total magnetic flux Φ_{TOT} and that XBP correspond to fluxes $\lesssim 10^{20}$ Mx (Golub *et al.*, 1977). The separation between the poles of the ER is an increasing function of age, with a growth rate of about 1 km/s.

XBP and active regions differ in such properties as latitude distribution, rotation rate, magnetic orientation, and solar cycle variation. Bright points have a more uniform latitude distribution than active regions. In fact, they can be found over the entire sun, even in coronal holes. ERs and short-lived (1–3 days) CaK plages have a similar distribution to that of XBP (Harvey and Martin, 1973). Magnetic pores, which live between 10 and 60 hours, are more broadly distributed in latitude than sunspots. The pores and CaK regions thus represent an intermediate form of emerging flux between the XBP and active regions. However, at least during Skylab, the XBP latitude distribution did show a peaking in the active region belts (latitude $<30°$).

Low-latitude XBP also tend to be highly nonuniform in longitude, like active regions, but during Skylab there was no obvious association with the active regions. At high latitude, XBP tend to be uniformly distributed in longitude.

Golub and Vaiana (1978) measured the rotation rates of regions living from one to seven days. The rate was found to vary with the lifetime (or size) of the region; e.g., XBPs rotate more slowly than short-lived active regions. The rotation rate for an XBP is similar to that of photospheric gas (i.e., 12.8°

day^{-1}), suggesting that shorter-lived features effectively "float" on the solar surface, whereas active regions are more deeply rooted. This in turn implies that the deeper layers of the sun rotate more rapidly than the surface gas.

XBP emission is typically less than one arcmin in diameter. Using Skylab X-ray and EUV images, Sheeley and Golub (1979) were able to resolve a bright point into two to three small loops, each 2500 km in diameter and 12000 km long. The flux loops emerging in bright points show no obvious correlation with the Ca network (Howard *et al.*, 1979; Harvey and Martin, 1973), unlike active regions (EFRs), which tend to emerge at the boundaries of supergranular cells (Bumba and Howard, 1965). However, this suggestion needs to be established with better statistics; if true, it again suggests that the active region flux comes from deeper layers than that of the shorter-lived features.

B. The thermal evolution of emerging flux regions. We have seen that in terms of primarily statistical properties the behavior of emerging flux regions is quite different over a spectrum whose opposite ends are characterized by bright points and active regions. Does this dichotomy exist in the physical plasma parameters of individual regions representative of these extremes?

Little and Krieger (1977) studied this question by attempting to predict the future development of an emerging region by studying the early evolution of its temperature and volume emission measure N_e^2V in soft X-rays. Because this work is unpublished but completed, I will discuss it in some detail here.

They selected a list of EFRs that were observed with the Skylab AS&E soft X-ray instrument based on two criteria: (1) the region was born on the observable disk, and (2) the time resolution was three to four hours for at least one to two days when soft (F3: 2–32, 44–54 Å) and hard (F1: 2–17 Å) filter pairs were available. Sixteen regions met these criteria and fell into two lifetime groups: short-lived ($\lesssim$4 d) and long-lived ($\gtrsim$10 d). The short-lived group was subdivided into those that were prominent enough to be given McMath plage numbers and those that were not (XBP). Table 7.1 lists these regions in terms of their lifetime, both as measured on CaK images and the soft X-ray images, and in terms of their derived temperature. Because the EFRs were unresolved by the instrument at birth, the temperature is the effective color temperature integrated over the entire region. Although the effects of scattering in the instrument were not fully corrected (deconvolved), the temperature was determined to be insensitive to deconvolution. The emission measure was more sensitive to deconvolution, but the relative values were useful; accurate densities could not be determined.

Figures 7.12 and 7.13 show examples of the long-lived and short-lived regions, respectively. As an example of the behavior of the long-lived regions, the time evolution of the temperature and emission measure of McMath region 12511 is shown in Figure 7.14. The region was detected at a threshold of about $N_e^2V = 4 \times 10^{46}$ cm^{-3} and grew rapidly with flaring (obvious flare data were excluded from the samples). Within a day, the value of N_e^2V became nearly constant at 10^{48} cm^{-3}. During emergence and

Table 7.1
Emerging X-Ray Flux Regions[a]

McMath No.	Lifetime: CaK Plage	Lifetime: X-Ray[b]	Derived T_e (10^6 K)	n[c]
		LONG-LIVED REGIONS		
12488	≥40 d	≥40 d	2.46	2
12504	≥40 d	≥40 d	2.41	3
12509	≥40 d	≥40 d	2.43±0.06	7
12511	≥40 d	≥40 d	2.48±0.07	16
12516	≥ 7 d	≥ 9 d	2.42±0.13	15
		SHORT-LIVED REGIONS		
12482	≥ 9 d	2 d	1.80±0.03	6
12483	5 d	2¾ d	1.85±0.06	7
12485	15 h	33 h	1.86±0.04	3
12486	27 h	20 h	—	
12487	2 d	39 h	1.82	2
12493	4 d	4½ d	1.98	2
XBP 1			1.85±0.04	3
XBP 2		d	1.87	1
XBP 3		d	1.92	1
XBP 7		27 h	1.80±0.04	4
XBP 8		42 h	1.89±0.09	5
XBP 9		d	1.84	1

[a]Adapted from Little and Krieger (1977).
[b]Lifetime from AS&E S-054 Filter 3 images.
[c]Number of observations.
[d]XBP observed on only one X-ray image pair.

later development, the temperature remained constant at about 2.45×10^6 K.

Figure 7.15 shows the morphological development and X-ray flux profile of short-lived McMath region 12483, and Figure 7.16 shows the evolution of its temperature and emission measure. Its brightness rise time was similar to the long-lived regions, but after reaching a maximum N_e^2V of 10^{47} cm^{-3} it began to decay quickly. This region also exhibited a constant temperature over its entire lifetime, but at a value of only 1.85×10^6 K. All the short-lived features had similar profiles, except that the XBP rise to peak brightness was faster than the time resolution, and the maximum N_e^2V was 5×10^{46} cm^{-3}.

We see from these results (Table 7.1) that there seem to be two distinct temperatures at which regions emerge; this temperature distinction permits us to predict whether a region will be short- or long-lived and to estimate what its maximum emission measure will be. The average temperature for long-lived regions was $2.45 \pm 0.04 \times 10^6$ K and for short-lived regions was $1.85 \pm 0.04 \times 10^6$ K, *with no overlap between the groups.* This result is remarkable because it implies that the energy balance and/or boundary conditions of the loops comprising the bright points is different from that of the loops initiating the growth of an active region. For instance, active region flux loops could be more deeply rooted in the convection zone than are those of bright points. Alternatively, the heating mechanism itself could be completely different for the two kinds of regions.

Wolfson *et al.* (1977) and Mosher (1980) have extended the Little and Krieger results for active regions using the Lockheed OSO-8 MXRH soft-X-

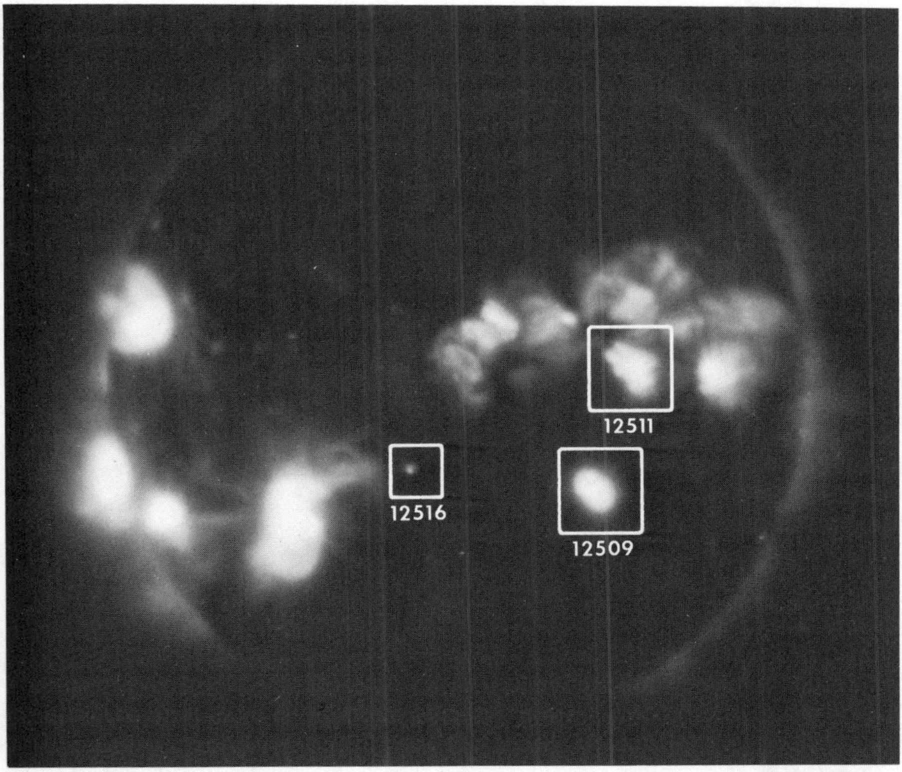

September 1 1814 UT

Fig. 7.12. Figures 7.12–7.16 are from Little and Krieger (1977): A short-exposure AS&E Filter 3 X-ray image from Skylab showing representative long-lived action regions studied from the time of their emergence.

ray experiment. The MXRH is not as sensitive as the Skylab AS&E instrument, so it could only measure temperatures and emission measures of brighter and therefore more developed regions (i.e., after the first sunspots had appeared). Wolfson *et al.* showed that the temperature of a newly born active region, though variable, always returned to a baseline value of about 3×10^6 K. They concluded, with Little and Krieger, that an increasing amount of material appears in such a region at a stable temperature. Mosher examined the emergence of six regions and found that, once the active region had grown sufficiently to produce spots, its temperature was in the range $2.5–3.5 \times 10^6$ K and could not be used to predict the region's further development. Again, short-lived regions were excluded from his sample.

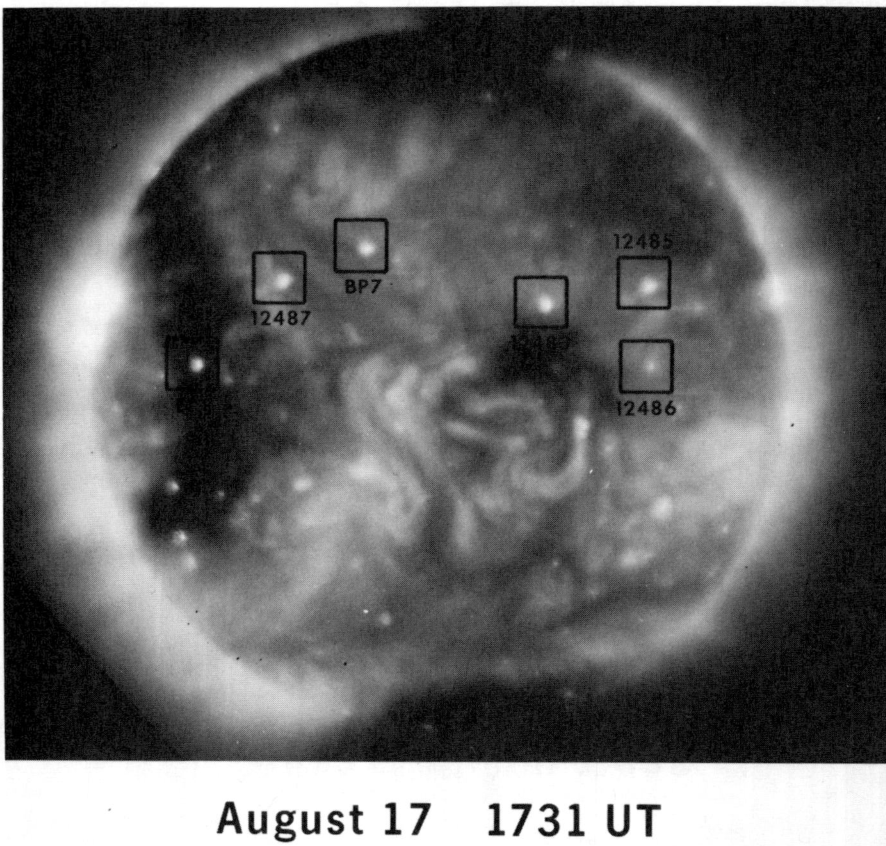

Fig. 7.13. A long-exposure Filter 3 X-ray image showing representative short-lived regions studied throughout their lifetimes.

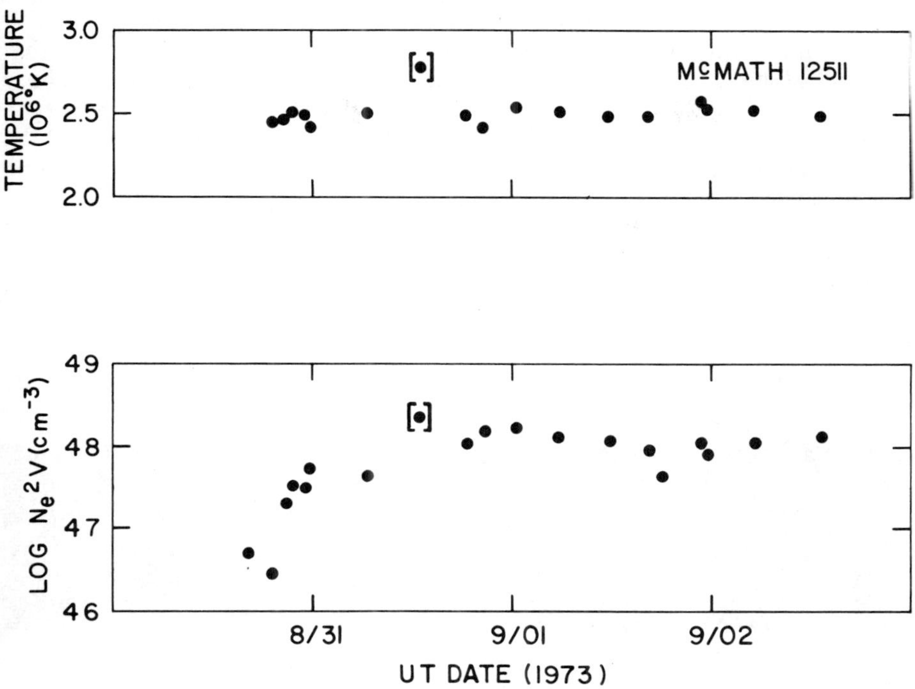

Fig. 7.14. The X-ray evolution in temperature and volume emission measure of long-lived active region McMath 12511. The brackets indicate a possible flare brightening of the region.

C. *The temporal behavior of bright points.* Sheeley and Golub (1979), Nolte *et al.* (1979), and Habbal and Withbroe (1980) have studied the temporal behavior of individual bright points. Nolte *et al.* examined integrated light curves of 17 individual XBP at 90-s time resolution and found evidence of steady-state heating as well as impulsive brightenings and rapid decays. They found that a significant number of XBP "flared" just before disappearing. Golub *et al.* (1974) estimated that about 10% of all XBP flare during their lifetime. Since XBP flares are associated with macrospicules (Moore *et al.*, 1977), or tiny prominence eruptions, these facts suggest that at least some XBP die following a catastrophic ejection of material.

Sheeley and Golub and Habbal and Withbroe found that the temporal behavior of a bright point was actually the average of the more rapid variations of individual loop structures. Even during periods of relatively constant integrated emission, individual loops were observed to sequentially form and disappear. These loops evolved on a time scale of about six minutes. Sheeley and Golub also observed these variations in a short-lived "active region" (lifetime 2.5 d) that had an arcade of tiny loops.

One of the most fundamental ways in which the short- and long-lifetime ends of flux emergence are different is in their variation over the solar cycle. XBP counts from Skylab and six AS&E rocket flights have been combined with sunspot numbers to reveal that the XBP and sunspot frequency distributions are anticorrelated (Davis *et al.*, 1977). Figure 7.17 shows that the XBP counts and the sunspot number R_Z vary almost 180° out of phase (Golub *et al.*, 1979). This demonstrates that the mechanism of emergence may be quite different from one part of the cycle to another.

The relative contribution of XBP and active regions to the total magnetic flux spectrum over a solar cycle can be estimated using the following assumptions: (1) all XBPs are physically the same, and (2) R_Z gives a relative indication of the amount of active region flux emerging throughout the solar cycle. Then, if the average XBP emerges with 3×10^{19} Mx of flux

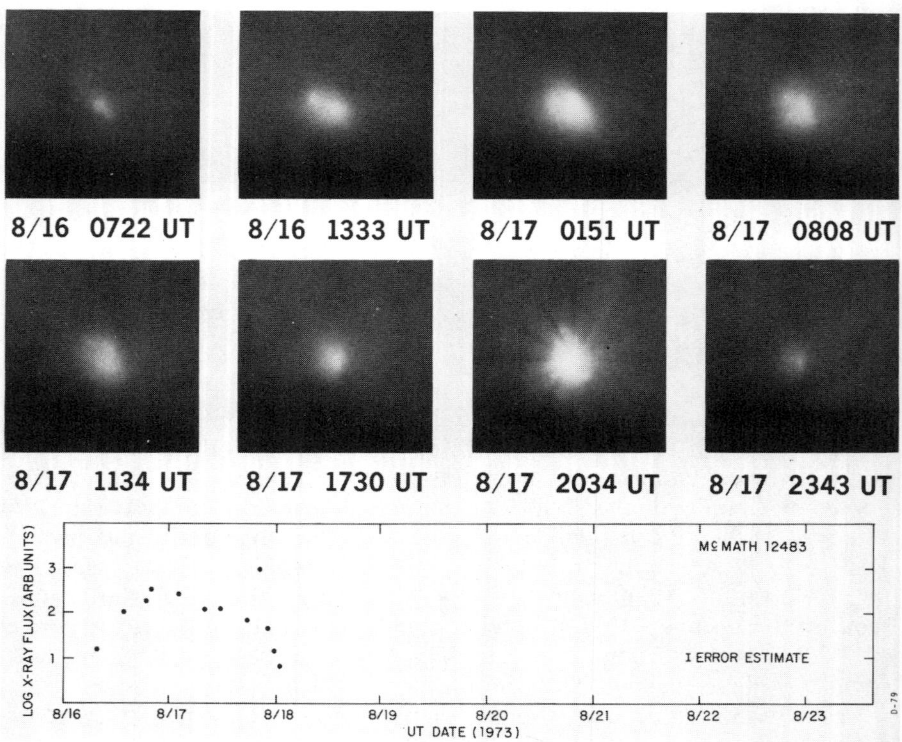

Fig. 7.15. The morphological development of short-lived McMath region 12483 during its lifetime. Each panel is 3 arcmin on a side. The bottom graph shows the evolution of the X-ray flux during this time. Note that the region flared within a few hours of its disappearance.

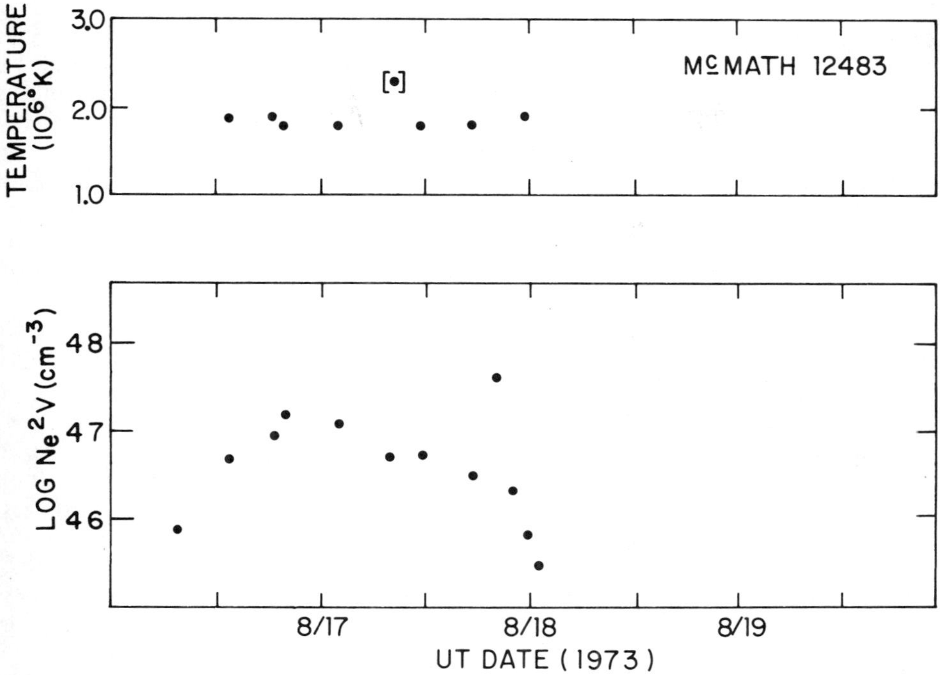

Fig. 7.16. The X-ray evolution in temperature and volume emission measure of the short-lived region shown in Figure 7.15.

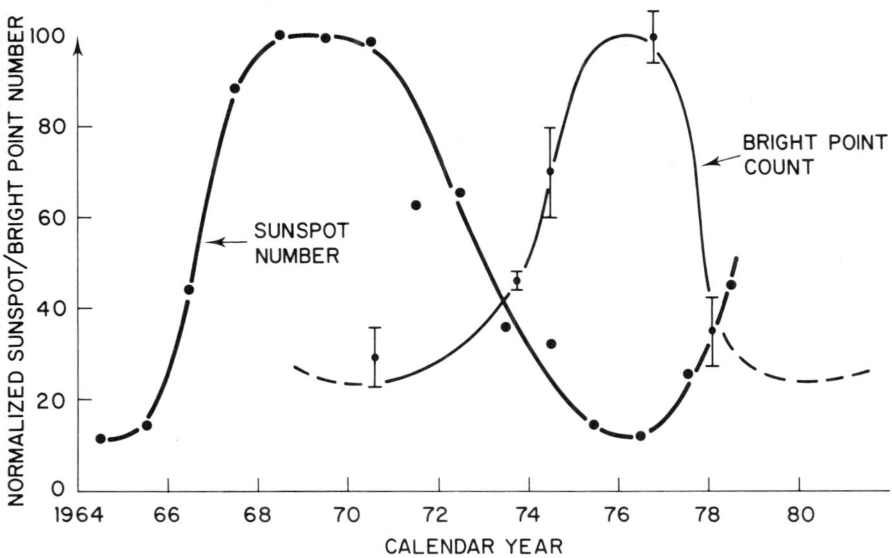

Fig. 7.17. Variation of the number of XBP with the solar cycle as represented by sunspot number R_Z. *Both parameters have been normalized to maximum values of 100 (from Golub* et al., *1979).*

(Golub *et al.*, 1977), 40% of the total magnetic flux in 1970 emerged as XBP, 90% in 1973, 95% in 1976, and 70% early in 1978. Bright points therefore are the dominant form of emerging magnetic flux through the declining, minimum, and rising phases of the solar cycle. This relation has been qualitatively confirmed for the current solar maximum by the observation of very few XBP on images obtained by an AS&E rocket flight in November 1979, as predicted by Figure 7.17. However, Sheeley (Chapter 2) points out that because of their large average flux ($\sim 10^{22}$ Mx) and long lifetime, active regions still dominate the total amount of magnetic flux on the sun except at solar minimum.

Finally, we can ask if other indicators of emerging flux fit into the pattern of the anticorrelation between XBP and active regions. Unfortunately, here the situation is confused. Since XBP are often identified with ER, we expect ER to follow the solar cycle variation of XBP. On the contrary, Martin and Harvey (1979) found that the number of ER varied *in phase* with the sunspot cycle, using data from 1970, 1973, and 1975–77. However, their original paper (Harvey and Martin, 1973) gave evidence that the number of ER peaked during the solar minimum of 1964–65. This conclusion was based on limited statistics at poorer spatial resolution. Their recent results also showed that the ER reached a minimum about one year before the sunspots did. They also found that short-lived (1–3 days) CaK plages seemed to be anticorrelated with sunspot number, like XBP. We have seen that such plages have physical characteristics similar to XBP (e.g., Little and Krieger). The studies of XBP, ER, and calcium plages have been done in different ways. The discrepancy in the patterns related to the solar cycle may be real or may be due to observational selection effects. If real, the differences may be related to the observations that ER characteristics of two cycles may coexist on the sun for a large fraction of a cycle (Martin and Harvey, 1979), and that XBP at least occasionally reveal a two-component latitude distribution. Future simultaneous observations at good time and spatial resolution using soft X-ray, optical, and magnetic field data are necessary to resolve these differences.

7.5 CONCLUSION

7.5.1 Summary

This chapter is an overview of our present understanding of the structure of active region loops as observed in the transition region and corona. The emphasis has been on the multispectral imaging of these "plasma loops," which act as tracers of magnetic flux loops, as observed with the Skylab instruments.

A major observational concern of the Workshop was to identify and classify active region loops using the Skylab data set. The key result of this effort was: The number of complete, identifiable plasma loops in an active region increases with the temperature regime in which the active region is viewed. This is related to the result that most of the material either in an individual loop or integrated over an entire active region is at coronal temperatures ($T_e > 10^6$ K). These "hot" loops are, to first order, isothermal and isobaric and are characterized by a relatively narrow temperature range of 2–3 × 10^6 K. There appear to be classes of cool loops whose emission peaks at temperatures characteristic of the chromosphere through the transition region. Loops or spikes with 10^5 K $< T_e <$ 10^6 K are often associated with sunspot umbrae, over which there is an absence of coronal material. The absence of observable radial temperature and pressure gradients and the lack of spatial association between cool and hot loops imply that any cool cores in those loops, if they exist, must be very thin. Therefore, based on thermal and temporal characteristics, there appear to be at least two distinct classes of active region loops: the umbral loops and the relatively stable hot loops. Intermediate between these may be classes of nonumbral cool loops that, because of their transient nature and restricted temperature range, have been infrequently observed.

The majority of hot loops in an active region form low-lying arcades crossing the neutral line. The brightest loops often end in sunspot penumbrae, may have coronal magnetic field strengths of about 300 G, and have been identified with the peak microwave emission in some active regions. Other classes of hot loops are associated with emerging flux, filaments, and interconnections to other active regions.

Empirical data have indicated that the temperature gradient at the base of hot loops must be steep, i.e., more than 2 × 10^{-2} K/cm. Compact, bright, cool elements exist at the base of hot loops, but vary on time scales short compared to their associated loops. The brightest loops are typically imbedded in strong and/or changing magnetic fields.

Individual hot loops evolve on long time scales of typically hours or days. Although their temperatures are relatively fixed at 2–3 × 10^6 K, the brightness of different hot loops varies by factors of two or three, implying variations in density between loops. Evidence has been found for steady-state heating of stable hot loops of all size scales. The sudden decay of such loops on the radiative time scale has been attributed to the sudden disruption of their heating mechanism. Cooler loops appear to evolve much faster, on time scales of minutes to tens of minutes. In the transition region, complete loops change by the proper motion of emission along the loop, whereas hot loops brighten and fade in place.

Finally, recent observations of the emergence of flux on the sun have revealed fundamental differences in the ends of the emergent flux spectrum as characterized by lifetime and size. These extremes, typified by XBP and ER (lifetime < 1 d) and active regions (lifetime > 4 d), differ in such properties as latitude distribution, rotation rate, magnetic orientation, solar cycle variation, and emergent temperature and peak emission measure. Because of their large numbers, the short lifetime-and-size end of the distribution dominates the emergent flux spectrum during the declining, minimum, and rising portions of the solar cycle. There is some evidence that the short- and long-lifetime ends of the distribution over the solar cycle are about 180° out of phase, implying that the mechanism of emergence varies systematically over the cycle. However, the total amount of flux emerging on the sun may be approximately constant over the cycle. Several lines of argument suggest that the bright points may essentially "float" on the surface of the sun, whereas active regions may be more deeply rooted in the convection zone.

7.5.2 Future EUV and Soft X-Ray Space Observations

Since most of the material in nonumbral active region loops is at coronal temperatures, and the characteristic radiation from the corona comes from resonance lines of the high-Z elements with wavelengths $\lesssim 60$ Å, it is clear that further progress in our understanding of active region loop structure must involve space instrumentation in the XUV and soft X-ray regime. I will briefly mention some spaceborne experiments in the EUV and soft X-rays being designed for the next generation of observations of active regions. Sounding rocket flights that test new methods of observing the sun give an idea of the kind of experiments which will be flown in the near future on Spacelab.

Such experiments include a spectrometer to measure coronal abundances, a combination soft X-ray telescope and spectrometer experiment to measure the energy balance of loops, EUV experiments to study the dynamics of active regions, and a Lyα coronagraph to measure temperatures and velocities in the corona. In the future, techniques will have to be developed to allow us to fully describe the thermal distribution of active region loops. This can be accomplished in part by the development of subarcsec imaging in the EUV and X-rays and by the development of high resolution spectrometers at the focal plane of telescopes.

Acknowledgements

The author benefited from his participation in the Skylab Solar Workshop on Active Regions, especially on the team working on the morphology and evolution of active regions. I thank A. Krieger, the chairman of that team, for his efforts on this chapter, R. Levine for a careful review of the final manuscript, and S. Kahler and B. Roberts for their comments. I am grateful to the following individuals for reports and/or figures used in the chapter: C.-C. Cheng (NASA/MSFC), K. Dere (NRL), L. Golub (CFA), B. Jackson (UCSD), J. Mariska (NRL), J. Mosher (Lockheed), N. Sheeley (NRL), and F. Orrall (Institute for Astronomy).

REFERENCES

Bohlin, J. D., Sheeley, N. R., and Tousey, R.: 1975, in M. J. Rycroft (ed.), *Space Research XV*, Akademie-Verlag, Berlin, p. 651.

Bonnet, R. M., Bruner, E. C., Acton, L. W., Brown, W. A., and Decaudin, M.: 1980, *Astrophys. J.* **237,** L47.

Bougeret, J. L. and Steinberg, J. L.: 1977, *Astron. Astrophys.* **61,** 777.

Bumba, V. and Howard, R.: 1965, *Astrophys. J.* **141,** 1492.

Cheng, C.-C.: 1980, *Astrophys. J.* **238,** 743.

Cheng, C.-C., Smith, J. B., and Tandberg-Hanssen, E.: 1980, *Solar Phys.* **67,** 259.

Davis, J. M., Gerassimenko, M., Krieger, A. S., and Vaiana, G. S.: 1975, *Solar Phys.* **45,** 393.

Davis, J. M., Golub, L., and Krieger, A. S.: 1977, *Astrophys. J.* **214,** L141.

Dere, K. P.: 1980, "XUV Structure of Solar Active Regions," Ph.D. Thesis, Catholic Univ. of America, Washington, D.C.

Dulk, G. A., Melrose, D. B., and Suzuki, S.: 1979, *Proc. Astron. Soc. Aust.* (in press).

Emslie, A. G. and Machado, M. E.: 1979, Center for Astrophysics Preprint No. 1034.

Foukal, P.: 1975, *Solar Phys.* **43,** 327.

Foukal, P.: 1976, *Astrophys. J.* **210,** 575.

Foukal, P.: 1978, *Astrophys. J.* **223,** 1046.

Foukal, P. V., Huber, M. C. E., Noyes, R. W., Reeves, E. M., Schmahl, E. J., Timothy, J. G., Vernazza, J. E., and Withbroe, G. L.: 1974, *Astrophys. J.* **193,** L143.

Gerassimenko, M., Nolte, J. T., and Petrasso, R. D.: 1976, *Solar Phys.* **48,** 121.

Gerassimenko, M., Solodyna, C. V., and Nolte, J. T.: 1978, *Solar Phys.* **57,** 103.

Gergely, T. E. and Erickson, W. C.: 1975, *Solar Phys.* **42,** 467.

Gergely, T. E. and Kundu, M. R.: 1975, *Solar Phys.* **41,** 163.

Gergely, T. E., Kundu, M. R., Golub, L., and Webb, D.: 1980, in M. R. Kundu and T. E. Gergely (eds.), *Radio Physics of the Sun*, D. Reidel, Dordrecht, Holland.

Glackin, D. L.: 1973, *Publ. Astron. Soc. Pacific* **85,** 241.

Golub, L.: 1980, *Proc. Roy. Soc. A* **297,** 595.

Golub, L., Davis, J. M., and Krieger, A. S.: 1979, *Astrophys. J.* **229,** L145.

Golub, L., Krieger, A. S., Harvey, J. W., and Vaiana, G. S.: 1977, *Solar Phys.* **53,** 111.

Golub, L., Krieger, A. S., Silk, J. K., Timothy, A. F., and Vaiana, G. S.: 1974, *Astrophys. J.* **189,** L93.

Golub, L., Maxson, C., Rosner, R., Serio, S., and Vaiana, G. S.: 1980, *Astrophys. J.* **238,** 343.

Golub, L. and Vaiana, G. S.: 1978, *Astrophys. J.* **219,** L55.

Habbal, S. R. and Withbroe, G. L.: 1980, *Solar Phys.* (in press).

Harvey, K. L., Harvey, J. W., and Martin, S. F.: 1975, *Solar Phys.* **40,** 87.

Harvey, K. L. and Martin, S. F.: 1973, *Solar Phys.* **32,** 389.

Hood, A. W. and Priest, E. R.: 1979, *Astron. Astrophys.* **77,** 233.

Howard, R., Fritzova-Svestkova, L., and Svestka, Z.: 1979, *Solar Phys.* **63,** 105.

Jackson, B. V.: 1979, Report to the Skylab Solar Workshop on Active Regions.

Krieger, A. S.: 1977, *Proc. of the OSO-8 Workshop*, Univ. of Colorado, Boulder, CO, p. 98.

Krieger, A. S., Vaiana, G. S., and Van Speybroeck, L. P.: 1971, in R. Howard (ed.), *Solar Magnetic Fields*, D. Reidel, Dordrecht, Holland, p. 397.

Kundu, M. R., Alissandrakis, C. E., Bregman, J. D., and Hin, A. C.: 1977, *Astrophys. J.* **213,** 278.

Kundu, M. R., Schmahl, E. J., and Gerassimenko, M.: 1980a, *Astron. Astrophys.* **82,** 265.

Kundu, M. R., Schmahl, E. J., and Rao, A. P.: 1980b, *Astron. Astrophys.* (in press).

Kundu, M. R. and Velusamy, T.: 1980, *Astrophys. J.* **240,** L63.

Levine, R. H.: 1977, in J. B. Zirker (ed.), *Coronal Holes and High Speed Wind Streams*, Colorado Assoc. Univ. Press, Boulder, CO, p. 103.

Levine, R. H. and Pye, J. P.: 1980, *Solar Phys.* **66,** 39.

Levine, R. H. and Withbroe, G. L.: 1977, *Solar Phys.* **51,** 83.

Little, S. J. and Krieger, A. S.: 1977, *Bull. AAS* **9,** 341.

Mariska, J. T., Feldman, U., and Doschek, G. A.: 1980, *Astrophys. J.* **240,** 300.
Marsh, K. A., Hurford, G. J., and Zirin, H.: 1980, *Astrophys. J.* **236,** 1017.
Marsh, K. A., Zirin, H., and Hurford, G. J.: 1979, *Astrophys. J.* **228,** 610.
Martin, S. F. and Harvey, K. L.: 1979, *Solar Phys.* **64,** 93.
Moore, R. L., Tang, F., Bohlin, J. D., and Golub, L.: 1977, *Astrophys. J.* **218,** 286.
Mosher, J. M.: 1980, Report to the Skylab Solar Workshop on Active Regions.
Nicholas, K. R., Bartoe, J.-D., Brueckner, G. E., and Van Hoosier, M. E.: 1979, *Astrophys. J.* **233,** 741.
Nolte, J. T., Solodyna, C. V., and Gerassimenko, M.: 1979, *Solar Phys.* **63,** 113.
Pallavicini, R.: 1980, in M. R. Kundu and T. E. Gergely (eds.), *Radio Physics of the Sun*, D. Reidel, Dordrecht, Holland.
Pallavicini, R., Vaiana, G. S., Tofani, G., and Felli, M.: 1979, *Astrophys. J.* **229,** 375.
Poletto, G., Vaiana, G. S., Zombeck, M., Krieger, A. S., and Timothy, A. F.: 1975, *Solar Phys.* **44,** 83.
Priest, E. R.: 1978, *Solar Phys.* **58,** 57.
Pye, J. P., Evans, K. D., Hutcheon, R. J., Gerassimenko, M., Davis, J. M., Krieger, A. S., and Vesecky, J. F.: 1978, *Astron. Astrophys.* **65,** 123.
Roberts, B. and Frankenthal, S.: 1980, *Solar Phys.* **68,** 103.
Rosner, R., Tucker, W. H., and Vaiana, G. S.: 1978, *Astrophys. J.* **220,** 643.
Schmahl, E. J. and Orrall, F. Q.: 1979, *Astrophys. J.* **231,** L41.
Sheeley, N. R.: 1980, *Solar Phys.* **66,** 79.
Sheeley, N. R. and Golub, L.: 1979, *Solar Phys.* **63,** 119.
Spruit, H.: 1977, *Proc. of the OSO-8 Workshop*, Univ. of Colorado, Boulder, CO, p. 83.
Stewart, R. T. and Vorpahl, J.: 1977, *Solar Phys.* **55,** 111.
Svestka, Z.: 1976, *Solar Flares*, D. Reidel, Dordrecht, Holland.
Van Hoven, G., Chiuderi, C., and Giachetti, R.: 1977, *Astrophys. J.* **213,** 869.
Van Speybroeck, L. P., Krieger, A. S., and Vaiana, G. S.: 1970, *Nature* **227,** 818.
Weart, S. R.: 1970, *Astrophys. J.* **162,** 987.
Webb, D. F. and Kundu, M. R.: 1978, *Solar Phys.* **57,** 155.
Webb, D. F. and Levine, R. H.: 1979, Report to the Skylab Solar Workshop on Active Regions.
Webb, D. F. and Zirin, H.: 1980, *Solar Phys.* (in press).
Withbroe, G. L.: 1975, *Solar Phys.* **45,** 301.
Wolfson, C. J., Acton, L. W., Leibacher, J. W., and Roethig, D. T.: 1977, *Solar Phys.* **55,** 181.

8. PHYSICS OF STATIC LOOP STRUCTURES AND SCALING LAWS

George L. Withbroe

8.1 INTRODUCTION

Observations of the corona made with high spatial resolution, particularly observations obtained with EUV and X-ray instrumentation capable of observing high-temperature radiation from the solar disk, indicate that the coronal structure in active regions consists of a series of magnetic flux tubes or loops filled with plasma at coronal temperatures. These loops, which have closed magnetic configurations (both ends of the loop rooted in the chromosphere), are the basic elemental components, or building blocks, making up the visible, EUV, and X-ray emitting corona in active regions. In order to understand the physical nature of the corona in these regions, it is necessary to understand the physical processes occurring in these elemental structures. What are the plasma heating and cooling mechanisms? What are the mechanisms for transporting material between the chromosphere and corona? Why do some loops appear to be dynamically and thermally stable while others are not? One method of achieving insights concerning these processes is through development of theoretical models that can account for the observed properties of active region loops. Quasistatic models, whose development has been stimulated by EUV and X-ray observations of active regions, are a class of models that appear to satisfactorily explain some of the observed properties of loops seen in active regions, quiet regions, and flares. This chapter discusses the theory of these models, some applications to observations, and some of the insights that they provide concerning coronal physics.

8.2 THEORY

The derivation of quasistatic loop models and their application to interpretation of EUV and X-ray observations have been discussed by Jordan (1975, 1976), Landini and Monsignori-Fossi (1975), Rosner *et al.* (1978), Craig *et al.* (1978), Withbroe (1978b), Hood and Priest (1979), Levine and Pye (1979), Machado and Emslie (1979), Vesecky *et al.* (1979), and others. The geometry is illustrated in Figure 8.1. The loop is assumed to be symmetrical, with the spatial coordinate s measuring distance (from one loop footpoint) along the

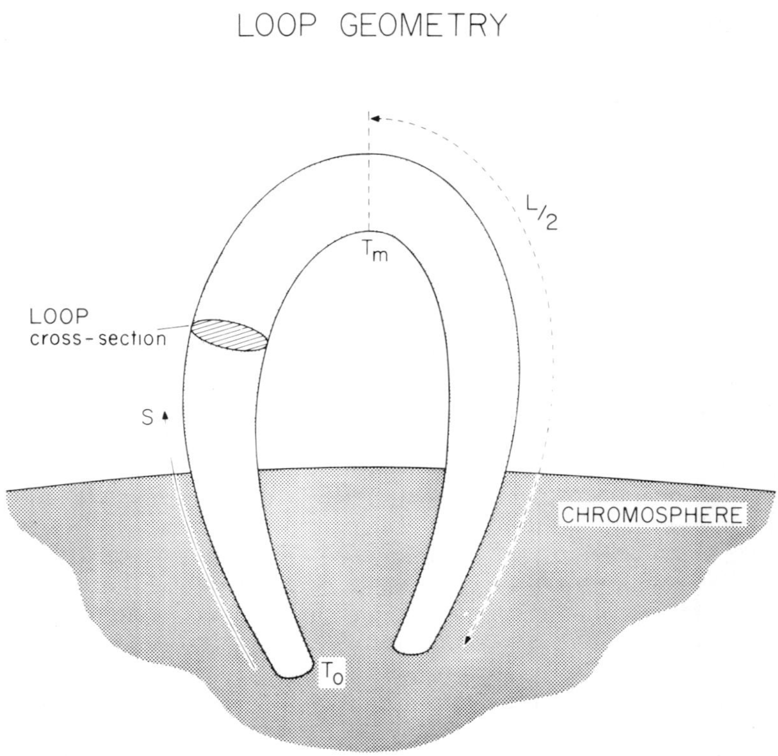

Figure 8.1. Schematic of a quasistatic loop.

loop parallel to the magnetic field defining the loop. The steady-state energy equation for a static plasma in a section of the loop with cross-sectional area *A(s)* is

$$\nabla \cdot \mathbf{F}_c = E_h(s) - E_r(s) \ , \tag{8.2.1}$$

where E_h is the energy input (erg cm^{-3} s^{-1}), E_r is the radiative loss (erg cm^{-3} s^{-1}), and F_c is the conductive flux (erg cm^{-2} s^{-1}). Because the thermal conductivity parallel to the field lines greatly exceeds that in the perpendicular direction (cf. Spitzer, 1962), the energy equation can be rewritten as

$$-\frac{1}{A(s)}\frac{d}{ds}\left[A(s)K_0 T^{5/2}\frac{dT}{ds}\right] = E_h(s) - n_e^2\,\Phi(T) \ , \tag{8.2.2}$$

where n_e is the electron density (cm^{-3}), $\Phi(T)$ is the radiative loss function (cf. Cox and Tucker, 1969; McWhirter *et al.,* 1975; Raymond *et al.,* 1976; Raymond, 1979), and K_0 is the coefficient of thermal conductivity (cf. Ulmschneider, 1970; Nowak and Ulmschneider, 1977). The plasma is assumed to be in hydrostatic equilibrium, with

$$\frac{dP}{ds} = -\varrho g_{11} \ , \tag{8.2.3}$$

where $P = P_g + P_v$, P_g is the gas pressure, P_v is the turbulent pressure (cf. Gabriel, 1976), ϱ is the density, and g_{11} is the component of gravity parallel to the magnetic field vector **B**.

The solution of the energy equation (Eq. 8.2.2) requires the specification of boundary conditions. We assume that the conductive flux at the chromospheric footpoints has a value of F_0. These footpoints are located at $s = 0$ and $s = L$, where L is the length of the loop measured from one footpoint up to the apex and down to the other footpoint. We assume that the footpoints are located at the height where the temperature is 3×10^4 K. At lower temperatures, the calculation of radiative losses becomes difficult because the primary sources of radiative loss are optically thick spectral lines and continua. The form of the radiative loss function used in Eq. 8.2.2 assumes an optically thin plasma. There are other advantages to restricting the model to temperatures above 3×10^4 K. In this temperature region, the plasma can be assumed to be fully ionized and the thermal conductivity varies as $T^{5/2} dT/ds$ (cf. Ulmschneider, 1970; Nowak and Ulmschneider, 1977). Since the loop is assumed to be symmetrical, the conductive flux equals zero at the apex of the loop, where $s = L/2$. We also assume that the maximum temperature in the loop T_m is located at the apex. These two conditions imply

$$\left[\frac{dT}{ds}\right]_{s = L/2} = 0 \ , \tag{8.2.4}$$

and

$$\left[\frac{d^2T}{ds^2}\right]_{s = L/2} < 0 \ . \tag{8.2.5}$$

An additional relationship between the loop parameters can be obtained by integrating the conductive flux equation

$$F_c(s) = K_0 T^{5/2} \frac{dT}{ds} \tag{8.2.6}$$

to obtain

$$L = 2 \int_{T_0}^{T_m} \frac{K_0 T^{5/2}\, dT}{F_c(s)} \ , \tag{8.2.7}$$

where T_0 and T_m are, respectively, the temperatures at the footpoints and the apex of the loop.

In order to illustrate some of the properties of quasistatic loop models, it is useful to consider solutions to the above equations derived for a simplified set of conditions. Since the coronal pressure scale height $H(=5000 T_m$ cm) is large compared to typical heights of active region loops, we assume that the electron pressure is constant along the length of the loop, with a value P. Because we are dealing with a fully ionized plasma with 10% helium, the gas pressure $P_g = P_e + P_H + P_{He} = 1.9P$. The heating function is assumed to have the form

$$E_h = AP^{\lambda}T^{\alpha} \ . \tag{8.2.8}$$

This equation is based on the discussion by Rosner *et al.* (1978), who derived values for λ between 1/2 and 7/6 and values for α between −1 and +2. These values are based on various heating mechanisms, including acoustic waves dissipated by viscous- or shock-wave damping, Alfven waves dissipated by anomalous conduction damping or mode conversion, and dissipation of electric currents by anomalous dissipation.

Under these conditions, the energy equation (Eq. 8.2.2) becomes

$$-dF_c = AP^{\lambda}T^{\alpha}ds - n_e^2\,\Phi(T)ds \ . \tag{8.2.9}$$

Integration of Eq. 8.2.9 after both sides of the equation have been multiplied by the conductive flux (Eq. 8.2.6) yields

$$F_c^2(T) = \frac{2AP^{\lambda}K_0}{(7/2+\alpha)}\, T_m^{\ 7/2+\alpha}\left[1-\left(\frac{T}{T_m}\right)^{7/2+\alpha}\right] - 2\left(\frac{P}{k}\right)^2 K_0 \int_T^{T_m} \Phi(T)T^{1/2}dT \ , \tag{8.2.10}$$

where $F_c(T_m) = 0$ from Eq. 8.2.4 and $n_e = P_e/kT = P/kT$. Equation 8.2.10 can be used to determine the conductive flux passing through the footpoints of the loop, $F_c(T) = F_0$. This allows Eq. 8.2.10 to be rewritten as

$$F_c^2(T) = F_0^2\left[1-\left(\frac{T}{T_m}\right)^{7/2+\alpha}\right] + 2\left(\frac{P}{k}\right)^2 K_0\left[\int_{T_0}^{T} \Phi(T)T^{1/2}dT - \left(\frac{T}{T_m}\right)^{7/2+\alpha}\int_{T_0}^{T_m} \Phi(T)T^{1/2}dT\right] , \tag{8.2.11}$$

where we have assumed $T_0/T_m \ll 1$.

The limiting case where $F_0 = 0$ is of interest. From the expression relating the loop length to the conductive flux (Eq. 8.2.7), one obtains

$$T_m = C_\alpha (LP)^{1/3} \ , \tag{8.2.12}$$

where C_α is a slowly varying function of α that has the approximate form,

$$\log C = 3.136 + 0.022\alpha \ , \tag{8.2.13}$$

for $-1.5 < \alpha < 3.0$. The limiting value for very large α is $\log C = 3.23$. Use of a different radiative loss function than that employed here (from Raymond, 1979) can result in a slightly different relationship between P, T_m, and L. For example, Craig *et al.* (1978) find $\log T_m \sim 4/13 \log (PL)$.

Figure 8.2 contains a plot of T_m as a function of LP for $\alpha = 0$ and a variety of values for F_0/F_{max}, where F_{max} is the maximum value that the conductive

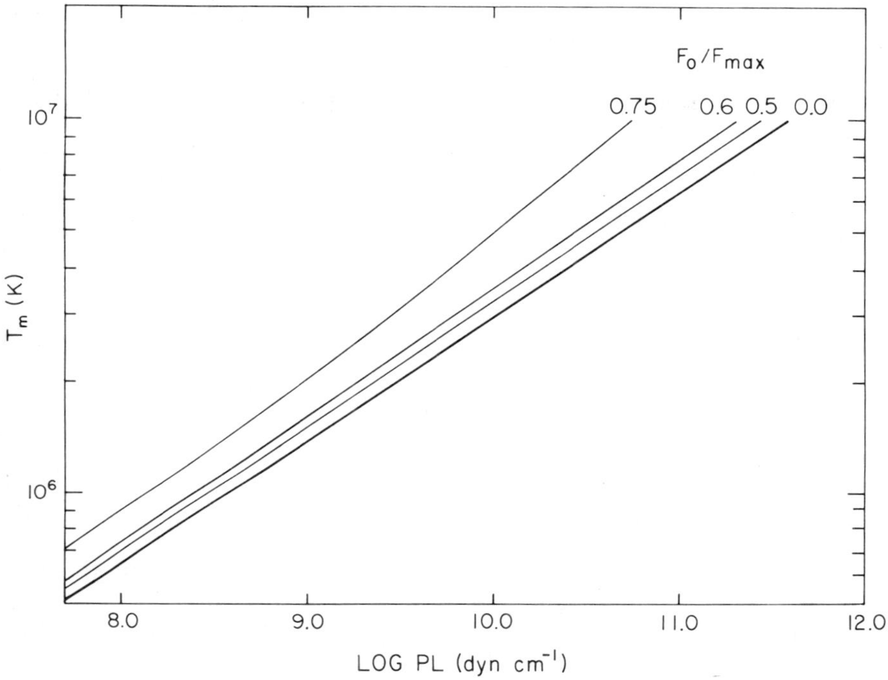

Figure 8.2. Variation of maximum loop temperature T_m as a function of the electron pressure P (dyne cm^{-2}) and loop length L (cm) for $\alpha = 0$ and several values of F_0/F_{max} (see text).

flux into the chromosphere can attain for a given loop length and temperature T_m. This value, which is approximately equal to $K_0 T_m^{7/2}/L$, can be obtained from Eq. 8.2.7 with the conductive flux given by Eq. 8.2.11 with $P = 0$. The line corresponding to $F_0/F_{max} = 0$ (see Eq. 8.2.12) is drawn as a heavy, solid line. Any loop with length, pressure, and maximum temperature falling to the right of this line is thermally and dynamically unstable. (This is discussed further below.) In principle, all loops with $0 \leq F_0 \leq F_{max}$ can exist; however, all of these need not be stable against perturbations to the heating function, temperatures, and/or densities (cf. Antiochos, 1979; Craig, 1979; Habbal and Rosner, 1979; Priest, this volume).

For a stable, static loop to exist, it is necessary that the conductive flux entering the upper chromosphere, F_0, be limited to values equal to or less than that which can be dissipated by radiation in chromospheric continua and lines. Empirical models of the chromosphere (cf. Vernazza *et al.*, 1973, 1979; Basri *et al.*, 1979) suggest that the conductive flux passing through the hydrogen Lyman alpha-emitting layers of the atmosphere (where $T = 2 \times 10^4$ K) to lower chromospheric layers is a negligible fraction of the energy radiated in Lyman alpha, the primary source of radiative loss in the upper chromosphere. This suggests that in a stable loop, the maximum value of the conductive flux entering the upper chromosphere is limited to $F_0 \leq F_{Ly\alpha}$ where $F_{Ly\alpha}$ is the Lyman alpha flux (cf. Machado and Emslie, 1979).

An upper limit to the Lyman alpha flux that can be produced by radiation of energy carried into the upper chromosphere by thermal conduction can be calculated using the optically thin approximation for the radiative loss function. Through use of procedures similar to those utilized in deriving Eq. 8.2.10, we obtain

$$F_0 = P\left[\frac{2}{K^2}\int_{T_c}^{T_0} \frac{K\Phi(T)dT}{T^2} - \frac{E_h(\text{upper chromosphere})}{P^2}\right]^{1/2} , \qquad (8.2.14)$$

where T_c is the temperature in the chromosphere below which no significant amount of energy is carried by conduction. The maximum value of F_0 that can be dissipated in the upper chromosphere is obtained from Eq. 8.2.14 with $E_h = 0$,

$$F_0 = 6 \times 10^5\, P \text{ erg cm}^{-2}\text{ s}^{-1} , \qquad (8.2.15)$$

where we have used values of K obtained from Nowak and Ulmschneider (1977) and the radiative loss function calculated by Raymond (1979).

If there is upper chromospheric heating by sources in addition to the heating by thermally conducted energy from the corona, then the source term E_h in Eq. 8.2.14 will reduce the maximum possible value of F_0 below that given by Eq. 8.2.15. In active regions, where typically $0.5 \leq P \leq 1.0$ dyne cm^{-2} s^{-1}, Eq. 8.2.15 suggests $F_0 \leq 6 \times 10^5$ erg cm^{-2} s^{-1}. Typical Lyα λ1216 fluxes (the major source of radiative energy loss in the upper chromosphere) are of the order of 2–3 times larger than this (e.g., Vernazza and Reeves, 1978). This suggests that in addition to the energy deposited from above by thermal conduction, the upper chromosphere in active regions is heated by energy transported upward from lower levels by radiation and/or nonradiative means (e.g., waves). However, given the approximations used in calculating Eq. 8.2.15, this problem should be investigated more thoroughly.

From the definition of $F_{max} \approx K_0 T_m^{7/2}/L$ above, Eq. 8.2.15, and Eq. 8.2.12, (which relates T_m and PL), we obtain

$$\begin{aligned} \frac{F_0}{F_{max}} &\approx \frac{6 \times 10^5\, PL}{K_0 T_m^{7/2}} \\ &\approx 4 \times 10^2\, T_m^{1/2} \\ &< 0.4, \quad \text{for } T_m > 10^6\, K. \end{aligned} \qquad (8.2.16)$$

This result, coupled with the probability that F_0 given by Eq. 8.2.15 is an upper limit (because of the likelihood that the upper chromosphere is optically thick and heated to some extent by energy carried up from below), suggests that F_0 is limited to a few tenths F_{max} or less. Because the relationship between the loop pressure, length, and maximum temperature is essentially independent of F_0/F_{max} in this case (see Figure 8.2), this means that for quasistatic loops with constant cross sections and $-1 < \alpha < 2$,

$$T_m = 1.4 \times 10^3 (PL)^{1/3} , \quad (8.2.17)$$

where we have ignored the small dependence on α (see Eqs. 8.2.12 and 8.2.13). This relationship is plotted (*heavy solid line*) in Figure 8.2.

The results given above have important implications concerning the flow of mass and energy in coronal loops and the flows between the loops and the chromospheric regions at the loop footpoints. All of the energy deposited in a coronal loop must be balanced by radiation from the plasma contained in the loop itself or by the chromospheric material at the loop footpoints. Thermal conduction simply provides a mechanism for transporting energy from one section of the loop to another (or to the footpoints of the loop); it does not dissipate energy. The total radiative loss from the plasma in the loop and the chromospheric material at the loop footpoints is proportional to P. Thus, any change in the rate of energy input must lead to a change in pressure if the loop is to balance the new rate of input by radiative losses. This suggests that the temperature-density structure of a coronal loop is controlled by the nonradiative energy deposition in the loop.

Consider a loop of length L and energy input E_1 (erg s^{-1}) in a static energy balance characterized by pressure P_1 and maximum temperature T_1. If the rate of energy input decreases to $E_2 < E_1$, then the loop will radiate more energy than it receives. The coronal plasma will become radiatively unstable, cool, develop a Raleigh-Taylor instability, and fall into the chromosphere. This lowers the pressure in the loop and in the upper chromospheric regions at the loop footpoints, which are in pressure equilibrium with the overlying coronal plasma. The lower pressure reduces the radiative losses, thereby enabling the loop to come into equilibrium again, with the radiative losses balancing the enerty input E_2. On the other hand, if the energy input increases with $E_2 > E_1$, then the plasma in the loop will heat up, increasing the maximum temperature and the downward conductive flux. Since the upper chromosphere is already radiating as much energy as it can, the increased energy input evaporates some chromospheric material, increasing the pressure throughout the loop and, hence, the radiative losses at all levels. This reduces the conductive flux and the rate of evaporation of chromospheric material. Equilibrium is achieved once the pressure has increased sufficiently that the radiative losses balance the new level of energy input E_2. If $E_2 > E_1$, then the new equilibrium values of T_m and P will be larger, with $P_2/P_1 \approx E_2/E_1$ and $T_2/T_1 \approx (E_2/E_1)^{1/3}$. (The precise relationship depends on the form of the heating function.) Thus, the effect of changing the rate of energy input is to move one up or down along the curve relating T_m, P, and L shown in Figure 8.2. One might expect that when the energy input changes there may be some oscillations around the equilibrium value of the pressure and temperature before stability is achieved. This has been demonstrated by Craig and McClymont (1979), who have investigated the dynamics of quasistatic loops subjected to variations in the energy input.

8.3 COMPARISON WITH OBSERVATIONS

Figure 8.3 contains a plot of T_m *vs* PL for the above theoretical model for $F_0 = 0$ (*heavy solid line*) along with empirical values (*points*) determined from several studies of X-ray and EUV-emitting loops (cf. Rosner *et al.*, 1978). The points in the lower half of the plot are from active regions, those in the upper half from flares. There is a substantial amount of scatter about the theoretical line, particularly for flares. Some of this is due most likely to errors in the determination of T_m, P, and L, which could easily amount to 0.1 dex or more for each parameter. For flares, it is possible that the loops may not have come into static energy balance. This may well explain the scatter to the left of the optimum range for the upper set of points. Another source of scatter may be due to the assumption that α is the same for all loops (cf. Eq. 8.2.12). The assumption that the loop cross section is constant along the length of the loop may also cause some of the scatter. It is likely that the loops increase with height because of divergence of the field lines defining the loop geometry (cf. Withbroe, 1978b; Vesecky *et al.*, 1979; Levine and Pye, 1979).

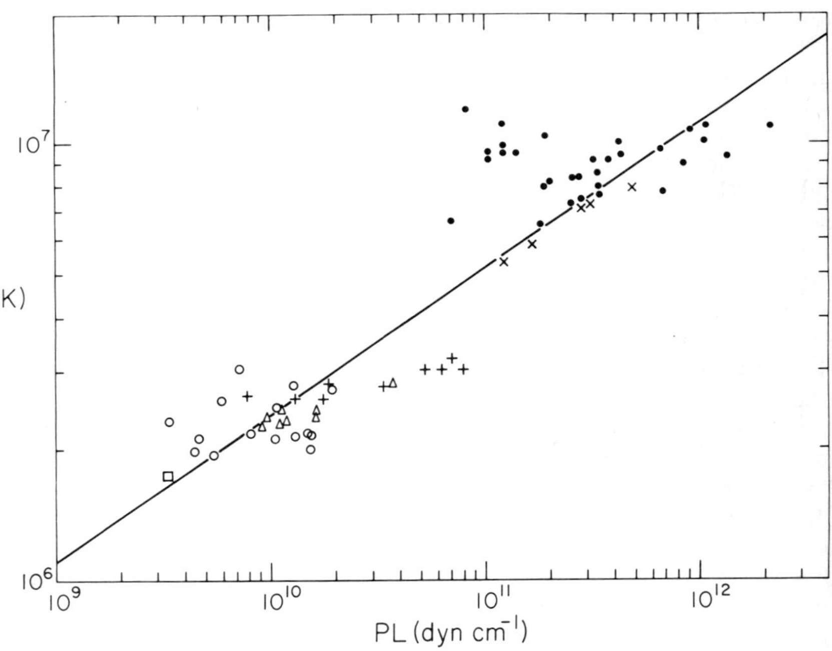

Figure 8.3. Plot (from Emslie and Machado, 1979) of maximum temperature T_m as a function of the loop pressure P (dyne cm^{-2}) and length L (cm). In this figure, the pressure is the total gas pressure $P_g = P_e + P_H + P_{He}$. The curve is the theoretical relationship (cf. Eq. 8.2.17). The empirical values are from Pallavicini et al., *1977* (filled circles); *Withbroe, 1978b* (diagonal crosses); *Pye* et al., *1978* (horizontal crosses); *Landini* et al., *1975* (triangles); *R. Chase (private communication,* circles*) and L. Golub (private communication,* open square*).*

Vesecky *et al.* have calculated the effects of field divergence in terms of a parameter Γ that is the ratio of the area at the apex of the loop to that at the footpoint. The effect of changing Γ for a particular loop model is shown in Figure 8.4. For a loop characterized by a given pressure P, maximum temperature T_m, and length L, the conductive flux near the loop footpoints increases with increasing Γ because of the channeling of the downward conductive flux produced by the converging (toward the loop footpoints) magnetic field defining the geometry of the loop. If T_m, L, and F_0 (the conductive flux into the upper chromosphere) are fixed, then P must increase as Γ increases in order that the radiative losses in the loop can reduce sufficiently the *total* amount of energy carried by conduction to the footpoints so that the energy *per unit area* F_0 does not change. This is why the curve relating P, T_m, and L shifts to the right as Γ increases in Figure 8.4.

Because of the scatter in the empirical data and the insensitivity of the theoretical scaling law (the relationship between P, T_m, and L) to variations in parameters such as F_0, α (the parameter describing the temperature sensitivity of the heating function), and Γ, comparisons between the theoretical scaling law and empirical data do not provide significant insights concerning heating mechanisms or the role of thermal conduction in heating the

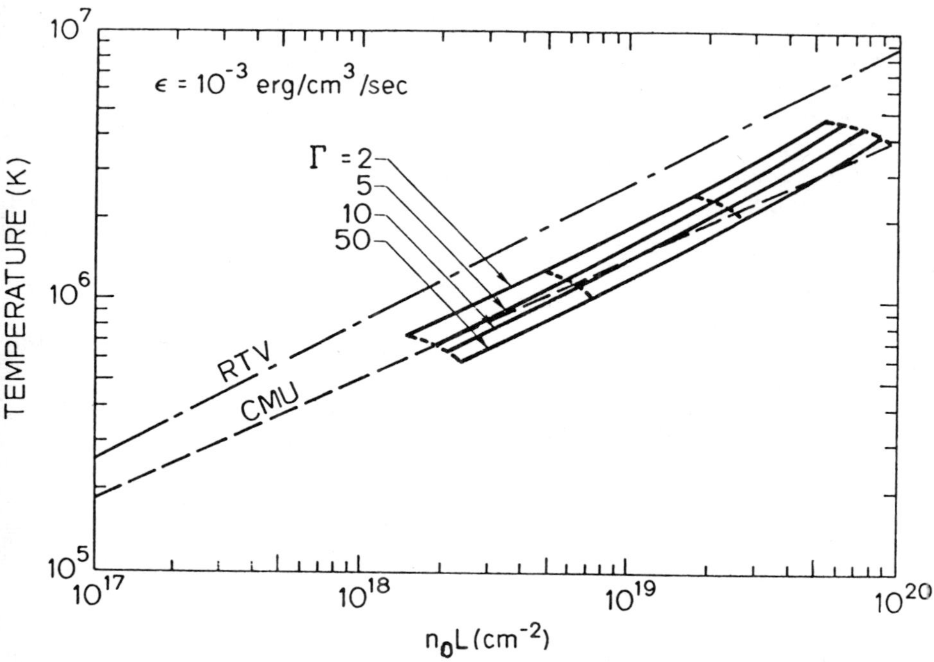

Figure 8.4. Plot (from Vesecky et al.*, 1979) of maximum loop temperature as a function of* n_0L, *where* $n_0 = P/kT_m$ *(cm^{-3}) is the electron density at the loop apex, and* L *is the loop length (cm). Curves are shown for several values of the area factor* Γ *with the dashed curve showing the behavior for constant* L. *The energy input* ϵ *is assumed to be uniform over the loop volume. The models of Rosner* et al. *(RTV) and Craig* et al. *(CMU) are shown for comparison. The differences between the latter curves result from differences in the radiative-loss functions utilized.*

upper chromosphere. On the other hand, the insensitivity to variations in parameters such as α and Γ implies that, if measurements of values of P, T_m, and L for a given loop depart significantly from the theoretical relationship, this is a good indication that the physical conditions in the observed feature are not consistent with those assumed in the theoretical quasistatic energy balance models. Given the uncertainties in the empirical values of P, T_m, and L ($\sim$0.1 dex) and the amount by which the theoretical curve in Figure 8.4 can be moved to the right (by about 0.3 dex for $\Gamma = 50$) and to the left (by about 0.2 dex, depending upon the assumed values of α and F_0), the empirical data are consistent with the predictions of the quasistatic model. However, this does not necessarily imply that these models provide a reliable representation of the physical conditions in loops. One needs additional empirical constraints.

Another empirical parameter that can be derived from measurements of EUV and X-ray emission lines is the differential emission measure $Q(T)$ defined by

$$Q(T)dT = n_e^2\, dV \ . \tag{8.3.1}$$

Craig and Brown (1976), Craig (1977), Withbroe (1975), and Levine and Pye (1979) have discussed limitations on empirical determinations of differential emission measures imposed by the use of spectral lines whose temperature-dependent contribution functions overlap. The characteristic width of a typical spectral line contribution function $f(T)$ is $\Delta \log T = 0.3$. Because of errors in intensity measurements and inaccuracies in atomic parameters, the best temperature resolution $\Delta \log T$ that can be achieved in determining empirical differential emission measures is about 0.3 dex. Methods of deriving differential emission measures from empirical data have been discussed by Batstone *et al.* (1970), Jordan (1975), Parkinson (1975), Withbroe (1975), Craig and Brown (1976), Craig (1977), Pye *et al.* (1978), Levine and Pye (1979), and others.

For a symmetric loop with cross-sectional area $A(s)$ and pressure P, the differential emission measure is

$$\begin{aligned} Q(T) &= 2A(s)n_e^2\left(\frac{dT}{ds}\right)^{-1} \\ &= \frac{2A(s)P^2K_0T^{1/2}}{k^2F_c(T)} \ , \end{aligned} \tag{8.3.2}$$

from the definition of differential emission measure and of the conductive flux (Eq. 8.2.6). The factor of two is due to the fact the loop has two legs.

Figure 8.5 contains a plot of $Q(T)$ as a function of $\log T$ calculated for quasistatic loops with constant cross sections, conductive flux into the chromosphere $F_0 = 0$, and various values of α, the parameter giving the temperature dependence of the heating function $E_h \approx T^\alpha$. All loops are assumed to have the same length. As is apparent, the shape of the curves of the differential emission measure are nearly identical. The only difference

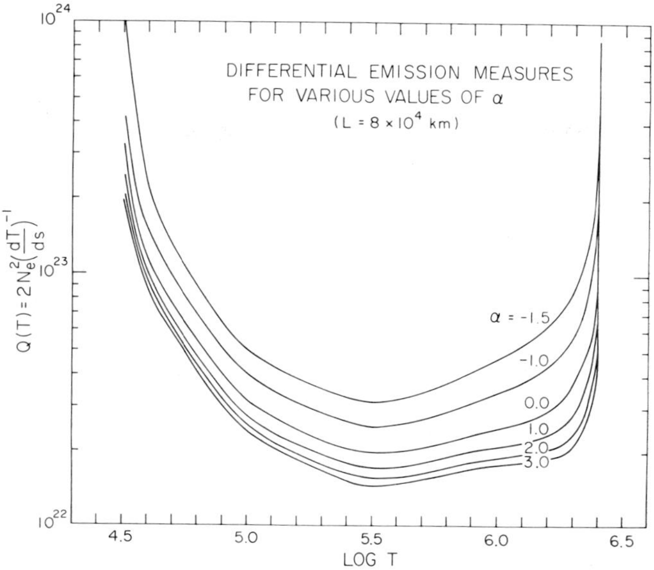

Figure 8.5. Variation of differential emission measure Q(T) as a function of log T for various values of α (the parameter controlling the temperature dependence of the heating function; see Eq. 8.2.8).

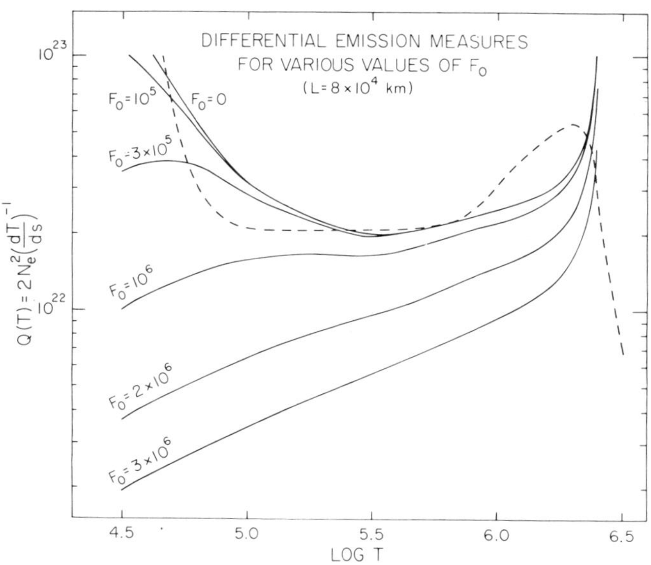

Figure 8.6. Variation of differential emission measure Q(T) as a function of log T for various values of F_0 (erg cm^{-2} s^{-1}), the conductive flux deposited by the loop in the upper chromosphere. For comparison, the maximum possible conductive flux F_{max} is 4×10^6 erg cm^{-2} s^{-1}. The parameter α is zero for these calculations. The dashed curve is based on empirical results obtained by Withbroe (1978a) and Levine and Pye (1979). The high-temperature part of this curve has a broad peak followed by a decline in emission measure because of the effects of the resolution imposed by the width in temperature over which emission lines used in deriving the curves are formed (see text). Additional factors affecting the shape of the empirical curve at high temperatures are the loop geometry and the fact that the emission originated in several loops.

is a relatively small vertical displacement caused by the slightly different pressures required to give the loops the same lengths. Given the insensitivity of the differential emission measures to changes in α, it is doubtful that one can measure loop lengths and differential emission measures with sufficient precision to determine α from empirical data.

Figure 8.6 shows the behavior of the differential emission measure for various assumed values of the conductive flux F_0 into the upper chromosphere. The corresponding value of F_{max}, the maximum conductive flux that can be deposited in the upper chromosphere for a loop of given length and maximum temperature, is 4×10^6 erg cm^{-2} s^{-1}. These curves demonstrate that the differential emission measure is sensitive to variations in F_0. Empirical differential emission measures (cf. Withbroe 1975, 1978a, 1978b; Levine and Pye, 1979) obtained from measurements of EUV and X-ray emissions from active regions indicate that the differential emission measure is essentially independent of temperature for $10^5 < T < 10^6$ K. In addition, empirical determinations of differential emission measures (cf. reviews by Jordan, 1975; Withbroe, 1978a) indicate that below 10^5 the differential emission measures increase with decreasing temperature. This behavior (*dashed curve* in Figure 8.5) implies that $F_0/F_{max} < 0.4$ as suggested in Section 8.2 during the discussion of the radiative dissipation of energy carried into the upper chromosphere by thermal conduction.

The differential emission measure close to the maximum temperature of the loop T_m is very sensitive to the assumed geometry in the loop. As a first approximation $s(T) \approx T^n$, where $n \approx 3$; hence, for a loop with cross-sectional area increasing with distance s from the footpoint, most of the increase in area and differential emission measure (which is proportional to the area) occurs near the temperature maximum. This is illustrated in Figure 8.7, which shows differential emission measures calculated by Vesecky *et al.* (1979) for loops characterized by a parameter Γ equal to the ratio of the cross-sectional area at the apex to that area at the footpoint. Because of this sensitivity to the variation of loop cross-sectional area, empirical differential emission measures can be useful probes of loop geometry (cf. Withbroe, 1978b; Levine and Pye, 1979).

8.4 CONCLUSIONS

On the basis of the results discussed above and in published papers on quasistatic loop models referred to earlier, it appears that these models provide a reasonable approximation to the physical conditions in coronal loops. The models predict values for parameters such as loop length, pressure, maximum temperature, and differential emission measure that are consistent with those observed. The models provide insights into mechanisms controlling the interchange of mass between the chromosphere and corona, and into the role that changes in the coronal heating in loops has in

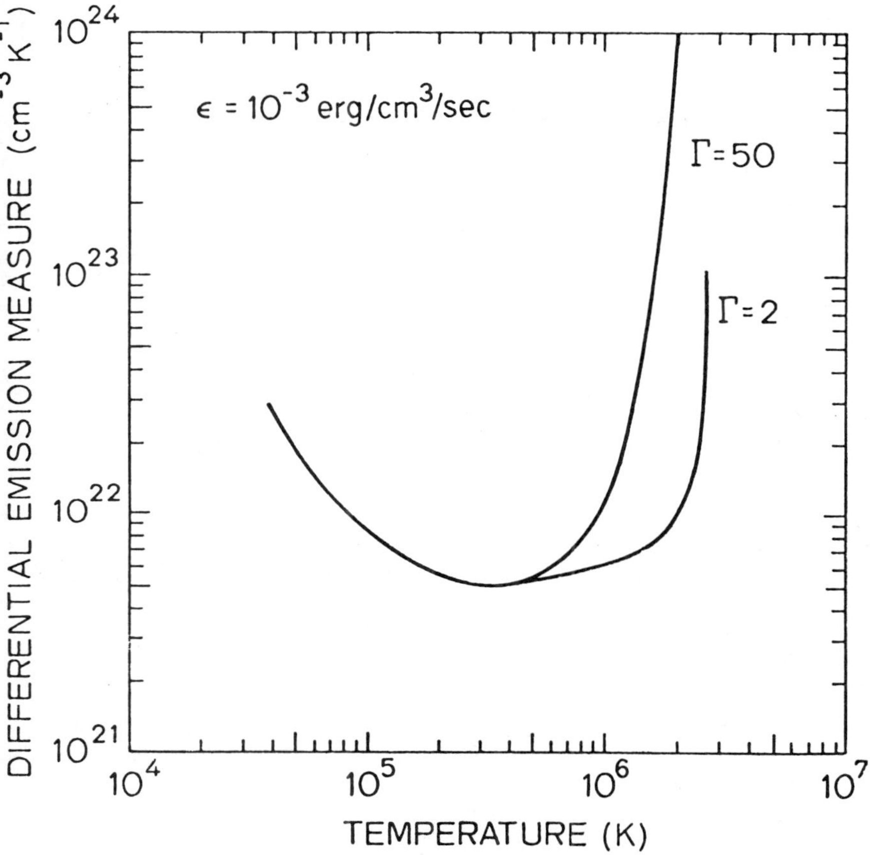

Figure 8.7. Plot (from Vesecky et al., 1979) of differential emission measure as a function of temperature for a loop of length L = 10^{10} cm, apex cross-sectional area 1 cm^{-2}, and area factors Γ of 2 and 50. The $\Gamma = 50$ curve has been multiplied by a constant factor to normalize it to the $\Gamma = 2$ curve at low temperature. The energy input ϵ was assumed to be constant over the loop volume.

initiating these flows. The models also demonstrate the insensitivity of the longitudinal temperature-density structure of loops to parameters, such as α, that characterize hypothetical steady-state heating functions. This is one of the most important results of the investigations into the properties of quasistatic models for coronal loops. In order to obtain empirical data placing significant constraints on possible mechanisms responsible for heating the coronal plasma in active region loops, it is necessary to measure consequences of these heating mechanisms other than that provided by the resultant longitudinal temperature-density structure. One must search for dynamical effects, mass motions, and signatures of nonthermal processes, as well as searching for temporal variations in the temperature-density structure.

REFERENCES

Antiochos, S. K.: 1979, Stanford Univ. Inst. for Plasma Res. Report No. 778.

Basri, G. S., Linsky, J. L., Bartoe, J. D. F., Brueckner, G. E., and van Hoosier, M. E.: 1979, *Astrophys. J.* **230,** 924.

Batstone, R. M., Evans, K., Parkinson, J. H., and Pounds, K. A.: 1970, *Solar Phys.* **13,** 389.

Cox, D. and Tucker, W. H.: 1969, *Astrophys. J.* **156,** 87.

Craig, I. J. D.: 1977, *Astron. Astrophys.* **61,** 575.

Craig, I. J. D.: 1979, presentation to Skylab Workshop on Active Regions.

Craig, I. J. D. and Brown, J. C.: 1976, *Astron. Astrophys.* **49,** 239.

Craig, I. J. D. and McClymont, A. N.: 1979, presentation to Skylab Workshop on Active Regions.

Craig, I. J. D., McClymont, A. N., and Underwood, J. H.: 1978, *Astron. Astrophys.* **70,** 1.

Emslie, A. G. and Machado, M. E.: 1979, Center for Astrophys. Preprint No. 1034.

Gabriel, A. H.: 1976, in R. M. Bonnet and P. Delache (eds.), *The Energy Balance and Hydrodynamics of the Solar Chromosphere and Corona,* G. de Bussac, Clermont-Ferrand, p. 375.

Habbal, S. R. and Rosner, R.: 1979, *Astrophys. J.* **234,** 1113.

Hood, A. W. and Priest, E. R.: 1979, *Astron. Astrophys.* **77,** 233.

Jordan, C.: 1975, *IAU Symp* **68,** 109.

Jordan, C.: 1976, *Phil. Trans. Roy. Soc.* **281,** 391.

Landini, M. and Monsignori-Fossi, B. C.: 1975, *Astron. Astrophys.* **42,** 213.

Landini, M., Monsignori-Fossi, B. C., Krieger, A. S., and Vaiana, G. S.: 1975, *Solar Phys.* **44,** 69.

Levine, R. H. and Pye, J. P.: 1979, *Solar Phys.* (in press).

Machado, M. E. and Emslie, A. G.: 1979, *Astrophys. J.* **232,** 903.

McWhirter, R. W. P., Thoneman, P. C., and Wilson, R.: 1975, *Astron. Astrophys.* **40,** 63.

Nowak, T. and Ulmschneider, P.: 1977, *Astron. Astrophys.* **60,** 413.

Pallavicini, R., Serio, S., and Vaiana, G. S.: 1977, *Astrophys. J.* **216,** 108.

Parkinson, J. H.: 1975, *IAU Symp.* **68,** 45.

Pye, J. P., Evans, K. P., Hutcheon, R. J., Gerassimenko, M., Davis, J. M., Krieger, A. S., and Vesecky, J. F.: 1978, *Astron. Astrophys.* **65,** 123.

Raymond, J. C.: 1979 (private communication).

Raymond, J. C., Cox, D., and Smith, B. W.: 1976, *Astrophys. J.* **204,** 290.

Rosner, R., Tucker, W. H., and Vaiana, G. S.: 1978, *Astrophys. J.* **220,** 643.

Spitzer, L.: 1962, *Physics of Fully Ionized Gases,* 2nd Ed., Wiley Interscience, New York.

Ulmschneider, P.: 1970, *Astron. Astrophys.* **4,** 144.

Vernazza, J. E., Avrett, E. H., and Loeser, R.: 1973, *Astrophys. J.* **184,** 605.

Vernazza, J. E., Avrett, E. H., and Loeser, R.: 1979, *Astrophys. J. Suppl.* (submitted).

Vernazza, J. E. and Reeves, E. M.: 1978, *Astrophys. J. Suppl.* **37,** 485.

Vesecky, J. F., Antiochos, S. K., and Underwood, J. H.: 1979, Stanford Univ. Inst. for Plasma Res. Report No. 740.

Withbroe, G. L.: 1975, *Solar Phys.* **45,** 144.

Withbroe, G. L.: 1978a, in *Proceedings of the OSO-8 Workshop,* Boulder, p. 2.

Withbroe, G. L.: 1978b, *Astrophys. J.* **225,** 641.

9. THEORY OF LOOP FLOWS AND INSTABILITY

E. R. Priest

9.1 INTRODUCTION

9.1.1 Active Regions Are Dynamic

A so-called "active" region has traditionally been regarded as evolving slowly through a series of essentially *stationary* states (apart from the occasional burst of activity during a solar flare). In the corona, the region consists largely of magnetic loops, and the modeling of the coronal plasma within each one has proceeded largely under the assumption of a static state; the density and temperature are then determined by the equations of energy balance

$$0 = \frac{d}{ds}\left(\varkappa_0 T^{5/2} \frac{dT}{ds}\right) - n^2 Q(T) + H \qquad (9.1.1)$$

and hydrostatic equilibrium

$$0 = \frac{dp}{ds} - nm_p g , \qquad (9.1.2)$$

where

$$p = 2nk_B T . \qquad (9.1.3)$$

(Here $\varkappa_0 T^{5/2}$ is the coefficient of thermal conduction parallel to the magnetic field, $Q(T)$ is the temperature-dependent part of the optically thin radiation, and H is an unspecified heating term.) The main assumptions, namely that flows are absent and that the plasma state is time-independent, are made on grounds of simplicity: one must walk before learning to run! Moreover, apart from the work of Vesecky *et al.* (1979) and Wragg and Priest (1980), it has been common practice to make the additional simplification that the loop pressure is uniform. [Even so, the solutions possess unusual properties, such as the little-understood feature of *nonequilibrium* (Hood and Priest, 1979).] A *scaling law* (Rosner *et al.*, 1978) may be derived simply by equating the orders of magnitude of each of the three terms in Eq. 9.1.1. When $Q(T)$ is proportional to $T^{-1/2}$, this law is

$$T_1 \sim (pL)^{1/3} \qquad (9.1.4)$$

for the summit temperature T_1 (in terms of the loop pressure p and length L); for a given heating H and loop length L, the loop pressure in turn adjusts itself to the value

$$p \sim H^{6/7} L^{5/7} . \quad (9.1.5)$$

(It may be noted that in this derivation nothing need be said about the boundary conditions.) A detailed discussion of static loop structures and scaling laws is given by Withbroe in Chapter 8 of this volume.

The *static* law (Eq. 9.1.4) for coronal loops gives a reasonable agreement with observation in that the predicted and observed lengths differ by only up to a factor of two. However, a cursory glance at time-dependent, ground-based observations shows a sunspot region to be truly *active*, and this is now confirmed by observations from space. The picture we now have is of *an active region that is dynamic, with continual activity in the form of a wide range of flows.* Some of these motions may be regarded as superposed on a background stationary state (e.g., small-amplitude waves), but others are highly nonlinear and *completely dominate* what is being observed (e.g., the transition region and possibly coronal loops). Many of the flows have not yet been observed enough for their physical properties and causes to be well understood; furthermore, the theoretical modeling of dynamic loops presents a much more difficult task than for static loops and has barely begun.

Ground-based observations have indicated the presence of many types of flow, as indicated schematically in Figure 9.1 and described in, e.g., Bruzek and Durrant (1977). (Note that the *dashed lines* are drawn to show rough

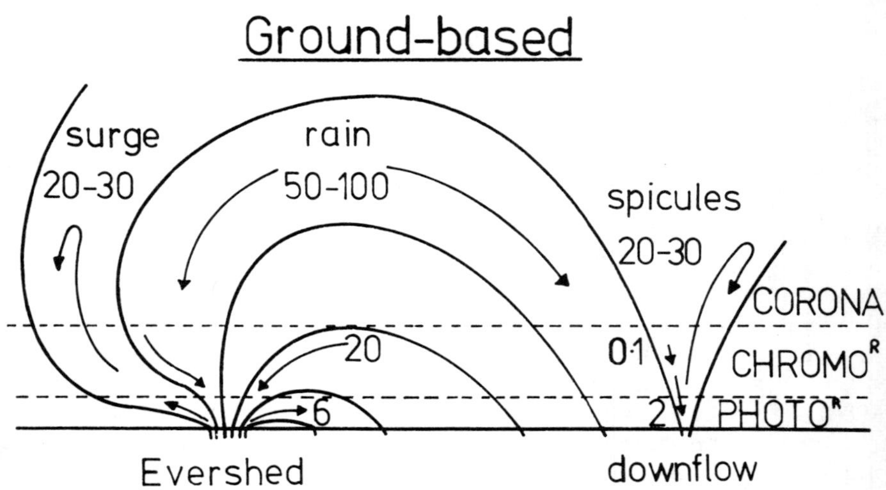

Fig. 9.1. A schematic active region, with preceding magnetic flux (left) *concentrated as a sunspot and following flux* (right) *more diffuse. Several types of flow* (seen from the ground) *are shown. The heavy-headed arrows indicate the flow directions, and the numbers give the typical flow speeds in km s*$^{-1}$.

levels in the atmosphere, but the idea of a plane-parallel atmosphere no longer seems completely relevant: for example, transition region material is often seen within loop structures rather than plane sheets and may be spread over a wide range of heights by spicular motions.) The types of flow are as follows:

(1) *Evershed flow* is a continuous *outward* motion at the photosphere of 6–7 km s^{-1} along the dark penumbral filaments of a well-established sunspot. *Inward* flow exists at higher altitudes (chromosphere) along *superpenumbral fibrils* and with typical speeds of 8–20 km s^{-1}. (A slower photospheric inflow is also present along the bright penumbral filaments.) A *moat* is often seen to form around a well-established sunspot; it is 10,000–20,000 km wide and shows a horizontal radial outflow of 1 km s^{-1}.

(2) *Downflow* is found at the *intense fields* of supergranule boundaries, which show up particularly well at the trailing edge of an active region. The speed is 1–2 km s^{-1} at the photosphere but only 0.1–0.2 km s^{-1} at the chromosphere. A slower upflow is found over larger areas around the centers of supergranule cells. Whereas the supergranular flow is largely horizontal at the photosphere (at 0.8 km s^{-1}), it becomes more isotropic at the middle chromosphere, where (at an altitude of 1400 km) the horizontal and vertical velocity components are both about 3 km s^{-1} (Section 4.2.1).

(3) *Surges* tend to occur especially at the leading edge of an active region. They are streams of plasma that are ejected upwards along slightly curved paths at typically 20–30 km s^{-1}, though occasionally a surge may move at 100–200 km s^{-1}, reach a height of up to 200,000 km, and last 10–20 min. The surge plasma either fades or returns back along the same path. It originates from a small flare-like brightening close to a sunspot.

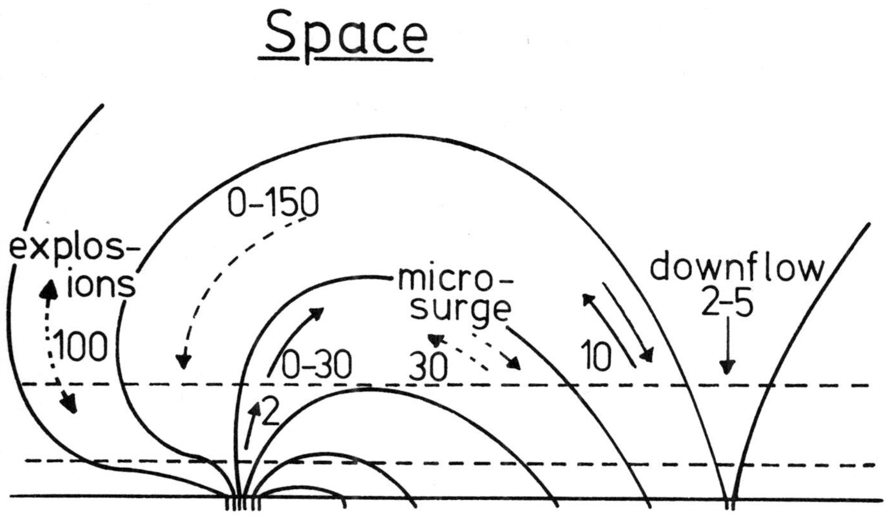

Fig. 9.2. Some of the space observations of flows in a schematic active region. Transient flows have dashed arrows, while large-scale, steady flows are indicated by solid arrows.

(4) *Spicules* are ejected from the low chromosphere (or below) at supergranule boundaries. They elongate over 5–10 min and reach a speed of 20–30 km s^{-1} and a height of typically 9000 km before fading in brightness or occasionally falling back.

(5) *Coronal rain* is cool plasma flowing down along curved paths at the free-fall speed of 50–100 km s^{-1}.

Very recently, observations from space on board OSO-8 and the HRTS (High-Resolution Telescope and Spectrograph) rockets have shown the existence (in the transition region) of new flows that are both fascinating and baffling (Figure 9.2). They may be divided into *transient* or *steady* motions, as follows:

(6) *Transient, small-scale, fast flows* have lifetimes of a minute or less and dimensions of only 1–2 arcsec. Over a sunspot, these flows are often *downward* at 0–150 km s^{-1}. Over a plage, *EUV bursts* or *microsurges* (Section 4.4.2) are seen in which the intensity increases by a factor between 5 and 20, and usually downflows of about 30 km s^{-1} are generated, although upflows are present in about a third of the cases. They occur at a rate of about 15/h in complex active regions and have lifetimes between about one and seven min. The downflows may represent coronal rain, whereas the upflows represent surges (Section 4.4.2). Occasionally, tiny *explosions* are found, with upward and downward motions of 100 km s^{-1}.

(7) *Steady, large-scale, slow flows* last for hours or more. They are generally upward over sunspots at speeds that increase from 1–4 km s^{-1} in the low and middle chromosphere to 10 km s^{-1} in the upper chromosphere and 20 km s^{-1} in the lower transition region. The flows are downward over the network at 2–5 km s^{-1} and both up and down over plage regions at typically 10 km s^{-1}. But even the conditions that give rise to upward rather than downward motions are as yet completely unknown; for instance, there is certainly no simple relation to the magnetic field strength or configuration.

More details and references for all these flows can be found in Chapters 4 and 5. Chapter 5 summarizes observations from two flights of HRTS with its $1'' \times 1000''$ slit. During the first flight, steady motions in an old quiet sunspot were studied. At transition region temperatures of typically 10^5 K, downflows at about 80 km s^{-1} were found for material cooling by radiation. At coronal temperatures (1.7×10^6 K), the flow over the spot was mostly upward at 0–20 km s^{-1} (with an average 2.5 km s^{-1}), whereas the flow over the surrounding plage was mainly downward at 0–35 km s^{-1} (average 13 km s^{-1}). The second HRTS flight investigated a complex, rapidly evolving active region in which a flare had recently started. At transition region temperatures, a downflow of 0–150 km s^{-1} over a light bridge was observed to persist for 260 s, and downflows of 10–60 km s^{-1} over the plage were found in $3'' \times 5''$ areas. Also, an explosive event lasted 20 s, with an upflow of up to 200 km s^{-1} and a downflow of 100 km s^{-1}; such explosions are common in the quiet sun, but cover less than 1% of the solar surface and have much smaller velocities. Chapter 4 reviews the results from ground-based data, Skylab, and OSO-8 for steady flows (supergranulation, Evershed, chromosphere, and transition region), quasisteady flows (fibrils, spicules, coronal rain, coronal loops), transient flows (surges, sprays, EUV bursts), oscillatory motions, and turbulent motions.

It is clear that we are only just beginning to determine the properties of these flows, and there are many unanswered questions about the relationship between the different flows seen at different heights or wavelengths. For instance, are the transient downflows related to coronal rain? Are the transient explosive events connected with microsurges, X-ray bright points, or magnetic reconnections? Is there a link between EUV microsurges and surges or spicules?

9.1.2 Causes of Flow

There is a wide variety of possible causes for the flow patterns that are found in active regions. Only a few can be mentioned here, and some are described in more detail in the rest of the chapter.

Many models for spicules and surges have been suggested. One possibility for *spicules* is that they are driven by *granular buffeting* (Roberts, 1979), where there is a resonance between the forcing granular motion at the edge of an intense magnetic flux tube and the vertical plasma motion within the tube. (An alternative possibility is a similar resonance, with the forcing by wave motions that originate within a supergranule cell and impinge on its boundaries.) Surges could be caused by the reconnection between *newly emerging* or *evolving* satellite flux and the ambient sunspot field, as in the *emerging flux model* (Heyvaerts, Priest, and Rust, 1977). Or they may be the result of *nonequilibrium* when the pressure within a closed magnetic structure exceeds a critical value.

Evershed flow may be a result of the interaction between the convection rolls and the magnetic field within the penumbra of a sunspot (Galloway, private correspondence). Another explanation is in terms of a *siphon flow* driven by a pressure difference that is maintained between the footpoints of a magnetic loop (Meyer and Schmidt, 1968). There are several possible reasons for such a pressure difference. For example, if the total pressure (plasma plus magnetic) is uniform at the photospheric base, regions of high magnetic field will have a low plasma pressure and vice versa. This would produce either an accelerated flow into strong field regions (such as sunspots or intense flux tubes) or a decelerated flow out of them. Another way of generating a footpoint pressure temporarily would be by photospheric motion. A converging plasma movement could compress the magnetic field and the plasma contained in it, and then the enhanced pressure could be slowly relieved by means of a flow. Similarly, a diverging flow could reduce the plasma pressure. Again, the effect of supergranular motion (typically ½ km s^{-1}) could be to drive a downflow of ½ km s^{-1} by viscous coupling in the intense tubes that make up the cell boundary and so maintain a pressure in those tubes, which may form loop footpoints. Yet another way of enhancing the pressure at a footpoint is by increasing the heating there (see Eq. 9.1.5).

Siphon flow may take place in coronal loops as well as the low-lying penumbral loops; indeed, it may give one explanation for the cool cores of some such loops. In addition to siphon flow, there are several other ways of explaining upflows and downflows, as follows.

An *upflow*, for example above a sunspot, may be driven by waves (either Alfven or magnetosonic) that are propagating upwards and dumping their momentum as well as their energy in the plasma. Alternatively, the upflow reported near a filament of 6–10 km s^{-1} (Lites *et al.*, 1976) may be part of an overall circulation within a magnetic arcade, driven by the downflow within the filament (Priest and Smith, 1979). Another mechanism for upflow is *evaporation*, which may be taking place in a mild way continually and not just during solar flares. Thus, as the magnetic configuration of the active region evolves, the plasma within each flux tube must respond and try to seek a new equilibrium state (if such exists). From Eq. 9.1.5, we see that increases in the length of a given magnetic loop or in the heating (which probably depends on the magnetic field) within the loop demands an increase in the pressure for a static, thermally isolated solution. This pressure increase occurs when extra material flows from below in a so-called "evaporation."

Downflow may be plasma that is returning to the chromosphere and photosphere from both spicules and surges. Alternatively, a downflow may be generated either in the photosphere or in the corona. One generation mechanism in the photosphere is the supergranular motion, already mentioned; another is the *intense magnetic field instability* (Parker, 1978; Webb and Roberts, 1978; Spruit, 1979). There are several ways of initiating a flow in the corona. For example, the opposite process to evaporation, *draining*, may occur when the length or heating of a loop decreases and the pressure falls (by Eq. 9.1.5). More dramatically, occasionally a loop may form a *cool core* (Jordan, 1975; Foukal, 1975), and the resulting excess material drains out of the loop (Levine and Withbroe, 1977). One way that cool cores may be produced is through the onset of *thermal nonequilibrium* (or, loosely speaking, thermal instability) when the loop pressure or length becomes too great or its heating too small (Priest, 1978; Hood and Priest, 1979). Once a cool core is produced, small-scale magnetohydrodynamic instabilities may drive a circulation of plasma from the ambient corona across the interface and into the core where it falls down (Priest, 1978).

There are many other ways in which flows may arise. For example, a host of wave motions is present in the solar atmosphere but is excluded from the subject of this chapter. Thermal instability may be continually occurring in loops, especially in the transition region, and may generate turbulent motions. Finally, as the large-scale magnetic field evolves in response to the motion of the photospheric footpoints through a series of largely force-free states, it may occasionally find that the threshold for the onset of a *magnetohydrodynamic instability* is passed; or, even, there may be no equilibrium at all, and a state of nonequilibrium ensues (Parker, 1972; Syrovatskii, 1978; Bobrova and Syrovatskii, 1979).

9.2 SIPHON FLOW

Interest in flows in coronal loops has been stimulated by the observation of large downflow velocities (up to 100 km s^{-1}) over spots, plages, and network from measurement of Doppler shifts in such lines as S IV and C IV (Brueckner, 1976; Bruner *et al.*, 1976; Doschek *et al.*, 1976). Furthermore, an analysis of Skylab observations has shown that ions formed at relatively low temperatures, such as C II, C III, O IV, O V, and O VI, are observable up to coronal heights of 60 Mm, which is much above their scale height (Foukal, 1975, 1976, 1978). More generally (Chapter 6), loops that are cooler than 2×10^6 K (Mg X) are found *to possess cool cores* at 5×10^5 K (Ne VII), which change in appearance over times of 15–30 min due to motions of up to 100 km s^{-1} along the field. By contrast, loops at 2.5×10^6 K (Fe XV) change much more slowly and seem to appear and fade in place.

One of the simplest places for the theorist to begin is with an analysis of steady siphon flow. Assuming that a pressure difference is maintained between the two footpoints of a magnetic loop, siphon flow is driven from the high-pressure to the low-pressure foot. In general, this flow is unsteady, but, if the footpoint pressures are constant and one waits several sound travel times L/c_s, a steady flow may be established. The properties of such a flow in coronal loops have been studied by Cargill and Priest (1980) and Noci (1980) as part of the Workshop activity, following previous work by Meyer and Schmidt (1968) for photospheric loops and Pikel'ner (1971) in connection with prominence formation. The plasma beta

$$\beta = \frac{2\mu p}{B^2}$$

is assumed to be much less than unity, the magnetic field constant in time, and the flow speed much smaller than the Alfven speed. Then the flow may be regarded as taking place in a rigid tube of given shape and cross section, with the notation indicated in Figure 9.3.

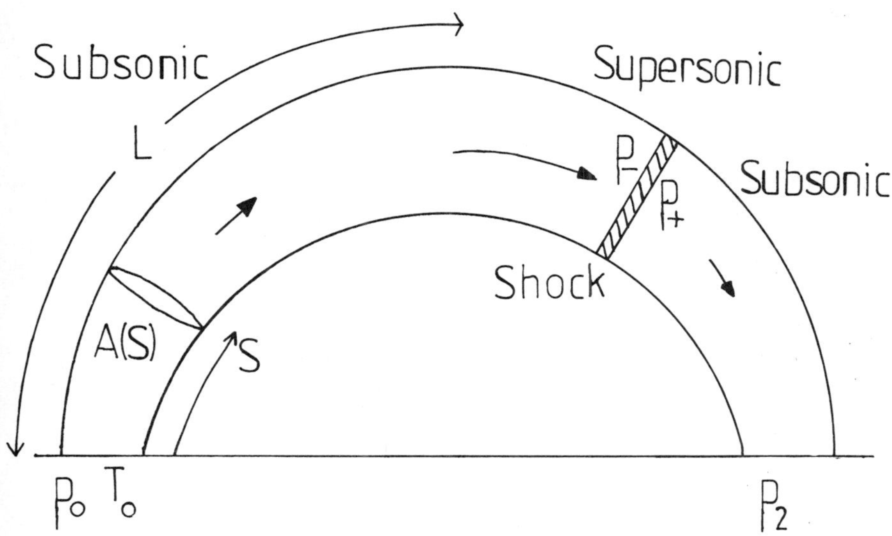

Fig. 9.3. The notation for siphon flow in a symmetric loop of length 2L, with the position from one footpoint measured by the coordinate s. The flow here is subsonic up to the summit and supersonic between the summit and a shock wave. (From Cargill and Priest, 1980.)

9.2.1 Equations for Steady Flow

The equations of continuity, momentum, state, and energy for steady, polytropic flow of a hydrogen plasma along a semicircular loop of length $2L$ are

$$nvA = n_0 v_0 A_0 \ , \tag{9.2.1}$$

$$m_p n v \frac{dv}{ds} = - \frac{dp}{ds} - m_p n g \cos \frac{\pi s}{2L} \ , \tag{9.2.2}$$

$$p = 2nk_B T \ , \tag{9.2.3}$$

$$p/n^{\alpha} = p_0/n_0^{\alpha} \ , \tag{9.2.4}$$

where p is the pressure, n the electron number density, T the temperature, α the polytropic index, m_p the mass of a hydrogen atom, and s the distance along the loop from a footpoint whose properties are denoted by subscript zero. Properties at the other footpoint are denoted by subscript two, and the problem is to determine the flow that results when p_0, T_0, and p_2 are prescribed. The assumption of polytropic flow is made on the grounds of simplicity; it allows analytical solutions. More recently, Cargill and Priest (1981) have solved the full energy equation numerically with radiation, conduction, and heating all present, and the results support the general features of the polytropic flow model.

After eliminating n, p, and T, the above set reduces to

$$\left(v - \frac{c_s^2}{v}\right) \frac{dv}{ds} = - g \cos \frac{\pi s}{2L} + \frac{c_s^2}{A} \frac{dA}{ds} \ , \tag{9.2.5}$$

where the polytropic sound speed c_s is given by

$$c_s^2 = \alpha p/(m_p n) \ .$$

Eq. 9.2.5 determines the flow speed $v(s)$, once the cross-sectional area $A(s)$ has been specified. It will be noted that the equation possesses a critical point, where the flow speed becomes sonic ($v = c_s$) and the right-hand side vanishes, just as in the solar wind equations. After imposing the condition that $v = v_0$ at $s = 0$, the solution to Eq. 9.2.5 may be found by a simple integration as

$$\frac{1}{2{c_{s0}}^2} (v^2 - v_0^2) + \frac{1}{\alpha - 1} \left[\left(\frac{v_0 A_0}{vA} \right)^{\alpha - 1} - 1 \right] = - \frac{2Lg}{\pi c_s^2} \sin \frac{\pi s}{2L} \tag{9.2.6}$$

when $\alpha > 1$, or

$$\frac{1}{2{c_{s0}}^2} (v^2 - v_0^2) + \log \frac{v_0 A_0}{vA} = - \frac{2Lg}{\pi c_s^2} \sin \frac{\pi s}{2L}$$

when $\alpha = 1$. The resulting density, pressure, and temperature in terms of v follow from

$$\frac{n}{n_0} = \frac{v_0 A_0}{vA} ,$$

$$\frac{p}{p_0} = \frac{n^\alpha}{n_0^\alpha} ,$$

$$\frac{T}{T_0} = \frac{n^{\alpha-1}}{n_0^{\alpha-1}} .$$

9.2.2 Solutions for Uniform Area

The simplest case, an isothermal loop ($\alpha = 1$) with a uniform area, illustrates the basic ideas. The critical point is located at the loop summit $s = L$, as indicated in Figure 9.4a, where the phase plane for solutions of Eq. 9.2.5 is shown.

For initial speeds v_0 slower than v_0^*, the flow is *subsonic* and symmetric about the top of the loop, so that the pressure ratio is unity. For $v_0 > v_0^*$, the results are unphysical, save for the purely supersonic flows. For $v_0 = v_0^*$, the flow becomes sonic at the loop summit.

When there is no pressure difference between the ends of the loop ($p_2 = p_0$), any of the subsonic flows may occur in this idealized model. However, an imposed pressure difference forces the flow to become *supersonic* beyond the loop summit and then to be decelerated (and heated) at a shock wave, which is located at some position on the downflowing leg (Figure 9.3). The effect of imposing different pressure differences is to change the location and strength of the shock wave and the values of the downstream speed v_2 (Figure 9.4b). For this particular case of a loop of half-length 50 Mm, supersonic flows are possible when the base upflow speed v_0^* is 0.34 c_{s0}; when the pressure ratio becomes less than 0.75, the shock moves below the footpoint and the model fails. For longer loops, one finds a lower base speed v_0^* and a wider range of allowable pressure ratios.

Since nvA is constant for the flow, the density has a profile that is essentially the inverse of the velocity; it decreases up to the summit for subsonic flow and right up to the shock wave for a shocked flow. If the flow is adiabatic rather than isothermal, this rarefaction of the plasma causes a fall in the summit temperature and pressure, too (e.g., Figure 9.6). For example, in a loop of half-length 50 Mm, *the effect of the flow is to reduce the temperature near the summit by typically a factor two and to reduce the pressure by up to a factor of ten.* This may possibly explain the *cool cores* that are found in some coronal loops.

The effect of a cross-sectional area that increases up to the summit and then decreases in a symmetric manner has also been studied. The sonic point remains at the top, but the speed there is reduced. If the area-increase or the loop length are too large, the subsonic solutions cease to exist.

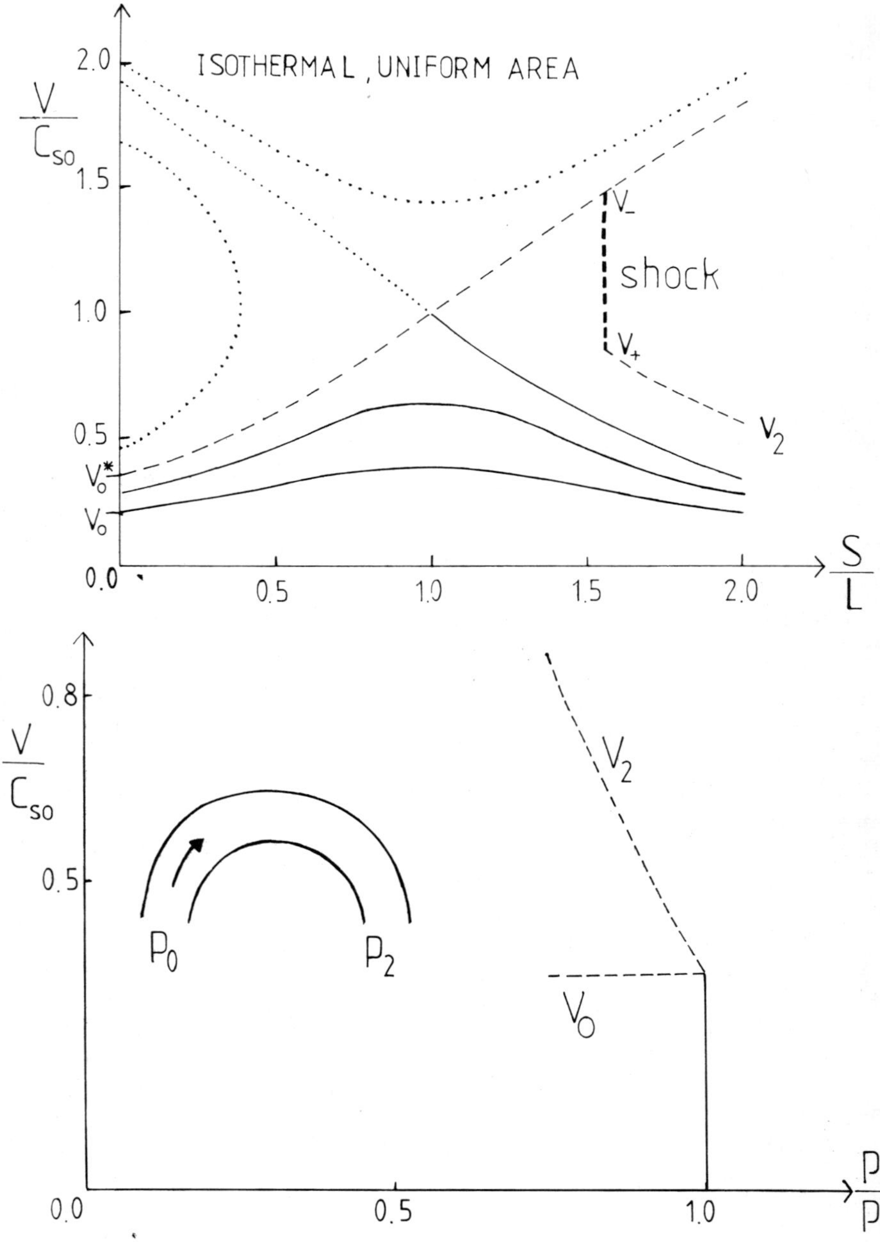

Fig. 9.4. Isothermal siphon flow in a loop of half-length 50 Mm and uniform cross-sectional area. (a) The flow speed v at a distance s along the loop. Dotted lines indicate unphysical or wholly supersonic solutions. The thick dashed line shows the shock transition from a speed v_- to v_+. (b) The footpoint flow speeds v_0 and v_2 as a function of the imposed pressure ratio p_2/p_0. The solid line denotes subsonic flows (for which $v_0 = v_2$), while the dashed lines denote flows containing shock waves (for which $v_2 > v_o$). (From Cargill and Priest, 1980.)

Furthermore, for a moderate area-increase, extra critical points appear and greatly complicate the topology. One effect of a large enough area-increase is to make the flow speed decrease to a minimum before increasing and passing through the sonic point. This effect can be quite substantial [and may partly explain the fall-off with height that Giovanelli (1977) observes in intense flux tubes near the photosphere]. For example, an area expansion by a factor of twenty in a loop of length 175 Mm gives a base speed v_0^* for adiabatic flow of 0.15 c_{s0} and a minimum speed of 0.06 c_{s0} at a distance of 40 Mm along the loop.

9.2.3 Loops with an Asymmetric Area

When the cross-sectional area $A(s)$ is not constant or symmetric about the loop summit, the critical (sonic) point of the flow equation (9.2.5) no longer occurs at the loop summit. Its location is determined by

$$\cos\frac{\pi s}{2L} = \frac{c_s^2}{gA}\,\frac{dA}{ds} \quad ;$$

thus for a *converging loop* ($dA/ds < 0$) the sonic point is situated on the downflowing leg, whereas for a *diverging loop* ($dA/ds > 0$) it is on the upflowing leg.

Consider a converging loop with a linearly decreasing area

$$A = A_0[1 - (s/4L)] \ ,$$

such that the area halves in going from one end to the other. (Similar results are produced for a more sophisticated profile with a difference in area between the ends and a maximum at the summit.) The resulting phase plane is shown in Figure 9.5a, again for a loop of half-length 50 Mm. The asymmetry can clearly be seen. It implies that for *subsonic solutions* the footpoint speeds v_0 and v_2 are no longer equal, which in turn means that the footpoint pressures are not the same. Indeed, evaluating Eq. 9.2.6 at the second footpoint and substituting

$$\frac{p_2}{p_0} = \left(\frac{v_0 A_0}{v_2 A_2}\right)^{\alpha}$$

gives

$$\frac{v_2^2 - v_0^2}{2c_{s0}^{\;2}} + \frac{1}{\alpha - 1}\left[\left(\frac{p_2}{p_0}\right)^{(\alpha-1)/\alpha} - 1\right] = 0 \quad \text{for } \alpha > 1$$

or

$$\frac{v_2^2 - v_0^2}{c_{s0}^{\;2}} + \log\frac{p_2}{p_0} = 0 \quad \text{for } \alpha = 1 \ .$$

Thus, the fact that v_2 exceeds v_0 implies that p_0 is greater than p_2 (and n_0 exceeds n_2), and the flow is *accelerated* into the converging loop. The way that the footpoint speeds vary with the pressure difference is given in Figure 9.5b. It can be seen that when $p_2/p_0 = 1$, the solution is *static*; i.e., there

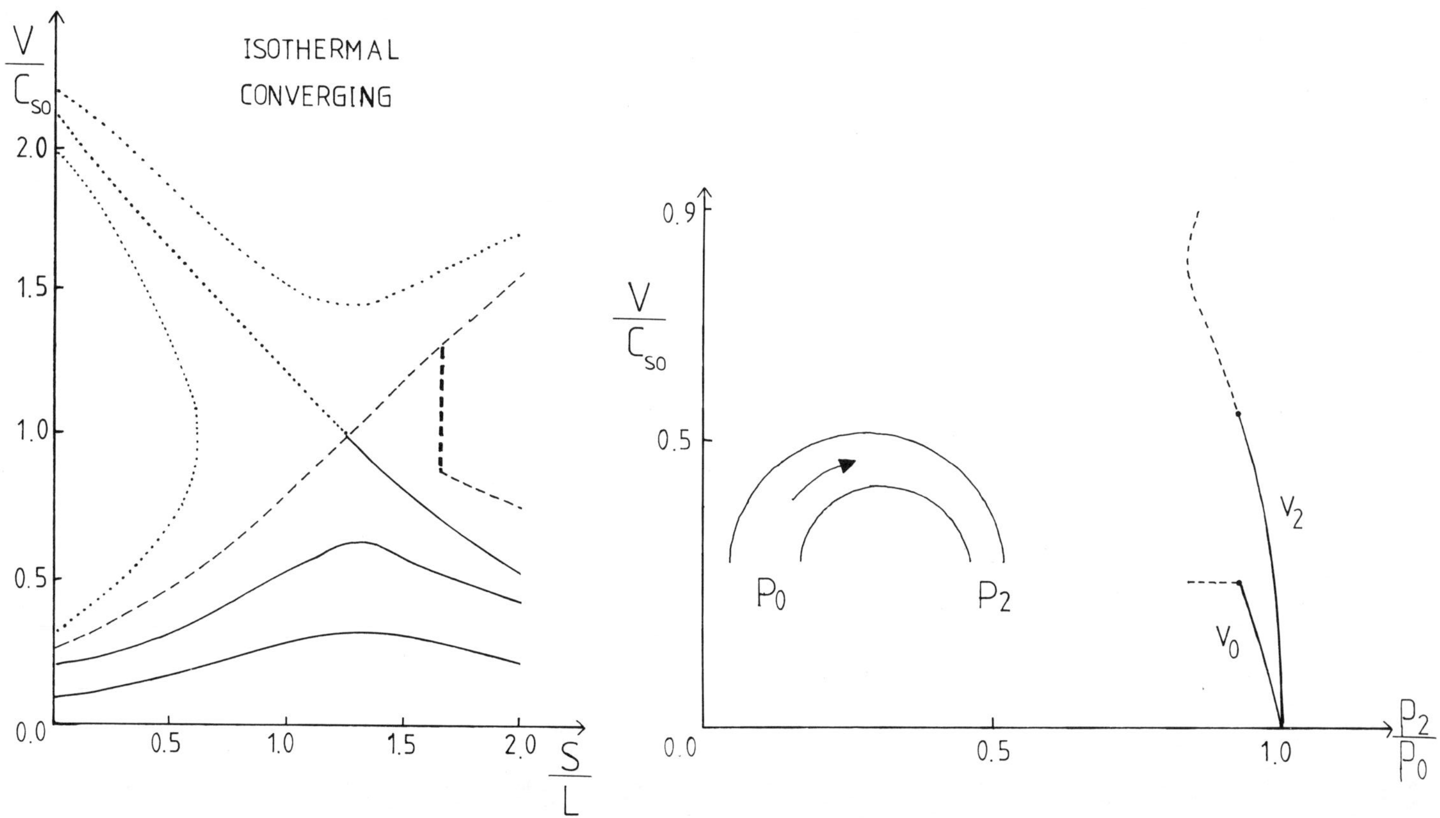

Fig. 9.5. Isothermal siphon flow in a converging loop of half-length 50 Mm and cross-sectional area that decreases by a factor of two from one footpoint to the other. (a) *The velocity profiles for subsonic* (solid) *and shocked* (dashed) *flow.* (b) *The footpoint flow speeds* v_0, *and* v_2 *as a function of the pressure ratio* p_2/p_0 *for subsonic* (solid) *and shocked* (dashed) *flow. (From Cargill and Priest, 1980.)*

is no flow. As p_2/p_0 decreases from unity, the solutions at first yield subsonic flow, and if p_2/p_0 decreases further the solutions become *shocked*.

The effect of increasing the convergence is to make the sonic point move further away from the loop summit and increase the range of pressure ratios p_2/p_0 that can generate subsonic flows. For adiabatic flow, the range of possible pressure ratios is far greater than for isothermal flow. *Diverging* loops give a phase plane for subsonic flows that may be obtained from Figure 9.5a by rotating the solutions about the summit $s = L$. All the subsonic flows now have v_2 smaller than v_0, so that p_2 exceeds p_0; the plasma moves in the direction of increasing pressure, and the flow is *decelerated* out of the diverging loop. For $p_2 > p_o$, there is also a shocked solution; but when $p_2 < p_0$ the flow is *accelerated,* and only shocked solutions are possible.

9.2.4 Polytropic Flow Model of Noci

Noci (1980) has studied the properties of the steady, siphon flow equations (9.2.1–9.2.4), with gravity decreasing like r^{-2} from the solar center and with the cross-sectional area A constant. He also puts $\alpha = 1.1$, since this value gives solar wind solutions in the inner corona that agree quite well with solutions to the full energy equation. Such a polytropic law (p/ϱ^{α} = constant) is commonly adopted as a rough approximation to the energy equation, especially when analytic solutions are sought.

The integral (Eq. 9.2.6) of the equation of motion is written in the form

$$\frac{1}{2}v^2 + \frac{2k_B}{m}\frac{\alpha}{\alpha - 1}T - \frac{GM_\odot}{r} = E \ , \tag{9.2.7}$$

where r is the heliocentric distance and the integration constant E is the total available energy. Whereas for solar wind solutions E is positive, here it can be negative, with a smallest value of $-GM_\odot/r_m$, where r_m is the heliocentric distance of the loop summit.

The Mach number is defined as

$$M = v/c_s \ ,$$

and, from Eqs. 9.2.7 and 9.2.1, it is determined (for a constant cross section) by

$$\left[M^{\frac{4}{1+\alpha}} + \frac{2}{\alpha - 1}M^{-2\left(\frac{\alpha-1}{\alpha+1}\right)}\right] f(s) = \left[M_0^{\frac{4}{1+\alpha}} + \frac{2}{\alpha - 1}M_0^{-2\left(\frac{\alpha-1}{\alpha+1}\right)}\right] f(0) \ , \tag{9.2.8}$$

where

$$f(s) = [E + GM_\odot/r(s)]^{-1} \ .$$

The subsonic solutions are defined by the parameters:

R — radius of the loop,
E — total energy,
v_0 — footpoint upflow speed,
p_0 — footpoint pressure.

For the shocked solutions, v_0 is determined by the extra condition that the solution pass through the sonic point; thus, the parameters are

$$R, E, p_0, p_2/p_0 \ ;$$

the last of these determines the position and strength of the shock (Section 9.2.2).

The variation of temperature along a typical loop is shown in Figure 9.6 for two values of E appropriate to coronal conditions. In each case, the temperature jump for five positions of the shock wave can be seen.

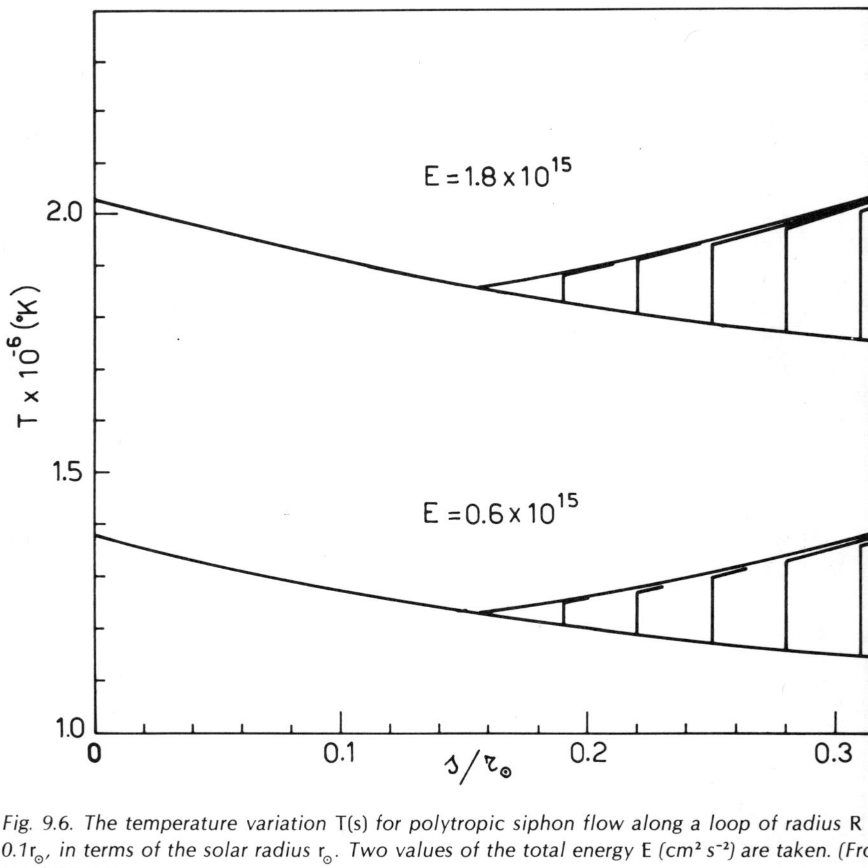

Fig. 9.6. The temperature variation T(s) *for polytropic siphon flow along a loop of radius* R = *0.1*$r_\odot$*, in terms of the solar radius* $r_\odot$*. Two values of the total energy* E ($cm^2 \, s^{-2}$) *are taken. (From Noci, 1980.)*

9.2.5 UV Emission

The density and velocity variations in a siphon flow may be large enough to give observable effects. In particular, Noci (1980) has investigated the effect on the UV brightness of the loop, taking into account lines of the ions O VI, Mg X, and Si XII, which are characteristic of the upper transition region and corona.

The width of the loop is chosen to be one-tenth of its radius. Variations across the loop and departures from ionization equilibrium are both neglected. For O VI and Mg X, the latter neglect is usually a fair approximation, because both lines have a rather flat population curve over the temperature range 1–2 $\times 10^6$ K and the temperature changes by only about 20% along the loop. However, Si XII has a population curve with a strong temperature dependence; thus, the results are inapplicable to loops of such a low density that the ionization or recombination times are comparable with the travel time between the footpoints.

Noci uses the standard expression for the intensity of radiation emitted in a coronal UV line (Pottasch, 1964; Withbroe, 1970) to obtain the results shown in Figure 9.7 for a loop at the disk center. The main effect of the flow is to cause a substantial drop in intensity of these lines, particularly in the "transonic" flows, where the shock has moved all the way down to the second footpoint. For *shocked flows* there is a *marked asymmetry* between the two sides of the loop, with a *sharp increase in intensity* at the shock. For *subsonic flows*, the intensity is symmetric about the loop summit and is much closer to the static value (shown *dashed*).

Figure 9.7b shows the effect of the parameter E on the intensity for transonic flow. [The loop temperature increases with E (Figure 9.6).] For constant base density n_0 (labeled n_1 in the figure), the differences in intensity between "hot" and "cool" loops is rather small for O VI, intermediate for Mg X, and large for Si XII. This clearly depends on the temperature dependence of the ionization equilibrium curve in the temperature interval considered (1–2 $\times 10^6$ K).

It is noticeable in Figure 9.7a that the intensities for static and subsonic loops do not differ considerably, while those for static and shocked loops do differ greatly in the downflowing leg, especially when the shock is close to the footpoint. This suggests a definite observable signature for supersonic flows. Indeed, it may well be that some loops (observed to extend from one footpoint and then disappear before reaching the second footpoint) possess this feature because of a subsonic–supersonic siphon flow.

9.3 EVAPORATION AND DRAINING

The magnetic field of an active region continually evolves because of the motion of its photospheric footpoints. If the configuration starts with one force-free equilibrium and the footpoints move to a new location, a new

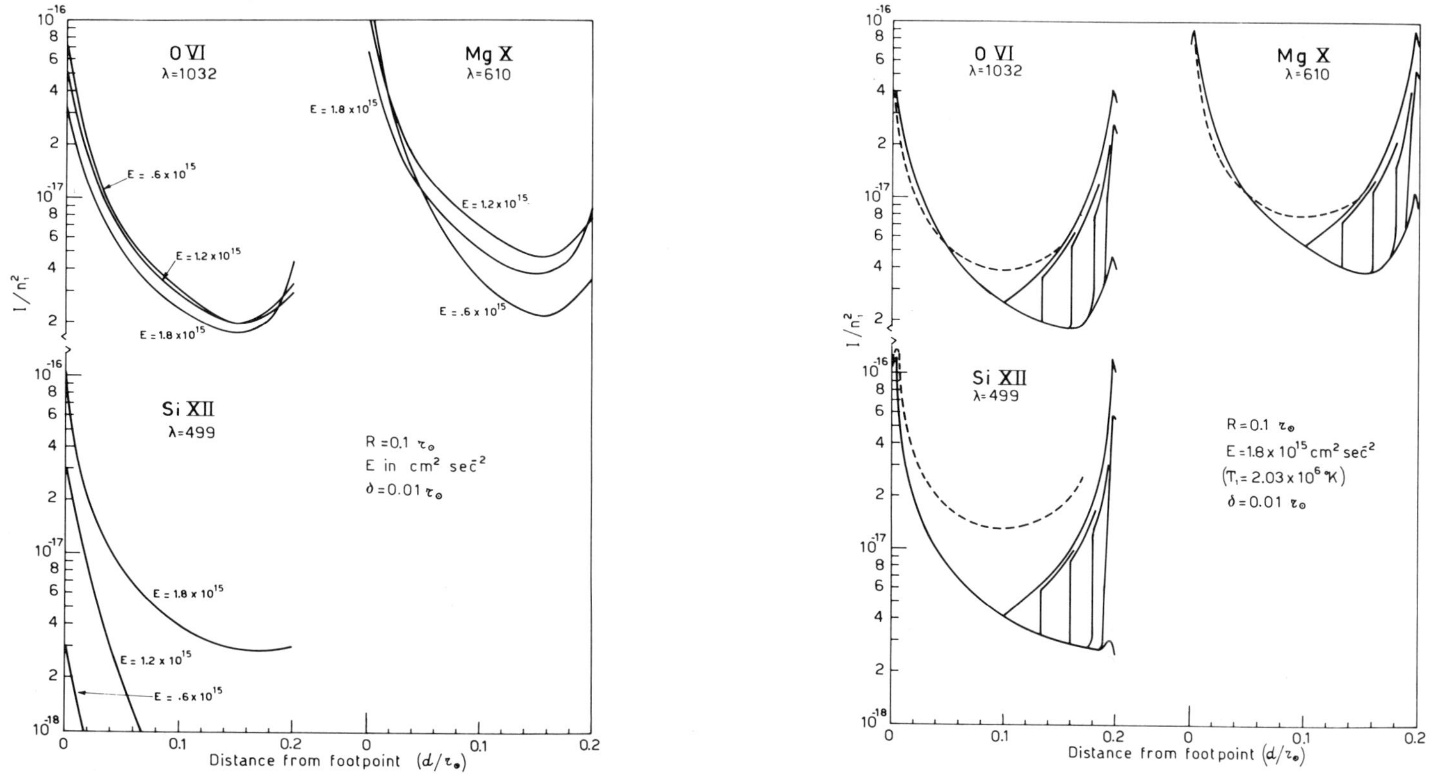

Fig. 9.7. Brightness variations I *along a loop containing siphon flow for three UV lines. The density at the upflowing footpoint is* n_1*, and* T_1 *(=2.03 × 10⁶ K) is the temperature there. The loop has a major radius of* $0.1r_{\odot}$ *and a minor radius of* $0.01r_{\odot}$. (a) *Flows driven by different pressure ratios for a total energy* $E = 1.8 \times 10^{15}$ *cm² s⁻². Dashed curves represent brightness variations for a static isothermal loop having the same footpoint temperature* T_1 *as the dynamic loop.* (b) *The effect of the energy parameter* E *on transonic flow. (From Noci, 1980.)*

magnetic equilibrium is set up comparatively quickly in the Alfven travel time—that is, provided the new equilibrium exists (Parker, 1972; Syrovatskii, 1978; Bobrova and Syrovatskii, 1979). Since the plasma pressure is much smaller than the magnetic pressure, the plasma contained within each flux tube then responds to the new magnetic equilibrium by seeking a new hydrostatic and energy balance; this adjustment takes place much more slowly, over the sound travel time. This means that plasma flows occur along field lines in response to the magnetic field evolution and in an attempt to establish equilibrium. Indeed, it may be that the corona never achieves such a state of rest but is continually prodded into activity by the restless movement of the field. In addition, there is the extra complication of magnetic nonequilibrium and instability possibly taking place. However, for the purpose of this section, let us suppose a *passive evolution of the magnetic field* through a series of equilibria and examine *the response of the plasma*.

Consider a loop in equilibrium (Figure 9.8) with a certain heating H and length L. If it is thermally isolated, the summit temperature may be written roughly according to Eq. 9.1.4 as

$$T_1 \sim (pL)^{1/3} ,$$

where the loop pressure is determined as

$$p \sim H^{6/7} L^{5/7} .$$

Thus the summit density scales as

$$n_1 \sim H^{4/7} L^{1/7} . \qquad (9.3.1)$$

Now, if the magnetic field changes in such a way that H or L *increases*, Eq. 9.3.1 implies that the new equilibrium possesses a higher density. To

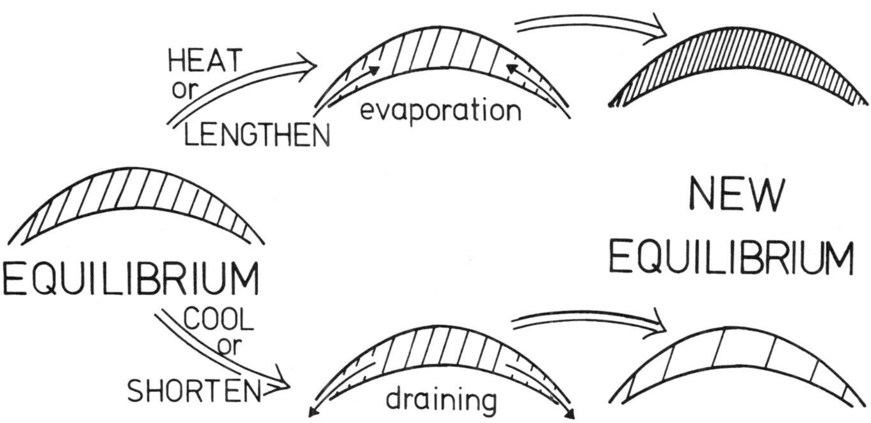

Fig. 9.8. The evolution of a coronal loop from one equilibrium to another by means of an evaporation or a draining, depending on whether the heating rate (or loop length) increases or decreases in value.

attain such an equilibrium, extra material must be brought up or *evaporated* from below along the loop, assuming it cannot travel across the field (Figure 9.8). Similarly, if the heating or length are reduced in value, there is too much material in the loop for equilibrium, and some of it must *drain* down until the pressure gradient offsets gravity and all the energy terms balance.

In the above simple description, some subtle responses have been omitted. For example, the "length" of a loop between the two footpoints (where the temperature gradient vanishes, say) will change as the heights of the footpoints respond. Also, the heating may itself depend on the new magnetic and plasma state and therefore is not free to be chosen independently. Furthermore, there are inadequacies in the scaling law (Eq. 9.1.4), which does not agree well with observations if only one class of loop is considered; also, Eq. 9.1.4 is based on an extremely simple energy balance, with an optically thin radiation law, and flows and turbulence excluded.

The transient flow that results during an evaporation or a draining is of course rather complex. It has been studied numerically during the Workshop by Krall and Antiochos (1980) and Craig (1980), as described below. Originally, these ideas were proposed in connection with flaring loops. [See Craig (1980) for a review of theoretical models for the thermal evolution of loop flares.] However, it is likely that transient flows are continually taking place in a mild way within normally evolving loops.

9.3.1 The Response of Active Region Loops to a Change in the Heating Rate

Krall and Antiochos (1980) have investigated the *response of the plasma to a change of heating rate* in a model active region loop. Soft X-ray images of the sun indicate little variation in emission from active region loops over periods of hours or days. This time scale is much longer than the radiative decay time of 10^3 s for plasma at a temperature 2–3.5 $\times 10^6$ K and density 10^9 cm^{-3} typical of such loops. It is therefore natural to assume that an approximate balance exists between the energy sources and sinks during a loop's lifetime.

Krall and Antiochos solve a set of time-dependent, single-fluid, one-dimensional equations along one field line of a dipole magnetic field. The energy balance includes thermal conduction, optically thin radiation $n^2Q(T)$, and a *spatially uniform* heating ϵ; the loop area $A(s)$ is allowed to vary with distance s along the loop, and the gravitational acceleration $g_\parallel(s)$ parallel to the loop is included. The equations of continuity, motion, and energy for $n(s,t)$, $v(s,t)$, and $T(s,t)$ are

$$\frac{\partial n}{\partial t} + \frac{1}{A(s)}\frac{\partial}{\partial s}(Anv) = 0 \ ,$$

$$n\left(\frac{\partial}{\partial t} + v\frac{\partial}{\partial s}\right)v = -\frac{1}{m}\frac{\partial p}{\partial s} + ng_\parallel(s) \ ,$$

$$3nk\left(\frac{\partial}{\partial t} + v\frac{\partial}{\partial s}\right)T + \frac{p}{A(s)}\frac{\partial}{\partial s}(Av) = \frac{1}{A(s)}\frac{\partial}{\partial s}\left(A\varkappa\frac{\partial T}{\partial s}\right) + \epsilon - n^2Q(T) \ ,$$

where

$$p = 2nkT \ .$$

The *boundary conditions* employed in solving these equations are of course most important. Krall and Antiochos assume that conditions of symmetry hold at the loop summit continuously. The base of the loop is taken at a constant chromospheric temperature $T_B = 3 \times 10^4$ K, and the boundary conditions there are rather subtle: since the chromospheric scale-height is much smaller than the loop height, the base is assumed to be rigid, except a mass flow (evaporation) through the base is allowed in response to an enhanced heat flux there. Furthermore, it is not practical to resolve numerically the temperature gradients in the transition region. This is a serious difficulty, because the temperature gradient near the base is very important; but an approximate analytic expression is used to jump across the transition region as follows: within the transition region, the changes in radiation, mechanical heating, and gravitational energy are neglected, and conservation of mass and energy gives

$$Anv = \text{constant},$$

$$\frac{\partial}{\partial s}\left[A\left(\frac{5}{2}pv - \varkappa\frac{\partial T}{\partial s}\right)\right] = 0 \ ,$$

where the total energy flux is set equal to zero under the assumption that the chromospheric velocity and heat flux are very small. These equations determine the velocity v_L and conductive flux F_L above the transition region, which are used as lower boundary conditions for the numerical integration in the corona. Two cases are considered. For *static conditions*, the heat flux is dissipated statically:

$$v_L(t) = 0$$

and

$$F_L(t) = \frac{-2\varkappa_0}{7A}\left(T_L^{7/2} - T_B^{7/2}\right) \Big/ \int_{s_L}^{s_B} \frac{ds}{A(s)} \ .$$

For *evaporation conditions*, the chromosphere can respond effectively to the heat flux only through mass motions; thus

$$v_L(t) = -\frac{2F_L}{5p_L}$$

and

$$F_L(t) = \frac{-2\varkappa_0}{5A} T_L(t)\left(T_L^{5/2} - T_B^{5/2}\right) \Big/ \int_{s_L}^{s_B} \frac{ds}{A(s)} \ .$$

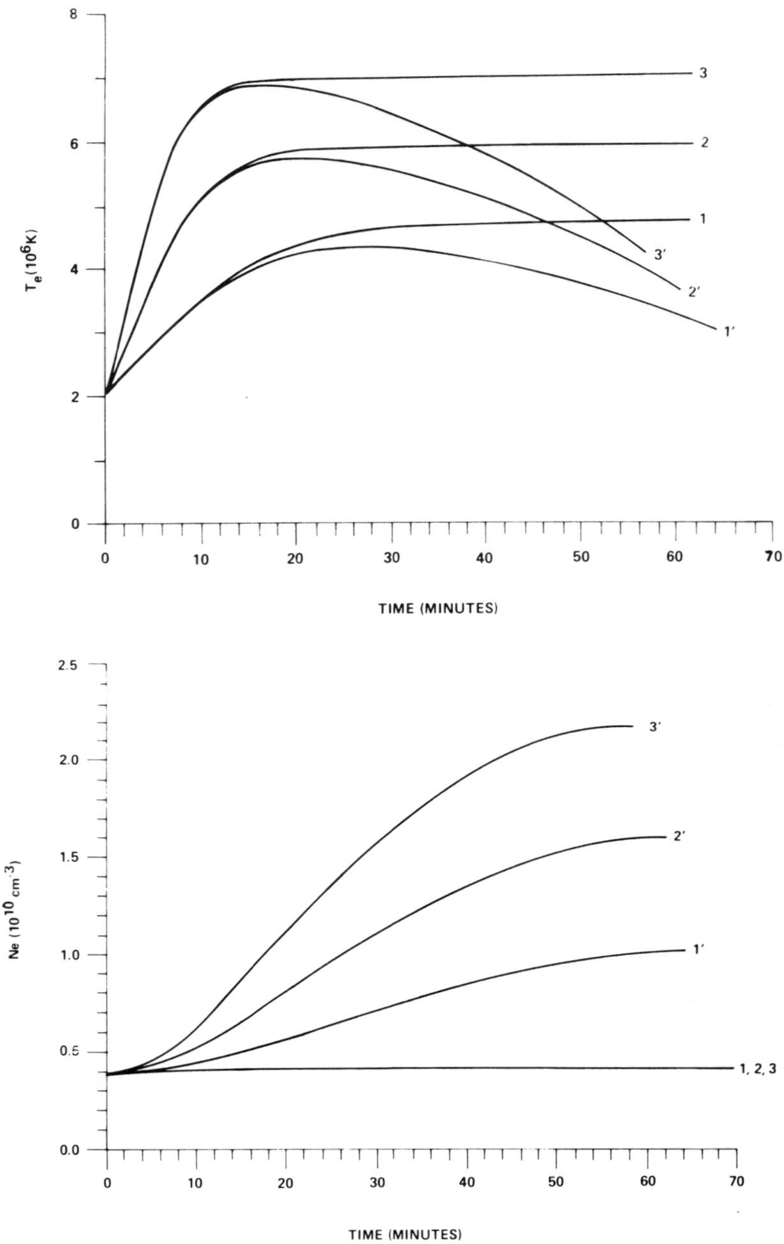

Fig. 9.9. (a) *Temperature and* (b) *density evolution at the summit of a coronal loop of length 5 × 10^9 cm, in response to an increased heating rate. Curves* 1, 2, *and* 3 *are for heating rates of three, six, and ten times the initial static value. Unprimed curves indicate static boundary conditions at the loop base, whereas primed curves are for evaporative conditions.* (c) *The integrated emission from the whole loop for a broadband aluminum channel of the proportional counter (8–12 Å) on the Skylab ATM X-REA instrument. (From Krall and Antiochos, 1980.)*

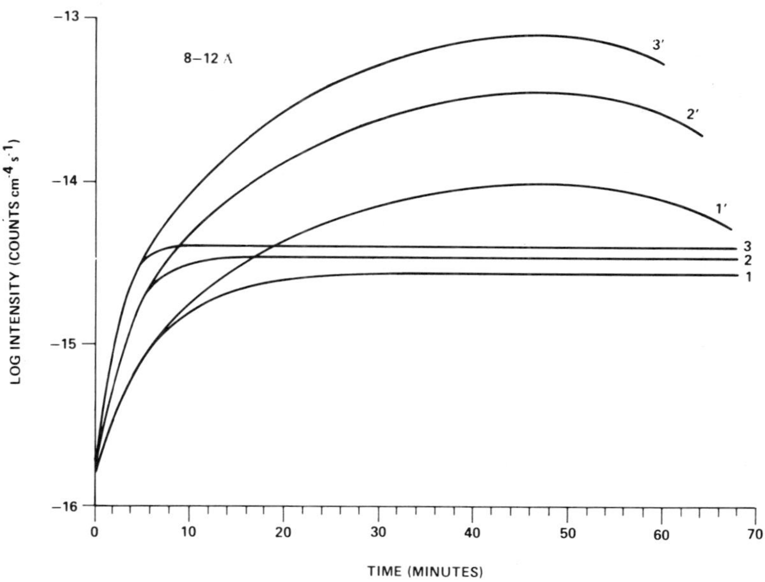

The initial state of the loop is *thermally isolated*; thus, the temperature gradient vanishes at the base and summit. The loop is also assumed to be in hydrostatic equilibrium.

A model loop 5 $\times$ 10^9 cm long with an initial summit density and temperature of 3.9 $\times$ 10^9 cm^{-3} and 2.2 $\times$ 10^6 K, respectively, was studied. The heating rate was gradually increased (by factors of three, six, or ten over a period of 2 min) in three numerical experiments. The resulting temperature and density evolution is shown in Figure 9.9 for the two types of base boundary condition. Curves *1*, *2*, and *3* result from *static conditions*, when the increased heat at the base is dissipated and lost from the system; curves *1′*, *2′*, *3′* are for *evaporative conditions*, when an upward mass flux is allowed to balance the enhanced heat load.

In the static case, no mass flow is allowed through the loop base. The density remains constant, while the temperature increases up to a new value at which the base conductive losses roughly balance the new heating rate. The effect of radiation is actually to enhance slightly the temperature rise. The mass flow is relatively minor. Because the final temperature gradient at the base is allowed to take any value, the new equilibrium is dependent on the time evolution and not just on the new heating rate.

In the evaporative case, the temperature begins to deviate significantly from the static case after about 15 min, when the conductive and evaporative energy fluxes become comparable with the heating rate. Subsequently, the density enhancement increases the radiative loss and causes the temperature to fall. Unfortunately, it was not possible to resolve temperatures below a few $\times$ 10^5 K and therefore the new equilibrium state could not be closely approached. However, this state is expected to depend solely on the new heating rate.

If some dissipation of the flux conducted to the chromosphere were allowed in the evaporative case, the resulting time profile would be expected to lie between the primed and unprimed curves. Also, a heating that is nonuniform would produce larger pressure imbalances and enhance the flows significantly; it is unclear to what extent this would alter the general time evolution calculated here. For longer loops, Krall and Antiochos find similar behavior to that of Figure 9.9, except that the time-scale is longer.

Figure 9.9c plots the calculated total emission of the loop as it would be measured by a broadband soft X-ray instrument on Skylab. It is roughly proportional to the square of the energy density. In the static case, with conduction the dominant loss mechanism, the emission curve resembles the curve for temperature squared. In the evaporative case, the base heat flux acts only to drive flows, with the total energy density held constant, and radiation is the only loss term. For the first five minutes, heating dominates either form of loss, and the boundary conditions make no difference. In the static case, the intensity levels off when conduction balances heating. In the evaporative case, the emission continues to rise as more material flows up. The increase in radiation is due mainly to the density enhancement; it is a cumulative result of the sustained conductive flux and produces the delay in the approach to a new equilibrium.

The above preliminary results indicate that, if significant changes in heating rates occur, the evolution of coronal loops may be affected significantly by mass and energy exchange with the chromosphere. In particular, evaporation may moderate temperature excursions within loops and temperature variations between them.

9.3.2 The Dynamic Approach to a Static Loop

Craig (1980) has employed a numerical code to follow the dynamic formation of a static loop. He solved the usual hydrodynamic equations of continuity, momentum, and energy for one-dimensional flow in the form

$$\frac{D\varrho}{Dt} = -\varrho\frac{\partial v}{\partial z}\,, \tag{9.3.2}$$

$$\varrho\frac{Dv}{Dt} = -\frac{\partial p}{\partial z}\,, \tag{9.3.3}$$

$$\frac{3}{2}\varrho\frac{D}{Dt}\left(\frac{p}{\varrho}\right) = \frac{\partial}{\partial z}\left(\varkappa\frac{\partial T}{\partial z}\right) - n^2Q(T) + H - p\frac{\partial v}{\partial z}\,. \tag{9.3.4}$$

The *gravitational acceleration* and viscosity are *neglected*, the *pressure is uniform*, and the radiation $n^2Q(t)$ is optically thin; the heating H (per unit volume) is proportional to density. However, the inclusion of gravity and nonuniform areas should not alter the *qualitative* behavior of the solutions.

The boundary conditions are those of symmetry at the loop summit together with a vanishing of the temperature gradient and plasma velocity at the base (always). This means that the loop is *isolated* from its surroundings, since *no mass or energy flux is allowed through the footpoints*. In other words, the loop is (rather unrealistically) given a closed base, so that the material initially present is only allowed to be redistributed within the loop, with no loss or addition of matter (unless the loop base is identified as some other point where $\partial T/\partial z$ vanishes within the domain of calculation); also, radiation is the only process by which energy can be lost, and this must be supplied solely by the heating as far as the loop as a whole is concerned However, it is difficult at present to see how to make the boundary conditions more realistic.

The initial state is a *uniform* plasma with the temperature 6×10^4 K and density 10^9 cm^{-3} such that heating and radiation balance exactly. The length of the loop is 3.3×10^9 cm. At $t = 0$, the equilibrium is disturbed by switching on an additional transient heating pulse H_t. In space, H_t has a Gaussian profile situated at the summit and having a half width of 5×10^8 cm that is much smaller than the loop length. In time, the pulse has a sinusoidal profile, increasing to a maximum of 10^{-2} erg cm^{-2} s^{-1} after 100 s and decreasing to zero at 200 s, after which it remains zero.

It is clear that there are (at least) two equilibrium solutions to Eqs. 9.3.2–9.3.4. The solutions both have

$$p = \text{constant}, \; v = 0;$$

one solution has T constant; the other has a profile that resembles that of the solar atmosphere, increasing from a minimum at the loop's base (where $\partial T/\partial z$ vanishes) to a maximum at its summit. The main object of the numerical experiment is to see how one equilibrium may evolve into the other impelled by a transient heat pulse; also, it is of interest to see how flows affect the relaxation and to see whether perturbations of finite amplitude can lead to thermal instability.

The results are shown in Figure 9.10. During the first minute, the plasma is heated at constant density. As soon as the temperature exceeds 10^5 K, the summit plasma heats rapidly due to radiative instability until conduction becomes effective at about 2×10^6 K. The enhanced summit pressure then drives a pressure wave down into the cooler material; this has the effect of transporting mass to the base to form a cool, dense region ($T \approx 2 \times 10^4$ K, $n \approx 10^{11}$ cm^{-3}). Figure 9.10a indicates that after 200 s the hot corona has formed, and the pressure pulse is propagating down; at the pulse, the temperature rises slightly, but then falls to chromospheric values as it cools because of the density enhancement. After 500 s, the loop structure is well developed, with coronal and chromospheric regions separated by a narrow transition layer 10^2-cm thick. Minor flows and counterflows then take place, the pressure wave reflecting off the ends of the loop, and the pressure gradually becoming more uniform (Figure 9.10b).

Figure 9.10c gives the evolution of coronal temperature and pressure at the loop summit. Quasistatic equilibrium is attained after about 30 min, by which time oscillations are at a low level and have a period of about 300 s, corresponding to the sound propagation time along the loop. The effect of an *extra injection* of energy at 1600 s (a "flare heating" experiment) is shown by the *dashed* curves. At first, the coronal pressure is enhanced and drives some plasma downwards; then there is a reflection off the dense chromo-

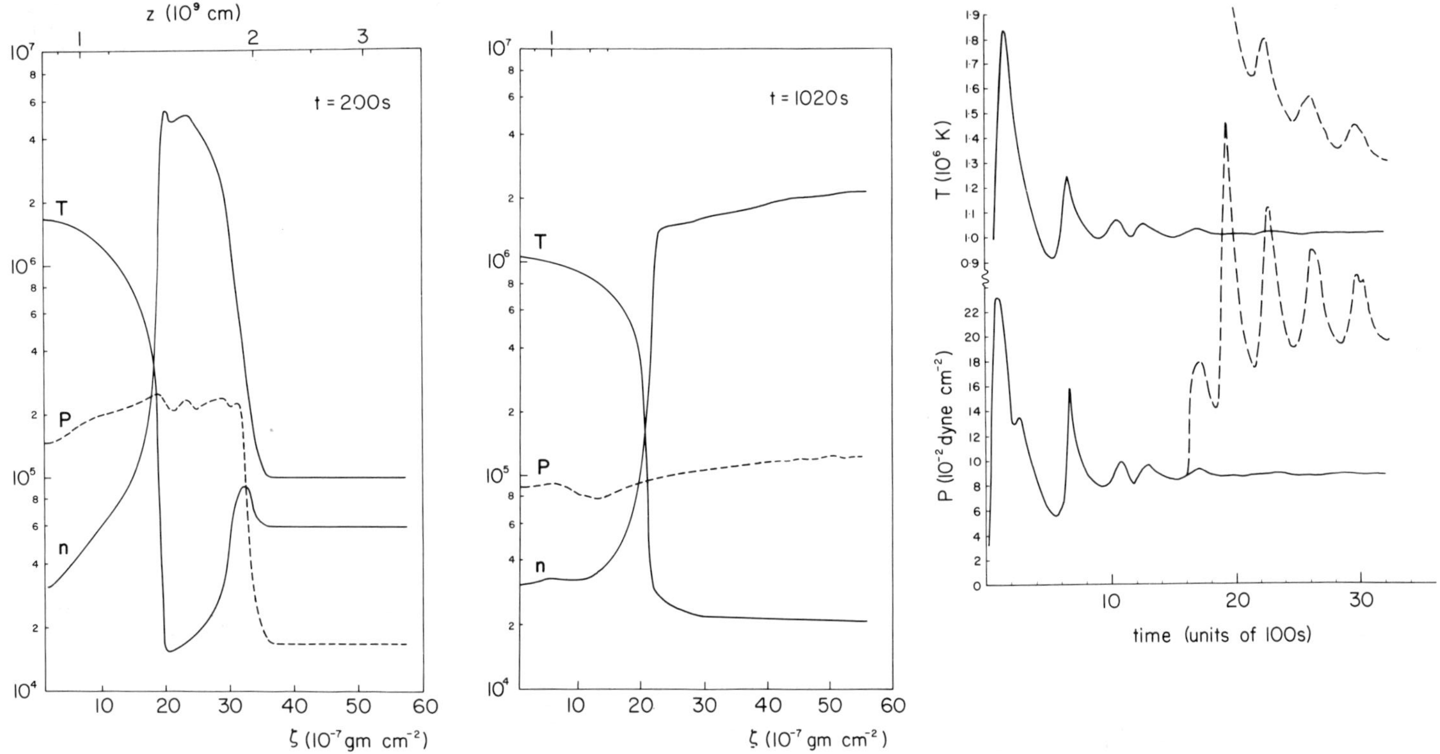

Fig. 9.10. The dynamic approach to static conditions in a coronal loop of length 3.3 × 10⁹ cm, with distance z *measured along the loop from the summit. Temperature* T *is in units of degrees K, density* n *in units of 10⁴ cm⁻³, and pressure* p *in units of 10⁻⁶ dyn cm⁻².* (a) *Profiles of* T, n, *and* p *after 200 s, when the corona has formed.* (b) *Profiles of* T, n, *and* p *after 1020 s, when the atmosphere is fully developed.* (c) T *and* p *at the loop summit as a function of time. The broken lines show the response to a second transient energy injection. (From Craig, 1980.)*

spheric material, and much more plasma flows back up into the corona (dynamic evaporation). The pressure and density achieve their maximum values at roughly the same time while the temperature is declining. The superposition of a hydrodynamic oscillation on the slower variation of temperature and pressure can clearly be seen. Although the evolution was not followed as far as a new equilibrium, the trend suggests that it differs from the previous equilibrium. The possibility of several equilibria is rather surprising. The new thermally isolated equilibrium possesses the same loop length, heating, and total mass, but it has a different pressure. Probably, the new solution is spurious, resulting from the detailed numerics of the transition region, which is not modeled adequately. [An integration of the equilibrium equations by Wragg (1981) has revealed (in most cases) just one solution in addition to the isothermal state.] However, there is no reason in principle why a steady, nonlinear oscillation should not exist as a solution—a point that evidently deserves further study. Such an oscillation could be driven by the radiation term and have a period equal to the radiative cooling time.

One difficulty with most numerical codes is to deal adequately with the very narrow transition region. Krall and Antiochos (1980) use an analytical approximation to the transition region structure as the boundary condition at the end of the loop. For the main calculation, Craig's scheme is numerically accurate; but for the flare heating experiment, it is inaccurate in the transition region, with only a few grid points in the steepest part, so that the region is not allowed to become as steep as it should; this influences the final state of the plasma, but does not affect the overall evolution and the main conclusions of the experiment. These conclusions are: a loop of fixed mass will relax on the conductive and radiative time-scales to an equilibrium; superimposed on the relaxation are oscillations at the sound travel frequency (typically 300 s!); and in an average flare heating process, mass is driven up into the corona, but the large transition region conductive fluxes have not yet been adequately modeled. Furthermore, the oscillations damp very slowly, so that in practice *a static loop may never exist*; rather, the plasma is likely to be in continual motion, with flows of 10^2–10^3 m s^{-1} through the transition region.

9.4 THERMAL STABILITY OF STATIC LOOPS

Many authors have recently set up *static* models for coronal loops, in which the pressure is usually assumed uniform and the temperature satisfies an energy balance equation of the form

$$0 = \frac{d}{ds}\left(\varkappa_0 T^{5/2}\frac{dT}{ds}\right) - n^2 Q(T) + H \ , \qquad (9.4.1)$$

where

$$n = p/(2k_B T) \ .$$

Thus, thermal conduction balances optically thin radiation n^2Q and heating, which is here assumed proportional to the density or uniform. The global stability of these models has been considered by Hood and Priest (1980a), Antiochos (1979), and Habbal and Rosner (1979), each in slightly different ways, which we outline below.

The equilibria for loops of length $2L$ and pressure p are obtained by solving Eq. 9.5.1 subject to the boundary conditions that

$$T = T_0 \text{ at the loop base } s = 0,$$

and

$$\frac{dT}{ds} = 0 \text{ at the loop summit } s = L. \tag{9.4.2}$$

This leads to a set of models depending on two parameters, pL and H/p, as obtained by Hood and Priest (1979). If, in particular, L and H are held fixed and p is altered, one finds a variation in summit temperature T_1 that is shown in Figure 9.11. The different solutions have different base conductive fluxes; but one model, known as a *thermally isolated* loop, has a vanishing base flux. This model depends on one parameter (pL) and leads to the scaling

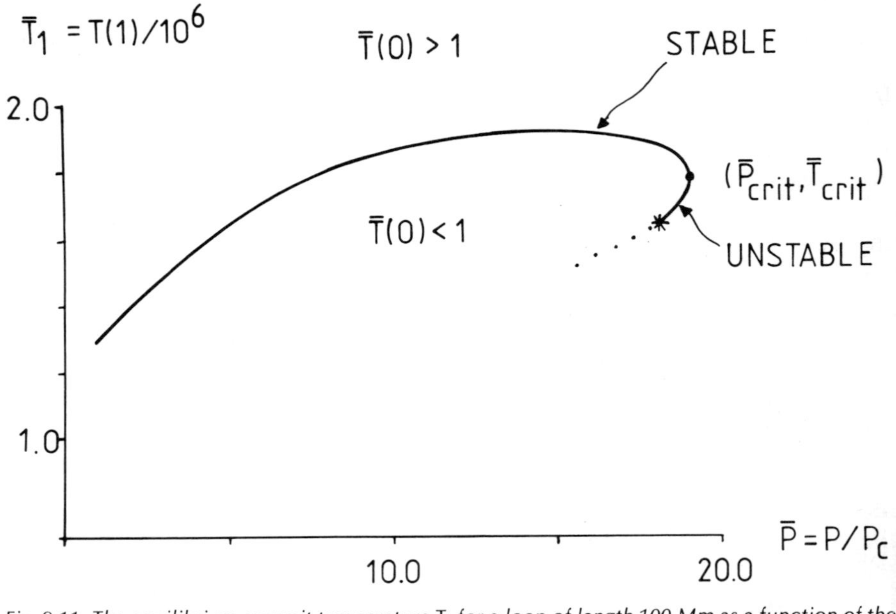

Fig. 9.11. The equilibrium summit temperature T_1 for a loop of length 100 Mm as a function of the pressure. The star shows the location of the thermally isolated solution. (From Hood and Priest, 1979.)

law, Eq. 9.1.4. Its location within the family of solutions of Figure 9.11 is denoted by a star.

It is well known that coronal plasma has a tendency to be *thermally unstable* (Parker, 1953; Field, 1965). This instability is due to the form of the radiative loss term in the energy balance. At constant pressure, the radiation is proportional to $T^{-2}Q(T)$, and for temperatures above about 2×10^4 K this is a decreasing function of temperature. Thus, a decrease in temperature leads to an increase in radiation, which in turn tends to cool the plasma and drive the instability. In many cases, *thermal conduction* may be so important that it provides stability by smoothing out any temperature perturbations more rapidly than they can grow. Indeed, it is only when the lengths of *current sheets* (Smith and Priest, 1977) or of field lines in *coronal arcades* (Priest and Smith, 1979) are large enough that thermal conduction may become ineffective and allow *prominences* to form. Within a thermally isolated loop, conduction and radiation are globally equal, and the question of stability is a delicate matter. The stability criteria of Field (1965) cannot be applied directly, since they were derived for a uniform medium. The thermally isolated loops considered here are nonuniform—indeed, their thermal structure is most important. This means that one needs to apply a normal-mode analysis to determine the global stability, as described below.

9.4.1 *Analysis of Hood and Priest*

Hood and Priest (1980a) consider the stability of static loops by applying a linear perturbation to the hydrodynamic equations. The fluid motions are assumed for simplicity to be directed along the magnetic field, and gravity is neglected. The time-dependent equations of motion, continuity, state, and energy for the density n^*, speed v^*, pressure p^*, and temperature T^* are then

$$mn^*\left(\frac{\partial v^*}{\partial t} + v^*\frac{\partial v^*}{\partial s}\right) = -\frac{\partial p^*}{\partial s} \,, \tag{9.4.3}$$

$$\frac{\partial n^*}{\partial t} + \frac{\partial}{\partial s}(n^*v^*) = 0 \,, \tag{9.4.4}$$

$$p^* = 2n^*k_B T^* \,, \tag{9.4.5}$$

$$\frac{n^{*\gamma}}{\gamma - 1}\left(\frac{\partial}{\partial t} + v^*\frac{\partial}{\partial s}\right)\left(\frac{p^*}{n^{*\gamma}}\right) = \frac{\partial}{\partial s}\left(\varkappa_0 T^{*5/2}\frac{\partial T^*}{\partial s}\right) - n^{*2}Q(T) + H \,, \tag{9.4.6}$$

where

$$Q(T^*) = \chi T^{*\alpha} \,.$$

Assuming that the perturbation is growing exponentially in time, the temperature is written in the form

$$T^*(s,t) = T(s) + T^{(1)}(s)e^{\omega t} ,$$

and similarly for the other variables, where $T(s)$ is the equilibrium state and ω is the growth rate. The resulting linearized equations are

$$mn\omega v^{(1)} = -\frac{dp^{(1)}}{ds} , \tag{9.4.7}$$

$$\omega n^{(1)} + \frac{d}{ds}(nv^{(1)}) = 0 , \tag{9.4.8}$$

$$\frac{p^{(1)}}{p} = \frac{n^{(1)}}{n} + \frac{T^{(1)}}{T} , \tag{9.4.9}$$

$$\frac{\omega}{\gamma - 1}\left(p^{(1)} - \frac{\gamma p}{n} n^{(1)}\right) + \frac{n^{\gamma} v^{(1)}}{\gamma - 1} \frac{d}{ds}\left(\frac{p}{n^{\gamma}}\right)$$

$$= \frac{d^2}{ds^2}(\varkappa_0 T^{5/2} T^{(1)}) - n^2 \chi \alpha T_0^{\alpha - 1} T^{(1)} - 2nn^{(1)} \chi T^{\alpha} . \tag{9.4.10}$$

A further simplification results if the instability time-scale (essentially the radiative time) is much longer than the time-scale L/c_s for sound wave propagation. Then sound waves are filtered out, and the perturbations are *isobaric*, with the integral of Eq. 9.4.7 reducing to

$$p^{(1)} \equiv 0 . \tag{9.4.7$'$}$$

After some manipulation, the equations reduce to one equation for the perturbed temperature $T^{(1)}(s)$, namely

$$\frac{d^2(\bar{T}^{5/2}\bar{T}^{(1)})}{ds^2} - \bar{p}^2\bar{L}^2\left((\alpha - 2)T^{\alpha - 3} + \frac{\bar{\omega}F}{\bar{T}\bar{p}}\right)\bar{T}^{(1)} = 0 , \tag{9.4.11}$$

where

$$F = 1 + \frac{\bar{T}}{\bar{T}^{(1)}} \frac{d\bar{T}}{d\bar{s}} \int_0^{\bar{s}} \frac{\bar{T}^{(1)}}{\bar{T}^2} d\bar{s} .$$

Dimensionless variables have been defined as follows:

$$\bar{p} = p/p_c, \bar{T} = T/T_0, \bar{s} = s/L, \bar{\omega} = \tau_{\text{rad}}\omega ,$$

where p_c is a typical coronal pressure,

$$\tau_{\text{rad}} = \frac{mc_p 2k_B}{\chi T_0^{\alpha - 2} p_c}$$

is the radiative time scale, and

$$\bar{L}^2 = \frac{L^2 p_c^2 \chi T_0^\alpha}{4k_B^2 \varkappa_0 T_0^{11/2}}$$

is a measure of the loop length.

Equation 9.4.11 must be solved subject to three boundary conditions and determines the value of the growth rate ω. If $\omega < 0$, the equilibrium state is stable; but if $\omega > 0$, it is unstable. The value $\omega = 0$ defines a state of *marginal* stability, separating stable from unstable regimes.

Hood and Priest point out that in order to deduce the nature of the stability, it is *necessary only to study the marginal case* $\omega = 0$, which leads to considerable simplicity. Thus, for instance, suppose a solution $T^{(1)}(s)$ of Eq. 9.4.11 is obtained with $\omega = 0$ that satisfies two of the boundary conditions, but does not satisfy the third, namely that $T^{(1)} = 0$ at $s = 0$, for example. If this marginal solution vanishes nowhere between $s = 0$ and $s = L$, then one can use the general *Sturm-Liouville* theory to deduce that a *negative* value of ω *would* make $T^{(1)}(0)$ vanish; in other words, the equilibrium would be stable.

Evidently, the nature of the stability depends critically on the boundary conditions that are imposed in solving Eq. 9.4.11. Consider first the case when the boundary conditions Eq. 9.4.2 are maintained for all time, so that the *fundamental* mode is *symmetric* and has

$$T^{(1)} = 0 \text{ at } s = 0 \ ,$$

and (9.4.12)

$$\frac{dT^{(1)}}{ds} = 0 \text{ at } s = L \ ,$$

together with a normalization condition

$$T^{(1)} = 1 \text{ at } s = L \ .$$

Then Eq. 9.4.11 may be solved and stability deduced.

However, when the summit temperature T_1 is known as a function of p, L, and H (and the boundary conditions Eq. 9.5.12 are used), the above elegant mathematical technique based on general Sturm-Liouville theory may be used to deduce a very general result. This is that *all equilibria lying on the upper part of the curve in* Figure 9.11 *are stable*, whereas *all those on the lower branch are unstable*. In particular, the *thermally isolated loop is unstable*. Furthermore, *if the base flux is increased from zero* so much that the equilibrium proceeds beyond the critical point onto the upper branch *then the loop is stable*. These general criteria may be verified for particular loops by solving Eq. 9.4.11 directly.

For the special case of *thermally isolated* loops, alternative boundary conditions to Eq. 9.4.12 may be adopted, since it is not clear which the sun actually imposes. For example, if the conductive flux remains zero at both feet, the *fundamental* mode is *antisymmetric* and has

$$\frac{dT^{(1)}}{ds} = 0 \text{ at } s = 0 \;, \qquad T^{(1)} = 0, \frac{dT^{(1)}}{ds} = 1 \text{ at } s = L \;. \tag{9.4.13}$$

Another possibility is that one foot maintains zero conductive flux, while the other keeps its temperature fixed, so that

$$\frac{dT^{(1)}}{ds} = 0,\ T^{(1)} = 1 \text{ at } s = 0 \;, \qquad T^{(1)} = 0 \quad \text{at } s = 2L \;. \tag{9.4.14}$$

The simplest thermally isolated loop is the *isothermal* solution $T \equiv T_0$, which may be analysed analytically, since the perturbation equation (9.4.11) reduces to

$$\frac{d^2\bar{T}^{(1)}}{ds^2} - \bar{p}^2\bar{L}^2[\alpha - 2 + (\bar{\omega}/\bar{p})]\bar{T}^{(1)} = 0 \;.$$

This has the solution

$$\bar{T}^{(1)} = A \cos k\bar{s} + B \sin k\bar{s} \;,$$

where

$$k^2 = \bar{p}^2\bar{L}^2[2 - \alpha - (\bar{\omega}/\bar{p})] \;.$$

When $\alpha > 2$, the loop is stable; when $\alpha < 2$ (i.e., $T > 1.5 \times 10^4$ K), the effect of applying the boundary conditions (Eqs. 9.4.12, 9.4.13, or 9.4.14) is to give instability when

$$16\bar{p}^2\bar{L}^2(2 - \alpha)/\pi^2 > \begin{cases} 4 \text{ for Eq. 9.4.12} \\ 4 \text{ for Eq. 9.4.13} \\ 1 \text{ for Eq. 9.4.14} \;. \end{cases}$$

Furthermore, it can be shown that $\bar{p}^2\bar{L}^2$ must certainly exceed these values to make the corresponding non-isothermal solution unstable. (This does not contradict the previous result that all thermally isolated loops are unstable, since it is impossible to have a thermally isolated loop when $\bar{p}^2\bar{L}^2$ is less than these values!) For *non-isothermal*, thermally isolated loops, a numerical solution of Eq. 9.4.11 was found to give *instability* for the boundary conditions Eqs. 9.4.12 or 9.4.13. The growth times were found to be bound below (for boundary conditions Eq. 9.4.12) by

$$\frac{\omega^{-1}}{\tau_{\text{rad}}} \gtrsim \begin{cases} 1.7 \quad \text{for } T_1 = 4T_0 \\ 200 \quad \text{for } T_1 = 25T_0 \\ 7{,}500 \text{ for } T_1 = 100T_0 \;. \end{cases}$$

Thus, for $T_1 = 100T_0$, this linear growth rate is rather slow, although, of course, nonlinear effects may give much faster rates.

Hood and Priest conclude that *thermally isolated loops are unstable* to perturbations that maintain the value of either the base temperature or the base conductive flux. Individual coronal loops may therefore be in a *dynamic state of ceaseless thermal activity*, unless some *stabilizing feedback mechanism* exists. Furthermore, a loop is shown to be *stable if its base conductive flux is large enough* that it lies on the upper of two equilibrium branches.

9.4.2 Analysis of Antiochos

Antiochos (1979) has independently considered the stability of static loop models. Like Hood and Priest, he concludes that the static models are unstable. The Hood and Priest method of studying marginal stability alone is simpler, but Antiochos is able to obtain more accurate values for the linear growth rate; he finds a much larger growth rate (for the case of a maintained base heat flux) than the lower bound given by Hood and Priest (for a maintained base temperature).

Antiochos sets up the linearized equations (9.4.7–9.4.10) for isobaric perturbations. He assumes as boundary conditions that the perturbations in velocity and heat flux vanish at the base, so that

$$v^{(1)} = \frac{d}{ds}(T^{5/2}T^{(1)}) = 0 \text{ at } s = L \ .$$

(For a thermally isolated loop, the latter condition reduces to

$$dT^{(1)}/ds = 0 \text{ at } s = L \ .$$

Also, it should be noted that the fundamental mode is antisymmetric, and the boundary conditions become just Eq. 9.4.13.) Since he wishes to consider perturbations that disturb only the coronal plasma and do not affect the plasma below the loop base (in the chromosphere), he does not allow any mass flux or change in the base heat flux. He reduces the equations to a single eigenvalue equation of Sturm-Liouville type for the perturbation velocity. It has the form

$$\frac{d}{dx}\left(y\frac{d\mu}{dx}\right) + Q(x)\mu + \lambda\mu = 0 \ , \qquad (9.4.15)$$

where

$$\mu(x) = y(x)^{5/14}\frac{dv^{(1)}}{dx} \ ,$$

$$Q(x) = \frac{1}{14}\frac{d^2y}{dx^2} - \frac{1}{y}\left(\frac{5}{14}\frac{dy}{dx}\right)^2 - y\left(\frac{dy}{dx}\right)^{-1}\frac{d^3y}{dx^3} \ ,$$

$$\lambda = -\omega\frac{5pL^2}{2\varkappa_0 T_1} \ ,$$

$$y(x) = \left(\frac{T(s)}{T_1}\right)^{7/2} ,$$

$$x = 1 - \frac{s}{L} .$$

Thus, $\mu(x)$ is the eigenfunction and λ is the growth rate measured in units of the conductive time-scale at the summit temperature.

In view of the complicated form of Eq. 9.4.15 and the fact that the equilibrium profile $y(x)$ is not itself in closed form, it is necessary to solve Eq. 9.4.15 numerically to deduce the growth rate. For the equilibrium state, the radiative loss function is taken in the form

$$Q(T) = \chi T^{-0.5} ,$$

and the loops are assumed to be *thermally isolated* with vanishing base conductive flux

$$\frac{dT}{ds} = 0 \text{ at } s = 0 .$$

Furthermore, the heating is allowed to depend on density and temperature like

$$H = H_0 n^{\gamma_1}/T^{\gamma_2} ,$$

where

$$\gamma = \gamma_1 + \gamma_2 ;$$

the three values 0, 1, and 2 are considered for γ. Also, three values for the ratio

$$\theta_b = T_0/T_1$$

of base-to-summit temperature are considered, namely 0.5, 0.1, and 0.02.

The first three eigenvalues λ_1, λ_2, and λ_3 are listed in Table 9.1. Since, in each case, the first eigenvalue (antisymmetric) is negative, we see that ω is positive and the *thermally isolated models are unstable*. This result also holds if exponents of 0 or −1 are used instead of −0.5 for the radiative loss; in fact, the growth rate increases as this exponent decreases. The growth rate also depends on the form of the heating; it decreases with decreasing γ, but even for large negative values of γ the equilibria are still highly unstable.

Table 9.1
First Three Eigenvalues for the Stability of a Thermally Isolated Loop

$\theta_b =$	0.5			0.1			0.02		
$\gamma =$	0	1	2	0	1	2	0	1	2
λ_1	-5.8	-7.6	-10.0	-79	-130.0	-260.0	-3800	-5800	-14,000
λ_2	-0.30	0.062	0.43	- 0.46	0.18	0.88	-0.39	0.28	1.1
λ_3	4.9	5.0	5.2	1.9	2.5	3.5	1.8	2.5	3.9

As the base temperature θ_b decreases, the growth rate increases rapidly, which suggests that the instability involves mainly the cool material in the transition region. Antiochos suggests that the *cool plasma near the base is responsible for the instability*, since its radiative time scale is much shorter than the coronal values. For example, in the case $\theta_b = 0.02$, a summit temperature 2×10^6 K, summit density 2×10^9 cm^{-3}, and loop length 4×10^9 cm give a growth time of only 0.1 s, which is close to the radiative cooling time for the base plasma. This exceeds the sound travel time of 0.01 s across a length-scale 2×10^4 cm characteristic of the perturbation; thus, the isobaric assumption remains valid. (It may be noted that the growth rate depends on the boundary conditions; the growth rate is much faster for the above case of a maintained base conductive flux than for a fixed base temperature treated by Hood and Priest.)

Figure 9.12a shows the spatial profile for the first (antisymmetric) unstable mode in the case $\gamma = 1$, $\theta_b = 0.02$. It can be seen that the perturbation is highly concentrated near the loop base, especially in the first scale-height, over only 10^{-5} of the loop length.

Antiochos also considers some models with a *nonvanishing base heat flux*. He simply starts with the thermally isolated loop when $\gamma = 1$ and $\theta_b = 0.02$, and then takes the loop base at successively greater heights and base temperatures θ_b. In other words, for the same equilibrium profile, the perturbation is imposed on successively smaller fractions of the loop. The results are given in Figure 9.12b for the first (antisymmetric) eigenvalue λ, as a function of θ_b. They agree with the conclusion of Hood and Priest that the *loop becomes stable if its base flux is large enough*. For this particular loop, stability occurs at a base temperature about 25% above the initial thermally isolated value and at a base heat flux that is about 13% of the maximum.

Antiochos concludes that the stability of a static loop model depends on the position of its base, namely the lowest point for which the model is valid. If the heat flux at the base is small enough, the loop is thermally unstable. He therefore suggests that the *lower transition region must be in a dynamic state* rather than a stationary one. The instability involves mainly the plasma near the base; he expects it to saturate rapidly and suggests that *static profiles for the lower transition region will definitely not be valid*, while those for the atmosphere above may still hold. He suggests that the lower transition region may have an oscillatory behavior, with periods of about the radiation time (<1 s) and wavelengths similar to the scale-height.

9.4.3 Analysis of Habbal and Rosner

Habbal and Rosner (1979) have considered the *local* stability of coronal loops, using the standard analysis of Field (1965). Thus, the equilibrium plasma is assumed to be essentially *uniform* (over a wavelength) and the perturbation temperature and heating function H are assumed to behave like

$$T^{(1)} = A \exp(\omega t + i2\pi s/\lambda) \ ,$$

$$H^{(1)} = B \exp(\omega t + i2\pi s/\lambda) \ ,$$

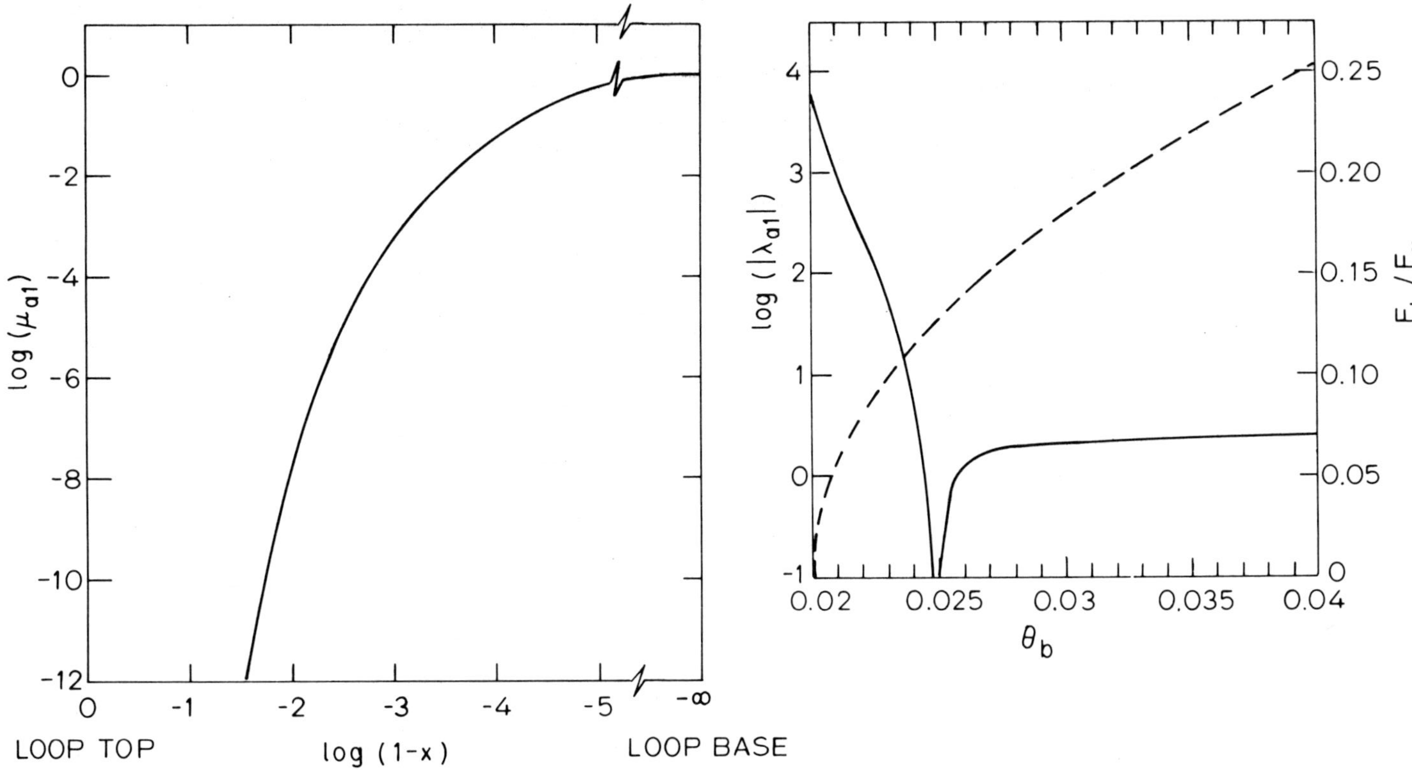

Fig. 9.12. The thermal instability of a loop of length L and base temperature T_0, which is a fraction θ_b of the summit temperature T_1. Radiation is proportional to $T^{-0.5}$ and heating is proportional to $n^{1-\gamma_2}/T^{\gamma_2}$ for any γ_2. (a) The profile $\mu(x)$ for the first unstable mode in a thermally isolated loop with $\theta_b = 0.02$, where $x = 1 - s/L$ and $\mu = L[T(s)/T_1]^{7/2}\, dv^{(1)}/ds$. The amplitude is normalized to unity at the base $x = 1$ ($s = 0$) of the loop. (b) The magnitude of the first eigenvalue λ_1 (solid curve) and ratio of base heat flux F_b to maximum heat flux F_m (dashed curve) as a function of base temperature θ_b. λ_1 is negative for $\theta_b < 0.025$ and positive for $\theta_b > 0.025$. $\lambda = \omega\tau_c$, where ω is the growth rate and τ_c the conduction time for coronal plasma. (From Antiochos, 1980.)

where λ is the wavelength, and ω is the growth rate of the perturbation.

First of all, they consider an *isobaric* perturbation caused by a fluctuation in the coronal heating rate. The resulting dispersion relation gives *instability* (Re $\omega > 0$) when the wavelength exceeds a critical value λ_c, namely,

$$\lambda^2 > \lambda_c^2 = \frac{5}{8\pi^2 \varkappa_0 T^{5/2}} \left[\frac{2}{5} Re \left(\frac{B}{A}\right) - \frac{H}{T} - \varkappa_0 T^{1/2} \left(\frac{dT}{ds}\right)^2 + \frac{\chi p^2}{2k_B^2} \left(\frac{5}{2} - \alpha\right) T^{\alpha-3} \right] , \tag{9.4.16}$$

where the radiation term has the form

$$n^2 Q(T) = n^2 \chi^2 T^\alpha .$$

From Eq. 9.5.16, it can be seen that conduction always stabilizes; radiation destabilizes when α is small enough; and heating stabilizes if Re $(B/A) \le 0$, so that the heating decreases with temperature. Of course, the analysis is valid only for wavelengths much less than a scale-height.

Next, an isothermal perturbation caused by a fluctuation in the chromospheric pressure is examined. In this case, instability results if

$$\text{Re}\,(B/A) < 2p\chi T^{\alpha-2}/4k_B^2 ,$$

or, in other words,

$$\frac{H^{(1)}}{H} < \frac{p^{(1)}}{p} .$$

In the simple case of heating that varies solely with pressure, like

$$H \sim p^m ,$$

instability occurs when

$$m < 1 .$$

Habbal and Rosner emphasize that in practice pressure variations at loop footpoints *would* be expected to cause coronal temperature fluctuations. Observationally, the variations in loop brightness are well-correlated with pressure variations, which are mainly a result of changes in density (Davis *et al.*, 1975). Furthermore, the temperature profiles for different solar features derived by Basri *et al.* (1979) (Figure 9.13) indicate that the height at which a given temperature is attained varies considerably from one feature to another; also, there are distinct plateaus at about 8000 and 20,000 K at which the conductive flux is small and (because the temperature rises so steeply above 8000 K) the coronal pressure for each feature is essentially the same as the pressure at 8000 K for that feature. In other words, the *coronal pressure is determined by the chromospheric pressure at about 8000 K* or, maybe, it is the coronal pressure that determines the chromospheric value! For bright plage regions, Basri *et al.* find a chromospheric pressure of 4 dyn cm^{-2}, while the coronal pressures from X-ray observations are 2–5 dyn cm^{-2} (Vaiana and Rosner, 1978). By comparison, the chromospheric and coronal pressures in quiet regions are both an order of magnitude lower. All this suggests that for atmospheric models one should *impose a vanishing temperature gradient at a fixed temperature* but allow the location or pressure to be free. However, such a base boundary condition at 2×10^4 K is inadequate for optically thin models, since the plateau is created by an optically thick effect, namely the leak of Lyman-α photons, and so the *optically thin models are not valid down as far as* 2×10^4 *K*! [For the optically thin equilibrium models, it is only Hood and Priest (1979) who have taken their base above 2×10^4 K.]

The *global stability* of thermally isolated loops is also considered, but, by comparison with the previous sections, Habbal and Rosner find such loops to be thermally stable. Rather than perform a full stability analysis (like Hood and Priest, and Antiochos), they consider stability against a specified form of temperature perturbation. Then, if it gives instability, one would expect the full system to be unstable; but if it gives stability, nothing can be said about the true stability, because it may be that a different form of perturbation grows. However, in one respect, they attempt to make their analysis more realistic than some others. They *allow the loop to respond to perturbations by moving its base* (where $T = T_0$ and $\partial T/\partial s = 0$) up or down in the atmosphere, which means that the loop length and base pressure change; this contrasts with Antiochos who keeps $\partial T/\partial s = 0$ at a fixed base, but (for a thermally isolated loop) it is equivalent to one of the sets of base conditions imposed by Hood and Priest (where T is maintained at a fixed level but $\partial T/\partial s$ is allowed to vary). Of course, one of the main difficulties is to know which boundary conditions occur on the sun for a time-dependent process! The following time-dependent energy equation is taken:

$$\frac{5}{2}\frac{p^*}{T^*}\frac{dT^*}{dt} - \frac{dp^*}{dt} = \frac{\partial}{\partial s}\left(\varkappa_0 T^{*5/2}\frac{\partial T^*}{\partial s}\right) - \frac{p^*}{4k_B^2 T^{*2}}Q(T^*) + H(p^*,T^*) , \qquad (9.4.17)$$

which is essentially the same as Eq. 9.4.6.

The equilibrium summit temperature T_1 is given by

$$T_1 = c_0 p^{1/\beta}$$

for a fixed length, and the heating function is assumed to be of the form

$$H = c_1 T^\gamma ,$$

where c_0, c_1, β, and γ are constants.

The base temperature T_0 is fixed at 2×10^4 K, at which the conductive flux is small (Figure 9.13), but the *physical location of the base is allowed to*

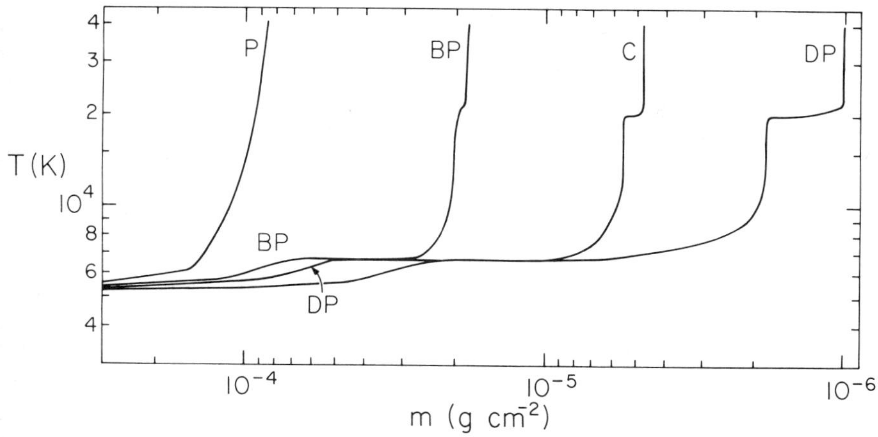

Fig. 9.13. The temperature in the lower solar atmosphere as a function of mass column density derived by Basri et al. *(1979) for several chromospheric features, namely, plage* (P), *bright point* (BP), *average quiet sun* (C), *dark point* (DP), *and the Vernazza* et al. *(1973) average quiet sun model* (VAL). *(From Habbal and Rosner, 1980.)*

vary as a result of the perturbation. In other words, the pressure is allowed to vary. The loop temperature is written in the form

$$T^*(s,t) = T(s) + T^{(1)}(s)e^{\omega t} \;,$$

as in Section 9.4.1, and the equilibrium profile $T(s)$ has a vanishing gradient $dT/ds = 0$ at $T = T_0$. But then, for simplicity, the approximation

$$T^{(1)}(s) = T_1^{(1)}$$

is adopted, namely that the temperature perturbation be *independent of space* and equal to the summit value. This is the crucial simplification in their analysis. It implies that the effect of the perturbation is to raise the temperature everywhere along the loop by the same amount—in particular, it means that the base temperature rises but its location remains the same, which is *inconsistent with the adopted base boundary conditions*. In other words, one cannot satisfy

$$\frac{dT^*}{ds} = 0 \text{ at } T^* = T_0$$

when

$$T^* = T(s) + T_1^{(1)}e^{\omega t}$$

and $T_1^{(1)}$ is constant. In order to allow the base location to vary, one needs to adopt a more general form for the perturbation.

With Habbal and Rosner's rough approximation, the integral of Eq. 9.4.17 over the loop reduces to

$$\omega = 2c_0^\beta c_1 T_1^{\gamma-\beta}[(\gamma - 2\beta + 5/2)(2 - 4\beta/7)^{-1}(\gamma + 7/2)^{-1}]$$

for the growth rate. (The alternative trivial solution $T_1 \equiv 0$ is the only one to resolve the above inconsistency.) Stability results (for $\beta < 3.5$) if

$$\gamma \leq 2\beta - 5/2 \ . \qquad (9.4.18)$$

For a given heating law, the value of γ is specified, and Eq. 9.4.18 determines the lowest β that gives stability. In particular, for acoustic wave heating, γ is 2, and therefore $\beta \geq 2.25$ gives stability; for acoustic shock heating, γ is $\beta - 1/2$, and therefore $\beta \geq 2$ gives stability; for current dissipation, γ is β, and therefore $\beta \geq 2.5$ gives stability; for Alfven wave or tearing mode heating, γ is $7\beta/6$, and therefore $\beta \geq 3$ gives stability. Because β is approximately 3 for the standard scaling law (which gives a reasonable fit with observations), Habbal and Rosner conclude that *all of the above heating mechanisms give stable loops* (for their perturbation alone), except that Alfven wave or tearing-mode heating gives neutral stability.

If the value $\beta \approx 3$ is regarded as given, Eq. 9.4.18 says that a heating mechanism must have

$$\gamma \leq 3.5$$

to give thermally stable loop structures.

9.4.4 Conclusion

Three different groups have independently asked the same question, namely, whether static loop models are thermally stable, and it is fascinating to see how the approaches differ. Indeed, in many respects they all complement (not compliment!) one another.

Habbal and Rosner give a local stability analysis that suggests the plasma is thermally *unstable* when the wavelength of the perturbation exceeds a critical value. However, they correctly point out that, for wavelengths comparable with the loop length, the local analysis is no longer valid and must be replaced by a global analysis in which the normal modes of the system are determined. They then attempt a global, essentially order-of-magnitude analysis and have the good idea of trying to allow the physical location of the loop base to move vertically. They consider only stability against a particular form of perturbation (one which is spatially uniform) and find that the system is stable *to such a perturbation.*

Antiochos (1979) and Hood and Priest perform a full global analysis (with more general perturbations than Habbal and Rosner) and reach the conclusion that *thermally isolated loop models are thermally unstable*. There are some differences in their approaches. Hood and Priest consider a wide class of equilibria (with an arbitrary base conductive flux) and establish a general result that *equilibria on an upper branch are stable*, whereas equilibria on a lower branch (including the thermally isolated loop) are unstable. They investigate the effect of several sets of boundary conditions. In particular, keeping the conductive flux or the temperature at the base fixed always gives instability. For the latter case (which, for thermally isolated loops, is equivalent to Habbal and Rosner's condition of fixing the temperature gradient at a fixed temperature), lower bounds on the growth-times are found that increase with the ratio T_1/T_0 of summit-to-base temperature. Antiochos concentrates on the case when the conductive flux is held fixed at the base and obtains the opposite trend for the variation of growth rate with T_1/T_0 because of his different boundary conditions; he goes further than Hood and Priest in obtaining accurate values for the growth-rate and profiles for the first unstable eigenfunction. The instability for this boundary condition is found to grow faster as T_1/T_0 is increased. The temperature perturbation is found to have a large amplitude, which is localized near the base (where the radiative time-scale is smallest). He deduces that the lower transition region is certainly unstable, but the atmosphere above may be well represented by a static model. However, the perturbation is not necessarily localized near the base, since with different boundary conditions the perturbation can be large right up the loop, even though driven near the base—the tendency for rapid growth near the base may affect the rest of the loop drastically. Furthermore, it would be interesting to see the effect on Antiochos' calculations of taking a radiative term proportional to T^{α} with an α that varies with temperature rather than being kept constant at, say, $-\frac{1}{2}$; for then an extra stabilizing tendency would exist near the base if the base temperature were low enough. Both Antiochos (1979) and Hood and Priest find that a loop becomes *stable if the perturbation is not allowed to reach down to the base* where the conductive flux vanishes but is constrained to exist only above a level where the temperature gradient is large enough; however, this does not nullify the main result that a perturbation that *does* extend right down to the base can be unstable.

In conclusion, the present limited stability analysis demonstrates convincingly that *thermally isolated loop models are linearly unstable*. Unless the atmosphere is able to provide some stabilizing feedback mechanism (for example, through an interaction with the surrounding atmosphere near the loop base or through forcing the loop base to move vertically in a certain way), the whole of a coronal loop may be in a *dynamic state of ceaseless thermal activity*. Much remains to be done in adding other effects (for instance, magnetic) to the stability calculations and in conducting numerical simulations that allow one to follow the nonlinear development, paying particular attention to the imposed boundary conditions.

9.5 DOWNFLOW AND UPFLOW IN INTENSE FLUX TUBES

It is now well established that supergranule boundaries are lined with tubes of intense magnetic field, which provide a natural channel linking the photosphere and chromosphere. Such intense flux tubes, where the field strength may exceed a kilogauss, can act as both a source and sink for material. They may be a source for spicules, providing tubes along which the plasma is ejected upwards by the squeezing and buffeting of granules on their sides (Parker, 1974; Roberts, 1979). Also, the intense tubes may provide a sink for returning spicular material and other plasma falling from the corona. The total mass involved in spicules is immense, greatly exceeding the amount of material escaping in the solar wind (Beckers, 1972; Pneuman and Kopp, 1977). This section describes three aspects of the theory of intense tubes that were extended during the Workshop: the creation of the tubes; dynamic influence of the tubes on the surrounding atmosphere; and finally the creation of upflows.

9.5.1 Intense Magnetic Field Instability

The properties of intense tubes, which comprise over 90% of the photospheric flux outside sunspots, have been reviewed by Stenflo (1976) and Harvey (1977). The tubes have field strengths in the range 1–2 kG, with diameters of 100–300 km, and they are located in the boundaries of supergranules, where strong, *steady downdrafts* occur at 0.5–2.0 km s^{-1}. The tubes also exhibit *outwardly propagating waves* (Giovanelli *et al.*, 1978), which have been modeled theoretically by Cram and Wilson (1975), Defouw (1976), and Roberts and Webb (1978).

An intriguing puzzle is—why does the sun arrange its surface field into intense, isolated flux tubes, in direct opposition to the expansive magnetic pressure? A possible answer is that convective motions provide the compression; but this does not seem to be able to give strong enough fields. The numerical modeling of Galloway (1977) does indicate intensification in excess of equipartition values, but it is uncertain to what extent this will still hold when the Boussinesq approximation is removed. An alternative approach has been adopted by Parker (1978), Webb and Roberts (1978), Spruit (1979), Spruit and Zweibel (1979), and Unno and Ando (1979), who consider the possibility of an *intense tube instability* within a partially formed tube, rather than convection around it. Both are driven by the superadiabatic temperature gradient that is present in the first few Mm below the solar surface.

Consider vertical motions in a slender vertical flux tube that is slowly diverging with altitude and has a radius much smaller than a scale-height. According to Roberts and Webb (1978, 1979) (see also Defouw, 1976), the equations of continuity, vertical momentum, transverse momentum, and isentropic energy are

$$\frac{\partial}{\partial t}\left(\frac{\varrho}{B}\right) + \frac{\partial}{\partial z}\left(\frac{\varrho v}{B}\right) = 0 \ , \qquad (9.5.1)$$

$$\frac{\partial v}{\partial t} + v\frac{\partial v}{\partial z} = -\frac{1}{\varrho}\frac{\partial p}{\partial z} - g \ , \qquad (9.5.2)$$

$$p + B^2/2\mu = p_e \ , \qquad (9.5.3)$$

$$\frac{\partial p}{\partial t} + v\frac{\partial p}{\partial z} = \frac{\gamma p}{\varrho}\left(\frac{\partial \varrho}{\partial t} + v\frac{\partial \varrho}{\partial z}\right) \quad . \tag{9.5.4}$$

These describe the nonlinear behaviour of vertical motions $v(z,t)$ in a *slender* tube, with an axial field $B(z,t)$, confined by an exterior gas pressure p_e. Equation 9.5.1 is obtained by eliminating (div **v**) between the equations of continuity and induction.

When flows and time-variations are absent, the gas is in hydrostatic equilibrium (denoted by subscript 0), with pressure balance maintained across the sides of the tube, such that

$$\frac{dp_0}{dz} = -\varrho_0 g, \quad \frac{dp_e}{dz} = -\varrho_e g, \quad p_0 + \frac{B_0^2}{2\mu} = p_e \ , \tag{9.5.5}$$

where p_0, ϱ_0, p_e, and B_0 are all functions of z.

For *linear* perturbations about this basic state, Webb and Roberts (1978) have supposed that

$$v(z,t) = \hat{v}(z)e^{i\omega t} \ .$$

If the exterior pressure perturbation is neglected, they show that the velocity amplitude $v(z)$ satisfies

$$\hat{v}'' + \left(\frac{c_T'^2}{c_T^2} - \frac{B_0'}{B_0} + \frac{\varrho_0'}{\varrho_0}\right)\hat{v}' + \left[\frac{\omega^2 - N_0^2}{c_T^2} - \left(\frac{B_0'}{B_0} + \frac{g}{c_0^2}\right)' - \left(\frac{B_0'}{B_0} + \frac{g}{c_0^2}\right)\left(\frac{c_T'^2}{c_T} - \frac{N_0^2}{g}\right)\right]\hat{v} = 0 \ , \tag{9.5.6}$$

where the primes denote differentiation with respect to z and

$$c_0(z) = \left(\frac{\gamma p_0}{\varrho_0}\right)^{1/2}, \ v_A(z) = \frac{B_0}{(\mu\varrho_0)^{1/2}} \ , \ c_T(z) = \frac{c_0 v_A}{(c_0^2 + v_A^2)^{1/2}}$$

are the sound, Alfven, and *tube* speeds, respectively, with

$$N_0^2 = -g\left(\frac{\varrho_0'}{\varrho_0} + \frac{g}{c_0^2}\right)$$

as the square of the Brunt-Väisälä frequency.

If it is assumed that external and internal temperatures are equal and γ is constant, Eq. 9.5.6 simplifies to

$$\hat{v}'' - \frac{1}{2\Lambda_0}\hat{v}' + \left(\frac{\omega^2 - N_0^2}{c_T^2} + (1 - \tfrac{1}{2}\gamma)\frac{N_0^2}{c_0^2}\right)\hat{v} = 0 \ , \tag{9.5.7}$$

where

$$\Lambda_0 = p_0/(\varrho_0 g)$$

is the temperature scale-height.

When either Eq. 9.5.6 or Eq. 9.5.7 is supplemented by two boundary conditions, we have an eigenvalue problem for the frequency ω, and it is of interest to determine what conditions give stability ($\omega^2 > 0$) and instability ($\omega^2 < 0$). The boundary conditions adopted here are that v vanish at two levels ($z = 0$ and $z = -d$), and the complications arise through the presence of nonconstant coefficients in the equations.

To illustrate the possibility of instability, consider a *local approximation* to Eq. 9.5.7 in which the coefficients are regarded as roughly constant. The solution is then (see Webb and Roberts, 1978)

$$\hat{v} \sim e^{z/(4\Lambda_0)} \sin n\pi z/d \ , \tag{9.5.8}$$

where n is an integer, and the frequency is given by

$$\omega^2 = \left(\frac{\gamma}{2c_0^2} + \frac{1}{v_A^2}\right) c_T^2 N_0^2 + \left(\frac{n^2\pi^2}{d^2} + \frac{1}{16\Lambda_0^2}\right) c_T^2 \ . \tag{9.5.9}$$

This dispersion relation shows that the plasma is stable for $N_0^2 > 0$. But, if the temperature stratification is superadiabatic, so that $N_0^2 < 0$, unstable modes may arise when the depth d is not too small; the time scale for growth is approximately

$$\tau \gtrsim \frac{(c_0/c_T)}{(\tfrac{1}{2}\gamma + c_0^2/v_A^2)^{1/2}|N_0|} \approx \frac{1}{g}\left|\frac{\Lambda_0}{\Lambda_0'}\right|^{1/2} \ . \tag{9.5.10}$$

For conditions typical of an intense tube ($c_0 = v_A$, $\gamma = 1.2$, $\Lambda_0 = 150$ km, and $\Lambda_0' = -1/4$), we find a time scale of 100 s for this convective instability.

From Eq. 9.5.9, the field strength that gives neutral stability ($\omega^2 = 0$) may be determined. With the above parameter values,

$$\beta_0 = 2\mu p_0/B_0^2$$

takes the critical value

$$\beta_c = \tfrac{1}{2}$$

for a neutral mode. The material is stable if the field strength is large enough (of the order of a kilogauss) that

$$\beta_0 < \beta_c \ .$$

The above local analysis serves only as a rough guide, because it ignores the detailed form of the scale-height $\Lambda_0(z)$. However, a full solution to

Eq. 9.5.6 with β not constant and T_0 not equal to T_e shows qualitatively the same features. For example, Spruit and Zweibel (1979) find $\beta_c = 1.8$. Thus, we conclude that *field strengths of* 1–2 kG *are required to make an intense tube convectively stable*, which is precisely in agreement with the observed values. (However, the effect of horizontal fluting instabilities remains to be analysed; perhaps they cause the tubes to disintegrate slowly over much longer time scales.)

Table 9.2 (from Webb and Roberts, 1978) shows how the critical field strength (and growth rate) varies with depth *d*. One can see that normally kilogauss field strengths are required for stability. However, Livingston *et al.* (1979) report that outside active regions the magnetic field of intense tubes may sometimes be lower than 1 kG. This may be explained (Roberts, private correspondence) if the depth *d* of such tubes is lower than normal. Thus, for example, if the depth is only 3 Mm (and $\Lambda_0' = -0.2$) the table shows that a tube is stable when its field exceeds 770 G. However, it is unclear, physically, why the tube depths should vary (Spruit, private correspondence).

Table 9.2a

The Critical Field Strength (in gauss) at $z = 0$ Necessary for Convective Stability in a Flux Tube of Given Depth d and Scale-Height Gradient Λ_0', for $\gamma = 1.2$ and $p_e(0) = 1.3 \times 10^5$ dyn cm^{-2}

	Depth *d*(km)				
Λ_0'	10^3	2×10^3	3×10^3	1.5×10^4	10^5
-0.15	stable	stable	stable	stable	stable
-0.2	490	670	770	1080	1280
-0.25	760	1040	1200	1750	unstable
-0.3	940	1290	1500	unstable	unstable

Table 9.2b

The Growth Rate (in s^{-1}), $i\omega_1$, of the Most Unstable Mode for Various Depths d and Scale-Height Gradients Λ_0' for $c_0 = v_A$, $\gamma = 1.2$, and $\Lambda_0(0) = 152$ km

	Depth *d*(km)				
Λ_0'	10^3	2×10^3	3×10^3	1.5×10^4	10^5
-0.15	stable	stable	stable	stable	stable
-0.2	stable	stable	stable	stable	1.2×10^{-3}
-0.25	stable	stable	3.2×10^{-3}	4.8×10^{-3}	4.8×10^{-3}
-0.3	stable	5.3×10^{-3}	6.7×10^{-3}	7.2×10^{-3}	7.2×10^{-3}

Solutions to the nonlinear set of equations (9.5.1–9.5.4) have not yet been investigated for the purpose of pursuing the nonlinear development of the instability; nor has the effect of an ambient convective cell been studied. However, Spruit (1979) has derived the possible equilibrium solutions to Eq. 9.5.5 that may be derived from the original equilibrium by an adiabatic displacement. Assuming equal interior and exterior temperatures initially, he shows that the vertical displacement $\xi(z_0)$ from a reference level z_0 is governed by an equation of the form

$$\xi'' = F(z_0, \xi, \xi') , \tag{9.5.11}$$

where the function F is nonlinear. This equation is solved numerically for the possible equilibria, subject to the boundary conditions that ξ vanish at two levels. It possesses nontrivial solutions only for $\beta_0 > 2.0$. In the collapsed state, the tube is cooler than the surrounding convection zone at low levels but hotter at high levels (above optical depth $\tau_{5000} = 1$). The field strength in the collapsed tube is enhanced principally in the first 1000 km below $\tau_{5000} = 1$.

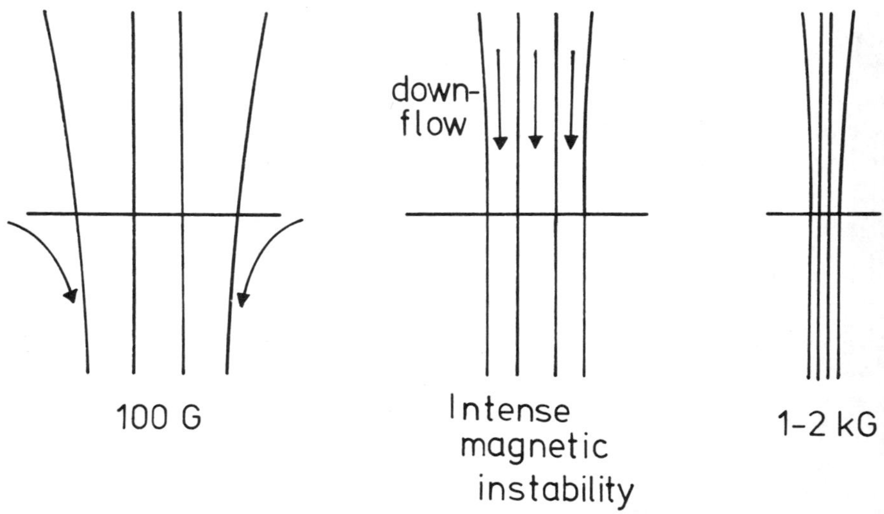

Fig. 9.14. The formation of kilogauss flux tubes by the intense magnetic field instability. Convective intensification is followed by instability, during which downflows are present, and the final state is convectively stable.

A general picture for the way the sun may assemble intense tubes from a diffuse magnetic field has been put forward by Parker (1978) and Spruit (1979), and mathematical aspects of it have been developed by Webb and Roberts (1978), Spruit (1979), and Spruit and Zweibel (1979). It is sketched in Figure 9.14 and may be described as follows. A diffuse vertical field is first swept into the downdrafts at the corners and boundaries of supergranules. [This process was described by Leighton (1963) and Parker (1963), but it cannot produce kilogauss fields; only equipartition magnetic pressures are likely, and these are at least an order of magnitude smaller than required values (Parker, 1976, 1977).] A flux tube of moderate field strength (say, several hundred gauss) extends across part of the superadiabatic convection zone (Parker, 1978) and is then subject to the *intense magnetic field instability* (elsewhere called the superadiabatic effect, or *convective instability*). On linear theory, this has two possible consequences (Spruit and Zweibel, 1979): if a downflow occurs, the tube cools, and its field strength increases until the field is so strong that it suppresses the motion; if an upflow takes place, the field weakens and disperses. Thus, a *concentration of the field is associated with a downdraft*. Once the field strength reaches a kilogauss or so, the instability switches off, and a *convectively stable intense tube* has been formed. Subsequently, it may be shuffled around by the flow or it may gradually disperse again due to, for instance, surface interchange instabilities or turbulent diffusion.

It should be noted that the above process is *transient* and thus should lead to *transient downflows* rather than steady, persistent flows; although it is possible that the observed downflows are in fact a spatial average over many intense tubes that are in the process of forming and evolving. The occurrence of the convective instability in intense tubes should have some effect on the atmosphere above; an attempt to model this effect is described in the next section.

9.5.2 Steady Flows

At photospheric levels, the steady downdrafts observed in the intense field boundaries of supergranules have flow speeds that vary between 2.5 km s^{-1} (Harvey and Hall, 1975) and 0.5 km s^{-1} (Giovanelli and Brown, 1977; Stenflo, 1973). The correlation between downflow and flux concentration appears in quiet region, network, and active region plages (Stenflo, 1976). Steady downflows of a few km s^{-1} are also found in the transition region over the network (Lites *et al.*, 1976; Doschek *et al.*, 1976).

Recently, Giovanelli and Slaughter (1978) have presented observations of the height variation of photospheric steady downflow in intense magnetic fields. The downward velocity increases rapidly with depth from 0 ± 0.1 km s^{-1} in Mg I 5183 Å (formed in the *nonmagnetic* atmosphere at about 840 km above $\tau_{5000} = 1$) to 0.6 km s^{-1} in C I (at $\sim$160 km) and 1.6 km s^{-1} at 0 km. By comparison, Stenflo (1973) finds an average downflow of less than 1 km s^{-1} and suggests that a substantial fraction of the mass downflux is occurring in the region just outside a magnetic element.

Theoretical models for steady flow along a flux tube have been set up by several authors. Parker (1977) argues that field concentration may be a result of such flow. Ribes and Unno (1976) analyse steady flow in a two-dimensional magnetic configuration typical of the chromosphere above a supergranule boundary; a later model (Unno and Ribes, 1979) includes compressive effects and gives a temperature excess similar to that observed in faculae, but the speed is rather too large.

Webb and Roberts (1978) consider steady downflow $v(z)$ in a flux tube of area $A(z)$ at photospheric levels. The flow satisfies the equations of continuity, flux, and motion, namely,

$$\varrho v A = \Phi \,, \tag{9.5.12}$$

$$BA = \Psi \,, \tag{9.5.13}$$

$$\varrho v v' = -p' - \varrho g \,, \tag{9.5.14}$$

where a prime denotes a z-derivative, Φ is the mass flux, and Ψ the magnetic flux; ϱ, v, A, B, p, and T are all functions of height z. [The equation of motion (9.5.14) along the field contains no term from the magnetic field, because the $\mathbf{j} \times \mathbf{B}$ force is directed across the field.] The gas law is written as

$$p = \varrho g \Lambda \qquad (9.5.15)$$

in terms of the temperature scale-height $\Lambda = RT/\mu g$. Combining Eqs. 9.5.12–9.5.15 gives

$$\left(1 - \frac{v^2}{g\Lambda}\right) \frac{v'}{v} = \frac{\Lambda'}{\Lambda} + \frac{1}{\Lambda} + \frac{B'}{B} \equiv \frac{\Phi'}{\Phi} \; , \qquad (9.5.16)$$

where

$$\Phi = B\Lambda \exp \left(\int \Lambda^{-1} \, dz \right) \; .$$

Equation 9.5.16 determines $v(z)$ for given temperature and magnetic field profiles. For a subsonic flow ($v^2 < g\Lambda$), the flow speed can only *decrease* with height if Φ decreases fast enough; for example, with an isothermal interior, the magnetic field would need to decrease faster than $\exp(-z/\Lambda_0)$ to make v decrease.

Equation 9.5.16 is supplemented by an equation of lateral pressure balance, namely,

$$p + (B^2/2\mu) = p_e \; , \qquad (9.5.17)$$

and so reduces to

$$\left(1 - \frac{v^2}{c_R^2}\right) \frac{v'}{v} = \frac{\Lambda'}{\Lambda} + \gamma g \left(\frac{1}{c_0^2} - \frac{1}{2c_e^2}\right) + \frac{g}{v_A^2} \left(1 - \frac{c_0^2}{c_e^2}\right) \; , \qquad (9.5.18)$$

where

$$c_R^2 = \frac{g\Lambda v_A^2}{g\Lambda + v_A^2} \; .$$

Webb and Roberts specify the flow $v(z)$ from the observationally determined profile of Giovanelli and Slaughter and deduce profiles for the interior temperature and magnetic field (Figure 9.15). One clear feature is the presence of an excess temperature (about 1000 K) in the first 300 km, and another is the very rapid fall in field strength to only 25 G at a height of 450 km (compared with 250 G in a static atmosphere). These results are consistent with the strong heating observed above intense fields (Stenflo, 1976); but, although a rapid fanning of flux tubes *is* observed (Frazier and Stenflo, 1972), little is known about the *exact* height variation of the field.

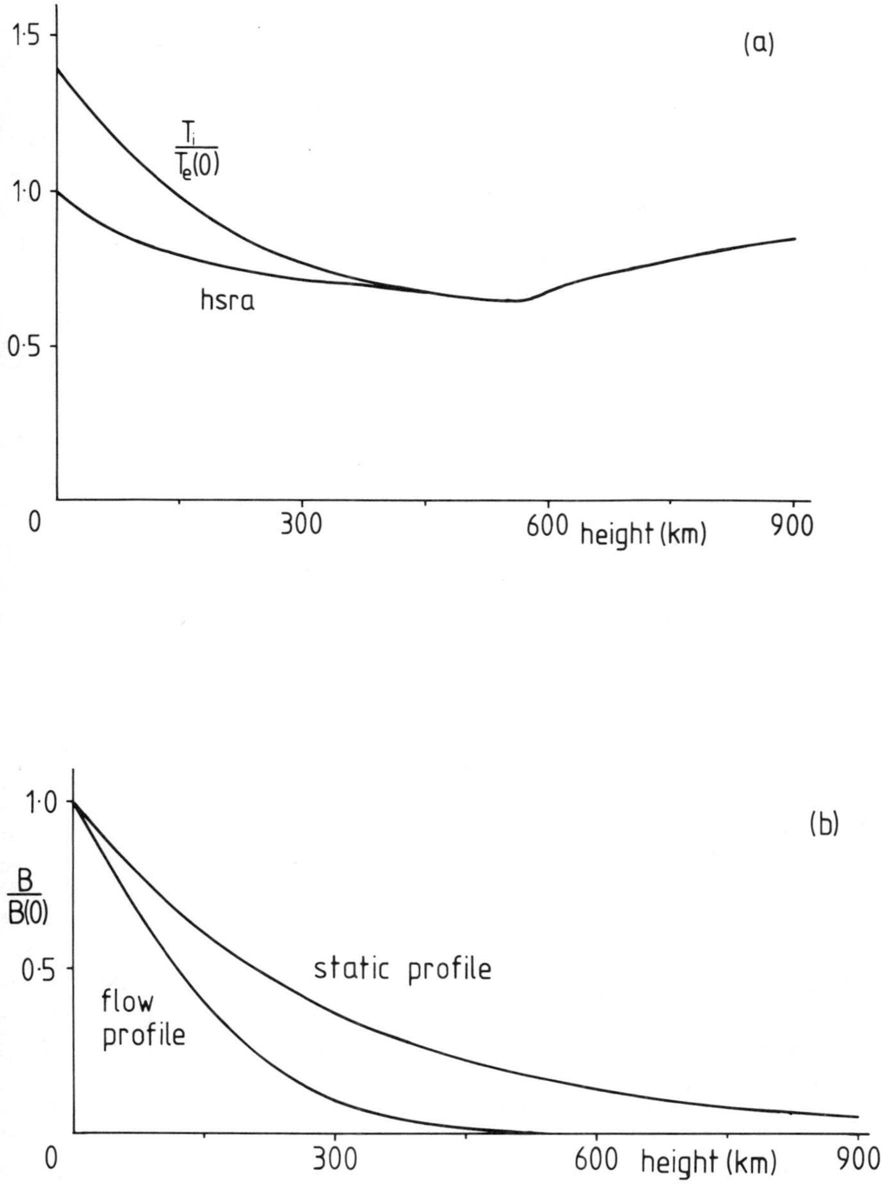

Fig. 9.15. Steady downflow in an intense flux tube: (a) *the temperature inside the tube;* (b) *the magnetic field strength;* (c) *a typical steady flow profile from the chromosphere up to the corona* (next page). *(From Webb and Roberts, 1980.)*

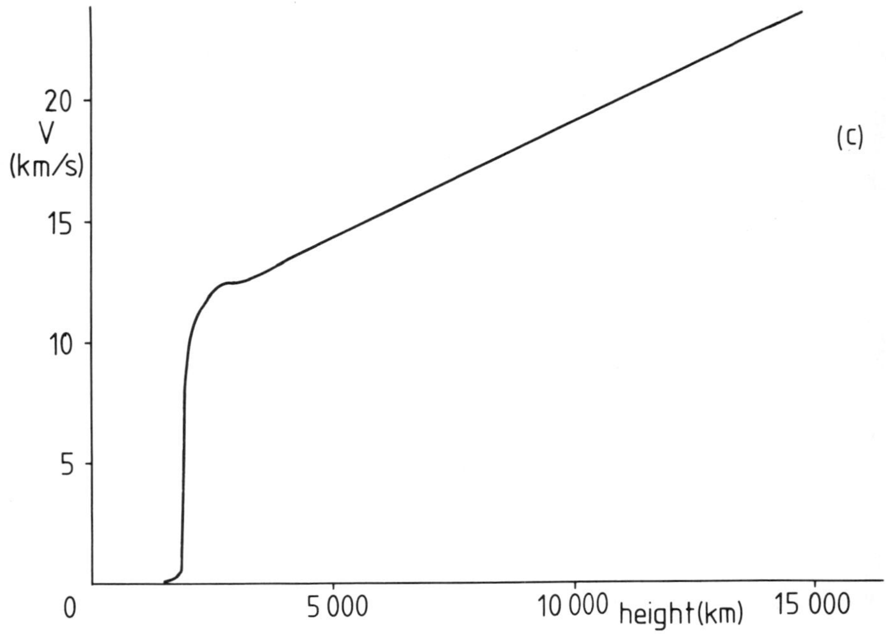

The conclusion from the analysis is that the *observed steady flow profile may be associated with an excess temperature and rapid decrease in field strength* over the first few hundred kilometers in altitude. Alternatives have been suggested, such as transverse inflow of neutral gas (Giovanelli, 1977), overstable oscillation or nonsteady flows (Spruit, 1979), or a finite-amplitude effect of convective instability (Unno and Ando, 1979). None of these possibilities has, however, been adequately explored and further investigation (both observational and theoretical) is necessary before a clear understanding of photospheric downflow can emerge.

Webb and Roberts also consider the possible effect on the corona of a small, steady photospheric flow. The observations from OSO-8 of Lites *et al.* (1976) indicate persistent downflows above the network, typical flow speeds being 2 km s^{-1} for the chromospheric line of Si II and 5 km s^{-1} for the transition zone line of Si IV. Webb and Roberts find from their model that a flow of only 0.1 km s^{-1} at a height of 1500 km can lead to speeds of 20–80 km s^{-1} in the corona, the value depending on the magnetic geometry (Figure 9.15c). For example, a flux tube that expands from a radius of 100 km at the photosphere to 5000 km in the corona gives a coronal speed of 26 km s^{-1}. However, it may well be that the photospheric flow is not steady at all (Spruit, 1979).

9.5.3 Spicule Generation in Intense Fields

The difficulty of explaining how the sun ejects matter in the form of spicules through the chromosphere and into the corona is a long-standing one, which has been reviewed recently by Athay (1976). (See also Beckers, 1972; Schmidt, 1974; Michard, 1974; Jefferies, 1974.) Moreover, the recent discovery that magnetic field strengths may be several kilogauss at supergranule boundaries needs to be incorporated in models of the origin of spicules. Indeed, Parker (1974) and Roberts (1979) have argued that spicules are closely related to intense flux tubes. The two are roughly comparable in number (10^5 spicules, 4×10^4 flux tubes), and both are concentrated in the network above supergranule boundaries.

Parker and Roberts have suggested that *spicules are the result of granular buffeting* on the sides of intense tubes. Roberts (1979) shows that such buffeting can drive a large-amplitude flow along the tube and, the narrower the tube, the larger the upflow that is generated. This is a *resonant effect*; for an isothermal atmosphere, the resonance occurs when the phase speed of the external driving oscillation approaches the natural speed c_T for longitudinal wave propagation in the tube. The effect of a single horizontal buffet is to increase the pressure inside the tube and force plasma upwards. The lifetimes of spicules (8–15 min) and granules (5–10 min) are comparable, but nonlinear effects such as the back reaction of the tube to compression would mean that the efficiency of the resonance is much less than 100%. Indeed, from the diameters of spicules (1000 km) and intense tubes (100–300 km), their densities (10^{-13} g/cm^3 and 10^{-7} g/cm^3), and the length of a spicule (10^4 km) by comparison with the depth of a granule (10^3 km), one may compare the masses and deduce that only a small fraction of the intense tube material needs to be ejected over a granular depth.

The resonance may be illustrated by considering the linear equation for isentropic wave propagation in a uniform magnetic field $B_0\hat{\mathbf{z}}$ threading a uniform atmosphere:

$$\frac{\partial^2}{\partial t^2}\left(\frac{\partial^2}{\partial t^2} - (c_0^2 + v_A^2)\nabla^2\right) \operatorname{div} \mathbf{v} + c_0^2 v_A^2 \frac{\partial^2}{\partial z^2} \nabla^2 \operatorname{div} \mathbf{v} = 0 \ , \qquad (9.5.19)$$

where ∇^2 is the Laplacian operator. For two-dimensional motions,

$$\mathbf{v} = (v_x(x), 0, v_z(x))e^{i(\omega t + kz)} \ ,$$

with no y-dependence. The transverse velocity amplitude $v_x(x)$ satisfies

$$\frac{d^2 v_x}{dx^2} - m_0^2 v_x^2 = 0 \ , \qquad (9.5.20)$$

where

$$m_0^2 = \frac{(k^2 c_0^2 - \omega^2)(k^2 v_A^2 - \omega^2)}{(k^2 c_T^2 - \omega^2)(c_0^2 + v_A^2)}$$

In an unbounded medium, Eq. 9.5.19 admits the usual magnetoacoustic wave solutions. For a column of magnetic field confined to $|x| < x_0$, Eq. 9.5.20 has (antisymmetric) solution

$$v_x = A_0 \sinh m_0 x, \; |x| < x_0 \; . \tag{9.5.21}$$

Suppose that an external oscillating pressure force

$$\delta p_e(z,t) = \tfrac{1}{2}\, \varrho_e v_e^2 \sin \omega t \cos kz \tag{9.5.22}$$

exists on the boundary of the tube; this may represent granular buffeting of a turbulent flow v_e with density ϱ_e. The buffeting will generate motions inside the tube. The constant A_0 is determined by constancy of total pressure (gas plus magnetic) and the longitudinal velocity v_z follows from the wave equation. On the axis ($x = 0$) of the tube

$$v_z(0) = \tfrac{1}{2}\, (\varrho_e/\varrho_0)(v_e/v_A)^2 c_T f(k) \cos \omega t \sin kz \;, \tag{9.5.23}$$

where

$$f(k) = \frac{\omega k c_T}{(k^2 c_T^2 - \omega^2) \cosh m_0 x_0} \; . \tag{9.5.24}$$

In the region $\omega < kc_T$, $f(k)$ has a maximum that grows as the width of the tube decreases; in a slender column, $v_z(0)$ may be large if $\omega \approx kc_T$.

The above analysis suggests that a spicule may be photospheric material that is driven up an intense tube by a resonant effect. As the material propagates upwards, it will presumably remain cooler than its surroundings because of the thermal insulation provided by the magnetic field; it will also expand somewhat along the field, but, guided by the field, it is likely to remain denser than the ambient chromosphere and corona.

However, three objections to the idea of granular buffeting have been raised by Spruit (private correspondence): the response of a flux tube to the wide spectrum of pressure fluctuations present in the photosphere would be far smaller than to a single frequency close to resonance; the inclusion of stratification in the analysis means that the resonance can operate in the low photosphere but not in the upper photosphere, where the granular frequencies lie below a cut-off and thus force the response to decay away; and the external pressure fluctuations cannot be acoustic in nature, since that would make the resonance take place only when $c_T > c_e$, which implies that $T_0 > T_e$, in contradiction with the photospheric observations. Evidently, much remains to be done in analyzing the effect of the intense field on the spicular process. Perhaps the spicule is driven in an intense tube by a nongranular resonance with the much larger scale magnetic disturbances that exist in the supergranule cell higher in the atmosphere. This would provide an accelerating force acting on the whole length of the spicule, producing a small pressure gradient along it (see Section 4.3.2).

9.6 MHD INSTABILITY OF LOOPS

Magnetohydrodynamic instabilities may occasionally produce dramatic events on the sun, such as erupting prominences and large solar flares. But they may also be occurring continually on smaller scales and produce a *general heating* of an active region. Because magnetic loops appear to form the building blocks of active regions, it is natural to consider the stability of such loops as isolated entities before investigating their interaction. Furthermore, because loops are observed to possess lengths that are at least ten times their widths, it is reasonable to model them by cylindrical structures.

Kruskal *et al.* (1958) and Shafranov (1957) considered the stability of a *cylindrical magnetic flux tube* with magnetic field components $(0, B_\theta(r), B_z(r))$ in terms of cylindrical polar coordinates r, θ, z. They found that an *infinitely long* tube is unstable to a *helical kink* perturbation

$$\underset{\sim}{\xi} = \underset{\sim}{\xi}(r)e^{i(\theta + kz)} , \qquad (9.6.1)$$

when

$$B_\theta/r + kB_z \geq 0 .$$

At equality, the wavenumber vector $(0, r^{-1}, k)$ is perpendicular to the equilibrium field, and the crests and troughs of the perturbation follow the field as it spirals around the axis.

A useful parameter in considering a tube of length $2L$ is

$$\Phi(r) \equiv \frac{2LB_\theta}{rB_z} ,$$

which represents the amount by which a field line is twisted about the tube axis in going from one end of the tube to the other. For such a finite tube, the Kruskal-Shafranov result states that, *for any twist* Φ, the tube is *unstable to helical kink perturbations* with wavenumbers

$$-k < \Phi/2L . \qquad (9.6.2)$$

The above result had several limitations, which demanded that a more detailed analysis be carried out. It applies to a simplified flux tube in which all the shear is concentrated in a thin sheet at a certain radius; whereas, in practice, for a coronal loop, one would expect Φ to vary continuously with radius. Furthermore, the following stabilizing effects were neglected:

(i) Line-tying. The photosphere is about a million times denser than the corona, which means that the footpoints of magnetic field lines are anchored by the dense plasma that is frozen to them. Any perturbations that begin in the corona would therefore be expected to die away as they propagate downwards. To analyze this effect in detail, one needs to include nonuniformities in the basic state along the loop, but, so far, line-tying has been

modeled simply by imposing the condition that the perturbation vanish at the ends of the loop. This should not be a bad approximation, since the photospheric density increase occurs over a relatively small height. The perturbation, Eq. 9.6.1, does not vanish for all Φ at any location z along the loop and so cannot model line-tying. However, Raadu (1972) has considered the effect of line-tying for a *force-free loop* by choosing a perturbation of the form

$$\underset{\sim}{\xi} = \cos\frac{\pi z}{2L}\left(\xi_r(r), -\frac{B_z}{B}\,\xi_0(r), \frac{B_\theta}{B}\,\xi_0(r)\right) e^{i(\theta + kz)} . \tag{9.6.3}$$

This vanishes at the ends $z = \pm L$ of the loop and is perpendicular to the equilibrium field. Following Bernstein *et al.* (1958) and Newcomb (1960), Raadu minimizes the energy of the system to determine the functions $\xi_0(r)$ and $\xi_r(r)$, and to find whether the perturbation makes the energy increase or decrease for given B_θ, B_z (and therefore Φ), and k. He finds that a loop becomes unstable to the kink perturbations (Eq. 9.6.3) when its twist exceeds a critical value, which is not determined precisely but lies somewhere in the range of 2.0π to 3.9π.

(ii) Pressure gradients. Suydam (1958) showed that a necessary condition for local stability in a cylindrical tube is that

$$\frac{dp}{dr} + \frac{rB_z^2}{8\mu\Phi^2}\left(\frac{d\Phi}{dr}\right)^2 > 0 . \tag{9.6.4}$$

If this is not satisfied, the tube is definitely unstable (in the absence of line-tying!); but if it is satisfied, it may be stable or unstable. Thus, a negative pressure gradient has a destabilizing influence, in the sense that, if dp/dr is sufficiently negative, Eq. 9.6.4 is violated. Since *some* loops are observed to possess strong radial pressure gradients (Jordan, 1975; Foukal, 1975), loop equilibria that are not force-free but magnetostatic, with

$$\nabla p = \mathbf{j} \times \mathbf{B} , \tag{9.6.5}$$

have been considered by Giachetti *et al.* (1977). They assumed that the magnetic field components have the form

$$\begin{aligned} B_\theta &= B_0(r)\, J_1(\alpha r) , \\ B_z &= B_0(r)\, J_0(\alpha r) , \end{aligned} \tag{9.6.6}$$

and then, after choosing a pressue profile that matches observations, they solved Eq. 9.6.5 to give $B_0(r)$. An undesirable feature of these solutions is that the field components reverse sign at the zeros of the Bessel functions J_0 and J_1.

(iii) Loop curvature. For a loop of major radius R_0, the analogous form of *Suydam's criterion* is *Mercier's criterion*. It states that a necessary condition for local stability is

$$\frac{dp}{dr}\left[1 - \left(\frac{2L}{R_0\Phi}\right)^2\right] + \frac{rB_z^2}{8\mu\Phi^2}\left(\frac{d\Phi}{dr}\right)^2 > 0 .$$

The curvature is thus a stabilizing influence, in the sense that a loop with dp/dr negative is stable if

$$R_0 < \frac{2L}{\Phi}$$

for all r.

In this section, we describe work by Hood and Priest (1980b) and Chiuderi *et al.* (1980) that has been undertaken as part of the Workshop activity on the magnetohydrodynamic stability of loops. Hood and Priest have demonstrated that probably the *main reason for the remarkable global stability of coronal loops is photospheric line-tying.* Loops become kink unstable when the twist Φ exceeds a critical value, which depends on the ratios of loop length to width and plasma pressure to magnetic pressure and also on the detailed transverse magnetic structure. A repeat of Raadu's analysis for a force-free field of uniform twist gives an upper limit on the critical twist of 3.3π, and an extension to more general magnetostatic equilibria with pressure gradients shows that the critical twist lies typically between 2π and 6π. When large loops become unstable, they may produce *solar flares*, but the less dramatic continual release of energy in smaller or weaker loops may provide *coronal heating*. Chiuderi *et al.* consider the effect of an external potential field on the stability properties of a cylindrical force-free flux tube. They find that (without line-tying) the flux tube is just as unstable as the constant-α case. Also, at wavelengths that are stable on this ideal MHD theory, the tube can be unstable to the tearing mode on time scales as small as 3000 s, even using a classical resistivity.

9.6.1 *Effect of Line-Tying and Pressure Gradients on Kink Instability*

For a cylindrically symmetric flux tube in equilibrium under a balance between a pressure gradient and a Lorentz force, the pressure and magnetic field components in the basic state satisfy

$$\frac{dp}{dr} = - \frac{d}{dr}\left(\frac{B_z^2 + B_\theta^2}{2\mu}\right) - \frac{B_\theta^2}{\mu r} . \tag{9.6.7}$$

Furthermore, a displacement $\underset{\sim}{\xi}$ from equilibrium causes a change in the energy (to second order) of

$$\delta W = \int \{\mathbf{Q}\cdot\mathbf{Q}/\mu + \mathbf{j}\cdot(\underset{\sim}{\xi}\times\mathbf{Q}) + (\nabla\cdot\underset{\sim}{\xi})\underset{\sim}{\xi}\cdot\nabla p + (\nabla\cdot\underset{\sim}{\xi})^2\gamma p\}\, dV , \tag{9.6.8}$$

where

$$\mathbf{Q} = \nabla \times (\underset{\sim}{\xi} \times \mathbf{B})$$

is the first-order change in the magnetic field. The stability of the equilibrium then depends on the sign of δW.

Hood and Priest (1980b) adopt a perturbation of the form of Eq. 9.6.3, which is perpendicular to the static field and is modulated by the factor cos $(\pi z/2L)$, so that it vanishes at the ends of the loop. It is also helically kinked in nature, with a wavenumber k along the loop. Then a minimization of Eq. 9.6.8 with respect to ξ_0 reduces the energy integral to one of the form

$$\delta W = \int \{F(d\xi_r/dr)^2 + 2H\xi_r d\xi_r/dr + G\xi_r^2\}\, dr \ , \tag{9.6.9}$$

where F, G, and H are functions of r that depend on the functions $p(r)$, $B_\theta(r)$, $B_z(r)$, and the wavenumber k. The function ξ_r, which in turn minimizes this integral, is given by the Euler-Lagrange equation

$$\frac{d}{dr}\left(F\frac{d\xi_r}{dr}\right) = \left(G - \frac{dH}{dr}\right)\xi_r \ , \tag{9.6.10}$$

subject to the boundary conditions

$$\xi_r = 1 \text{ and } \frac{d\xi_r}{dr} = 0 \text{ at } r = 0 \ .$$

The solution may then be inserted into Eq. 9.6.9 to determine whether δW is positive or negative and thus whether the equilibrium is stable or unstable to the perturbation. However, Newcomb (1960) established that a necessary and sufficient condition for stability is that ξ_r remain positive for all r, and this is the test that is employed in practice.

Consider first the *force-free* field with *uniform twist*, having

$$B_z = \frac{B_0}{1 + r^2/b^2} \ , \quad B_\theta = \frac{B_0 r/b}{1 + r^2/b^2} \ ,$$

such that the twist

$$\Phi \equiv \frac{2LB_\theta}{rB_z} = \frac{2L}{b}$$

in independent of r. B_z decreases monotonically from B_0 on the axis $r = 0$ to zero at infinity, while B_θ increases from zero to a maximum of 1/2 B_0 at $r = b$ and then decreases to zero at infinity slower than B_z. Using this field, Hood and Priest integrate Eq. 9.6.10 for a range of values of wavenumber k and twist Φ; the results are shown in Figure 9.16a. Below the *dashed* line,

$$-k < b^{-1} \ ,$$

or, since $\Phi = 2L/b$,

$$-k < \Phi/2L \ .$$

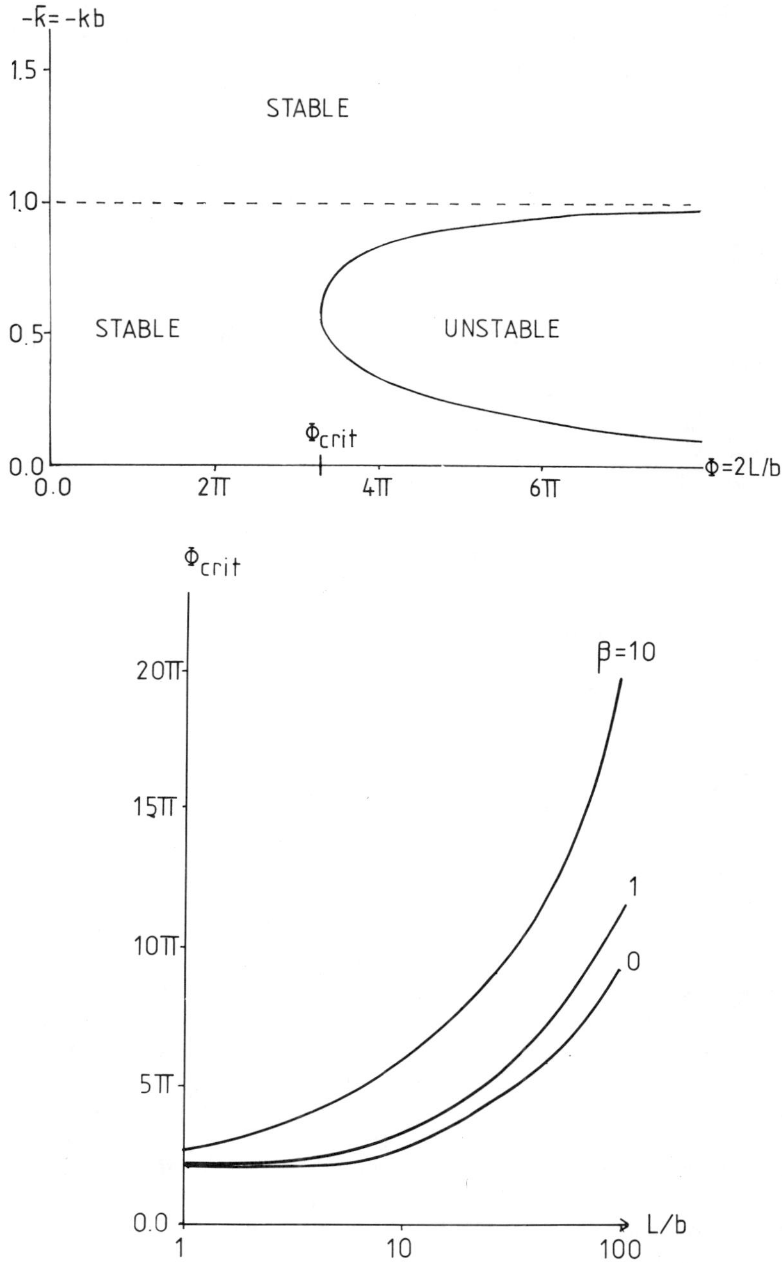

Fig. 9.16. Critical conditions for the onset of kink instability due to the perturbation (Eq. 9.6.3) in a coronal loop of length 2L. (a) *Stability diagram for an axial wavenumber* k *in a force-free field of uniform twist* Φ. (b) *The critical twist* Φ_{crit} *at* r = b *in a magnetostatic field as a function of the aspect ratio* L/b *for various values of* $\beta \equiv 2\mu p_c/B_0^2$ *where* p_c *is the ambient coronal pressure and* B_0 *the magnetic field strength on the axis. (From Hood and Priest, 1980b.)*

Thus, the Kruskal-Shafranov condition (Eq. 9.6.2) is recovered for instability. The extent to which the solid curve lies below the dashed line in the figure therefore illustrates the additional stabilizing effect of line-tying: for small amounts of twist, the loop is stable, but, when the twist exceeds

$$\Phi_{\text{crit}} \cong 3.3\pi \ ,$$

a range of wavenumbers appears over which the loop is kink unstable. Hood and Priest (1980b) stress that they have studied stability to just this particular perturbation (Eq. 9.6.3). However, they are also able to derive the sufficient condition for stability to the ($m = 1$) kink that Φ be less than 2π; in other words, if all kink perturbations are allowed, the instability onset will occur somewhere in the range

$$2\pi \leq \Phi \leq 3.3\pi \ .$$

Other equilibria are also considered. For example, if the axial field is

$$B_z = \frac{B_0}{1 + r^2/b^2}$$

and the twist Φ is uniform (but not necessarily equal to $2L/b$), we have

$$B_\theta = \frac{B_0 \Phi r}{2L(1 + r^2/b^2)}$$

and Eq. 9.6.6 determines the pressure as

$$p = p_c + \frac{B_0^2[(b\Phi/2L)^2 - 1]}{\mu(1 + r^2/b^2)^2} \ ,$$

where p_c is the ambient coronal pressure at infinity. The stability results indicate that for small wavenumbers the loop is more stable than the Kruskal-Shafranov limit (Eq. 9.6.2) predicts, whereas for large wavenumbers it is less stable. At a given small wavenumber, the stabilizing effect of line-tying dominates and makes the critical twist for instability bigger than the Kruskal-Shafranov value ($-2Lk$). On the other hand, for large wavenumbers, line-tying becomes relatively unimportant; however, according to the Suydam criterion, the pressure profile for $\Phi > 2L/b$ has a negative gradient and a destabilizing influence, which makes the loop unstable for smaller twists than predicted by Kruskal and Shafranov.

Another equilibrium of interest has a uniform axial component

$$B_z = B_0$$

together with a variable twist

$$\Phi(r) = \frac{\Phi(0)}{1 + r^2/b^2} \ ,$$

while the pressure is

$$p = p_c + \frac{B_0^2 b^2 \Phi(0)^2}{\mu(1 + r^2/b^2)^2 4L^2} \ .$$

Again, for small k the loop is more stable than the Kruskal-Shafranov limit, and for large k it is less stable. For each value of the aspect ratio L/b, there is *a critical twist* Φ_{crit} (at $r = b$, say) beyond which the loop becomes unstable at some wavenumber. As L/b increases, so does Φ_{crit} (Figure 9.16b), which may seem surprising at first. However, there are two factors to consider. The pressure profile is destabilizing and so, in order to obtain the same pressure profile, a loop that is twice as long would need to have a twist that is twice as large. By contrast, the (weaker) effect of line-tying is to make a loop that is twice as long possess a critical twist that is less than twice as large. Figure 9.16b shows the effect of $\beta \equiv 2\mu p_c/B_0^2$ on Φ_{crit}: as β is increased, the pressure rises in proportion; but its gradient remains the same, which increases the size of the positive-definite term in Eq. 9.6.8 and thus makes the equilibrium more stable. For an aspect ratio of $L/b = 10$, typical of active region loops, it can be seen that the smallest twist (at $r = b$) to give instability is 2.8π at $\beta = 0$ and about 3.3π at $\beta = 1.0$. For these magnetostatic (rather than force-free) equilibria, the values for Φ_{crit} are upper bounds, since only perturbations perpendicular to the magnetic field are considered and more general displacements may produce instability at lower thresholds.

9.6.2 *Stability of a Constant-α Force-Free Field Surrounded by an External Potential Field*

For the particular case of a force-free field, Chiuderi *et al.* (1980) have extended the previous work of Giachetti *et al.* (1977) by removing one of its main weaknesses, namely that the field possesses reversals as one moves out from the axis. The method is the same as that described above, in which an Euler-Lagrange equation (9.6.10) is solved for the radial perturbation ξ_r. However, a different basic field is adopted and the modulation of $\underset{\sim}{\xi}$ that models line-tying is absent, with the result that Eq. 9.6.10 exhibits regular singular points at the radii, where

$$\mathbf{k} \cdot \mathbf{B} = 0 \ ,$$

$$\text{or} \quad B_\theta/r + kB_z = 0 \ . \tag{9.6.11}$$

This complicates the analysis, and the singular points subdivide the radial interval $(0, \infty)$ into separate subintervals, whose stability can be studied independently.

Voslamber and Callebaut (1962) considered a constant-α force-free field without line-tying ($\alpha = -B_\theta^{-1}\, dB_z/dr$). If it extends to infinity in the radial direction, it is unstable for all wavenumbers $k \le 0.272\alpha$. But, if the field is restricted to a radial distance $r \le 3.176\alpha^{-1}$, it is stable; thus, in laboratory plasmas, stability is achieved by surrounding a plasma column by a perfectly conducting wall within this radius. Such a wall is of course not present on the sun, and Chiuderi *et al.* consider an alternative possible stabilizing influence to line-tying, namely the presence of a potential field beyond some radius. In their equilibrium field, $\alpha(r)$ takes a constant value α_0 for $r \leq r_0$ and then decreases to zero for $r > r_0$ over a distance δ. This has the mathematical advantage of reducing the number of singular points in the Euler-Lagrange equation and the physical advantage of removing many of the unwanted field reversals. Specifically,

$$\alpha \equiv -\frac{1}{B_\theta}\frac{dB_z}{dr} = \begin{cases} \alpha_0 , & r \leq r_0 \\ \tfrac{1}{2}[1 + \cos(\pi(r - r_0)/\delta)], & r_0 < r < r_0 + \delta \\ 0 , & r > r_0 + \delta . \end{cases} \tag{9.6.12}$$

Figure 9.17a shows how the range of unstable wavenumbers k (*shaded*) varies with r_0 in the particular case when $\alpha_0\delta = 2$. Regardless of the value of r_0, the flux tube is always unstable to some range of k. When $\alpha_0 r_0$ is small, there are no reversals in the field, and the unstable wavenumbers are negative, with at least $-0.5 < k/\alpha_0 < 0$; but, for large enough $\alpha_0 r_0$, B_z shows a reversal and a range of positive, unstable wavenumbers appears; at the same time, the range of negative, unstable k decreases and eventually disappears, only to reappear when a reversal in B_θ appears. The conclusions are therefore that these more realistic fields with fewer field reversals are *no more stable than the constant-α* case and that the most unstable region is that of low wavenumbers (for which line-tying is most effective, as discussed in the previous section).

Chiuderi *et al.* also consider the onset of *tearing-mode instability* in their structure. At each radius r_s, this is driven by finite resistivity effects localized in a narrow layer, with wavenumber given by

$$\mathbf{k} \cdot \mathbf{B}(r_s) = 0 ,$$

the same as Eq. 9.6.11. Furth *et al.* (1963) have shown that an estimate for the growth rate can be obtained from the ideal magnetohydrodynamic equations, with the neglect of resistivity making the ideal solution discontinuous at r_s. The growth time is

$$\tau = H \tau_R^{3/5} \tau_A^{2/5} ,$$

where

$$\tau_R = L^2/\eta$$

is the resistive decay time for an overall length scale L and magnetic diffusivity η; also,

$$\tau_A = L(\mu\varrho)^{1/2}/B$$

is the Alfven transit time. The constant of proportionality H is proportional to $(\Delta')^{-4/5}$, where Δ' is the jump in $\xi_r^{-1}\, d\xi_r/dr$ at $r = r_s$ and depends only on the

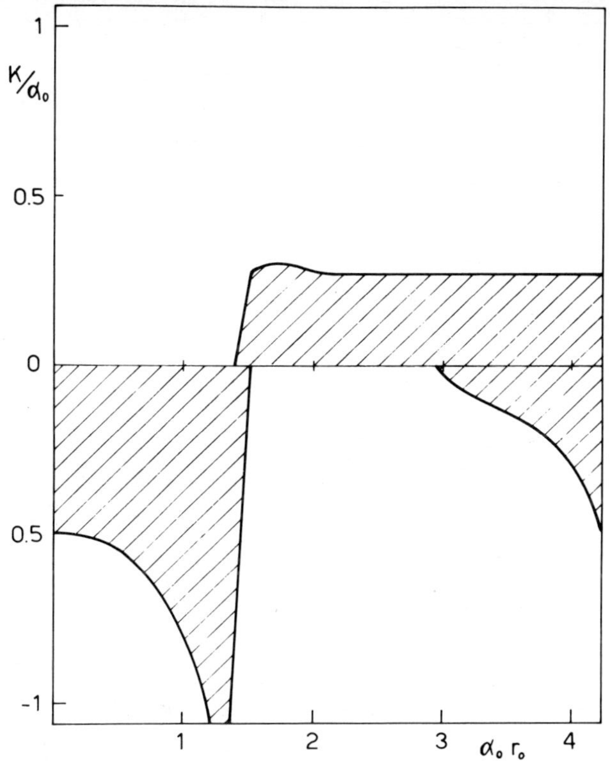

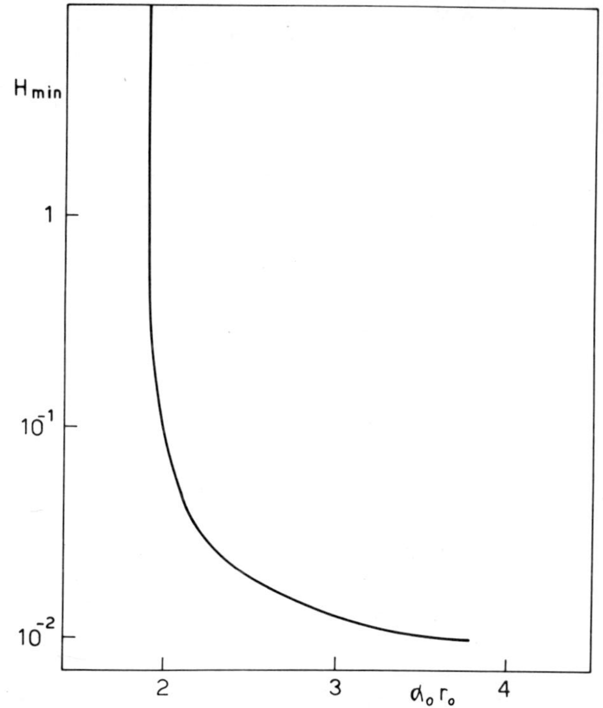

Fig. 9.17. The instability of a force-free flux tube of constant $\alpha\,(=\alpha_0)$ surrounded by a potential field at a radius $r_0 + 2/\alpha_0$. (a) The range of kink unstable wavenumber k (shaded) as a function of r_0. (b) The normalized growth time $H_{min} = \tau/(\tau_R^{3/5}\tau_A^{2/5})$ for the fastest growing tearing mode as a function of r_0, where τ_R and τ_A are the diffusion and Alfven travel times. (From Chiuderi et al., 1980.)

details of the equilibrium field. For typical coronal values

$$T = 1.5 \times 10^6 \mathrm{K},\ n = 3 \times 10^{14} m^{-3},\ B = 10^{-3}\ \text{tesla (10 gauss)},\ L = 10^5 \mathrm{m}\ ,$$

the tearing mode growth time is 3.3×10^5 s, using the classical diffusivity. Since this is much longer than τ_A, only values of k that are stable within the framework of ideal theory are considered. It is found that the smallest such wavenumber (just above the ideal stability boundary) is not necessarily resistively unstable, especially when the resistive layer is located in the region where the field is potential. Figure 9.17b gives the behavior of H_{min} (for the fastest growing model) as $\alpha_0 r_0$ varies. When $\alpha_0 r_0$ is small, H_{min} is of order unity or larger, since the lowest k's are resistively stable, and higher wavenumbers grow slowly; when $\alpha_0 r_0$ is larger than three, H_{min} is much smaller, and the growth time falls to a value of 3000 s, which is relevant for flares. The conclusion is then that flux tubes that are stable for ideal theory may be resistively unstable, with time-scales close to flaring times, even using classical values for the resistivity.

9.7 CONCLUSION

At one time, the solar atmosphere was believed to be plane-parallel, with the magnetic field playing an unimportant role. Then it was realized that the magnetic field dominates the behavior of many of the features in the atmosphere; still, the plasma was thought to be largely in a static state. Now, however, the picture of an active region is of a very dynamic environment, with continual motion of many different types.

It is clear that we are only just beginning to understand the wide variety of flows that are present. Observers have provided us with a fascinating glimpse of some of the motions above the solar surface. Many more observations are urgently needed to determine the conditions for the occurrence of each type of flow, the relationship between the different types, and their link with the magnetic configuration.

The theory for both steady and transient flows is only in its infancy, and this chapter provides a summary of some of this preliminary theory. It is not meant to be comprehensive, but represents the interests only of those in the Dynamics Group of the Workshop. The main results from the work are listed below:

(1) *Siphon flow* may be generated if a pressure difference is maintained between the footpoints. For small pressure differences, the flow is subsonic; but for moderate values, it may become supersonic at the summit and be slowed down by a standing shock wave. The presence of *cool cores* and the apparent appearance of only half a loop may be due to such flows.

(2) The evolution of active region magnetic loops will lead to continual *evaporation* and *draining* of the plasma contained in them. In particular, an increase in heating rate can produce evaporation and signifi-

cant mass motions. Also, a loop of fixed mass will relax to a static equilibrium over the conductive and radiative time-scale with superimposed oscillations at the sound travel frequency.

(3) Static models for thermally isolated loops are *thermally unstable*. Unless the atmosphere can provide some stabilizing mechanism coronal loops must be in a *dynamic state* of thermal activity.

(4) Kilogauss photospheric fields may be formed by an intense magnetic field instability, with an associated transient downflow. Photospheric downflows of only 1 km s^{-1} may produce flows at coronal levels of 20–80 km s^{-1}.

(5) A magnetohydrodynamic stability analysis of a loop suggests that the main reason for magnetic stability may be *line-tying* of loop footpoints in the dense photosphere. (Radial pressure gradients can also have a stabilizing effect.) When the twist in a loop is too large (for example, 3.3π for a uniform twist), the loop becomes *kink unstable*. Furthermore, the addition of a potential field beyond a certain radius does not make the loop more stable.

Acknowledgements

I wish to thank: F. Orrall for running the Workshop with such sensitivity; the theoretical members of the Dynamics Group (and others), namely, S. Antiochos, C. Chiuderi, I. Craig, S. Habbal, A. Hood, K. Krall, G. Noci, B. Roberts, R. Rosner, and H. Spruit, who have supplied me with work summaries, which I have closely followed; the other half of the Dynamics Group, namely, G. Athay, G. Brueckner, J. Cook, P. Lemaire, B. Lites, and also H. Zirin, who have fascinated and puzzled us with their observations and stimulated our interest in the subject.

REFERENCES

Antiochos, S.: 1979, *Astrophys. J.* **232,** L125.

Athay, R. G.: 1976, *The Solar Chromosphere and Corona: Quiet Sun*, D. Reidel, Dordrecht, Holland.

Basri, G. S., Linsky, J. L., Bartoe, J.-D. F., Brueckner, G., and Van Hoosier, M. E.: 1979, *Astrophys. J.* **230,** 924.

Beckers, J. M.: 1972, *Ann. Rev. Astron. Astrophys.* **10,** 73.

Bernstein, I. B., Frieman, E. A., Kruskal, M. D., and Kulsrud, R. M.: 1958, *Proc. Roy. Soc. (London)* **A244,** 17.

Bobrova, N. A. and Syrovatskii, S. I., 1979: *Solar Phys.* **61,** 379.

Brueckner, G. E.: 1976, *Bull. Am. Astron. Soc.* **8,** 345.

Bruner, E. C., Chipman, E. G., Lites, B. W., Rottman, G. J., and Shine, R. A.: 1976, *Astrophys. J.* **210,** L97.

Bruzek, A. and Durrant, C. J.: 1977, *Illustrated Glossary for Solar and Solar-Terrestrial Physics*, D. Reidel, Boston.

Cargill, P. J. and Priest, E. R.: 1980, *Solar Phys.* **65,** 251.

Cargill, P. J. and Priest, E. R.: 1981 (submitted).

Chiuderi, C., Einaudi, G., Ma, S., and Van Hoven, G.: 1980 (in preparation).

Craig, I. J. D.: 1980, in E. R. Priest (ed.), *Solar Flare Magnetohydrodynamics*, Gordon and Breach.
Cram, L. E. and Wilson, P. R.: 1975, *Solar Phys.* **41,** 313.
Davis, J. M., Gerassimenko, M., Krieger, A. S., and Vaiana, G. S.: 1975, *Solar Phys.* **45,** 393.
Defouw, R. J.: 1976, *Astrophys. J.* **209,** 266.
Doschek, G. A., Feldman, U., and Bohlin, J. D.: 1976, *Astrophys. J.* **205,** L177
Field, G. B.: 1965, *Astrophys. J.* **142,** 531.
Foukal, P. V.: 1975, *Solar Phys.* **43,** 327.
Foukal, P. V.: 1976, *Astrophys. J.* **210,** 575.
Foukal, P. V.: 1978, *Astrophys. J.* **223,** 1046.
Frazier, E. N. and Stenflo, J. O.: 1972, *Solar Phys.* **27,** 330.
Furth, H. P., Killeen, J., and Rosenbluth, M. N.: 1963, *Phys. Fluids* **6,** 459.
Galloway, D. J.: 1977, Ph.D. Thesis, Cambridge University.
Giachetti, R., Van Hoven, G., and Chiuderi, C.: 1977, *Solar Phys.* **55,** 371.
Giovanelli, R. G.: 1977, *Solar Phys.* **52,** 315.
Giovanelli, R. G. and Brown, N.: 1977, *Solar Phys.* **52,** 27.
Giovanelli, R. G., Livingston, W. C., and Harvey, J. W.: 1978, *Solar Phys.* **59,** 49.
Giovanelli, R. G. and Slaughter, C.: 1978, *Solar Phys.* **57,** 255.
Habbal, S. R. and Rosner, R.: 1979, *Astrophys. J.* **234,** 1113.
Harvey, J. W.: 1977, *Highlights of Astronomy* **4,** Part II, p. 223.
Harvey, J. W. and Hall, D.: 1975, *Bull. Am. Astron. Soc.* **7,** 459.
Heyvaerts, J., Priest, E. R., and Rust, D. M.: 1977, *Astrophys. J.* **216,** 123.
Hood, A. W. and Priest, E. R.: 1979, *Astron. Astrophys.* **77,** 233.
Hood, A. W. and Priest, E. R.: 1980a, *Astron. Astrophys.* (submitted).
Hood, A. W. and Priest, E. R.: 1980b, *Solar Phys.* **64,** 303.
Jefferies, J. T.: 1974, *IAU Colloq.* **56,** p. 71.
Jordan, C.: 1975, in S. R. Kane (ed.), 'Solar Gamma, X, and EUV Radiation', *IAU Symp.* **68,** p. 109.
Krall, K. R. and Antiochos, S. K.: 1980 (in preparation).
Kruskal, M. D., Johnson, J. L., Gottlieb, M. B., and Goldman, L. M.: 1958, *Phys. Fluids* **1,** 421.
Leighton, R. B.: 1963, *Ann. Rev. Astron. Astrophys.* **1,** 19.
Levine, R. H. and Withbroe, G. L.: 1977, *Solar Phys.* **51,** 83.
Lites, B. W., Bruner, E. C., Chipman, E. G., Shine, R. A., Rothman, G. J., White, O. R., and Athay, R. G.: 1976, *Astrophys. J.* **210,** L111.
Livingston, W. C., Harvey, J. W., and Heasley, J. N.: 1979 (in preparation).
Meyer, F. and Schmidt, H. U.: 1968, *Z. Angew. Math. Mech.* **48,** 218.
Michard, R.: 1974, *IAU Colloq.* **56,** p. 3.
Newcomb, W. A.: 1960, *Annals of Physics* **10,** 232.
Noci, G.: 1980, *Solar Phys.*
Parker, E. N.: 1953, *Astrophys. J.* **117,** 431.
Parker, E. N.: 1963, *Astrophys. J.* **138,** 552.
Parker, E. N.: 1972, *Astrophys. J.* **174,** 499.
Parker, E. N.: 1974, *Astrophys. J.* **189,** 563; **190,** 429.
Parker, E. N.: 1976, *Astrophys. J.* **204,** 259.
Parker, E. N.: 1977, *Ann. Rev. Astron. Astrophys.* **15,** 45.
Parker, E. N.: 1978, *Astrophys. J.* **221,** 368.
Pikel'ner, S. B.: 1971, *Solar Phys.* **17,** 44.
Pneuman, G. W. and Kopp, R. A.: 1977, *Astron. Astrophys.* **55,** 305.
Pottasch, S. R.: 1964, *Space Sci. Rev.* **3,** 816.
Priest, E. R.: 1978, *Solar Phys.* **58,** 57.
Priest, E. R. and Smith, E. A.: 1979, *Solar Phys.* **64,** 267.
Raadu, M. A.: 1972, *Solar Phys.* **22,** 425.
Ribes, E. and Unno, W.: 1976, *Astron. Astrophys.* **53,** 197.
Roberts, B.: 1979, *Solar Phys.* **61,** 23.
Roberts, B. and Webb, A. R.: 1978, *Solar Phys.* **56,** 5.

Roberts, B. and Webb, A. R.: 1979, *Solar Phys.* **64,** 77.
Rosner, R., Tucker, W. H., and Vaiana, G. S.: 1978, *Astrophys. J.* **220,** 643.
Schmidt, H. U.: 1974, *IAU Colloq.* **56,** p. 34.
Shafranov, V. D.: 1957, *J. Nuc. Energy* II **5,** 86.
Smith, E. A. and Priest, E. R.: 1977, *Solar Phys.* **53,** 25.
Spruit, H. C.: 1979, *Solar Phys.* **61,** 363.
Spruit, H. C. and Zweibel, E. G.: 1979, *Solar Phys.* **62,** 15.
Stenflo, J. O.: 1973, *Solar Phys.* **32,** 41.
Stenflo, J. O.: 1976, 'Energy Balance and Hydrodynamics of the Solar Chromosphere and Corona', *IAU Colloq.* **36,** 143.
Suydam, B. R.: 1958, *Int. Conf. on Peaceful Uses of Atomic Energy—2nd*, Geneva, p. 187.
Syrovatskii, S. I.: 1978, *Solar Phys.* **58,** 89.
Unno, W. and Ando, H.: 1979, *Geophys. Astrophys. Fluid Dynamics* **12,** 107.
Unno, W. and Ribes, E.: 1979, *Astron. Astrophys.* **73,** 314.
Vaiana, G. S. and Rosner, R.: 1978, *Ann. Rev. Astron. Astrophys.* **15,** 363.
Vernazza, J. E., Avrett, E. H., and Loeser, R.: 1973, *Astrophys. J.* **184,** 605.
Vesecky, J. F., Antiochos, S. K., and Underwood, J. H.: 1979, *Astrophys. J.* **233,** 978.
Voslamber, D. and Callebaut, D. K.: 1962, *Phys. Rev.* **128,** 2016.
Webb, A. W. and Roberts, B.: 1978, *Solar Phys.* **59,** 249.
Webb, A. W. and Roberts, B.: 1980 (in preparation).
Withbroe, G. L.: 1970, *Solar Phys.* **11,** 42.
Wragg, M. and Priest, E. R.: 1980 (submitted).
Wragg, M.: 1981, Ph.D. Thesis, St. Andrews University.

10. MECHANISMS OF ENERGY SUPPLY

Joseph V. Hollweg

10.1 INTRODUCTION

This chapter summarizes some current thinking about the physical mechanisms that may be responsible for the energy supply of solar active regions, including the propagation of the energy from the solar convection zone, where it is presumed to originate as convective motions, and the ultimate dissipation of that energy into heat and mass motion of the overlying atmospheric layers. This chapter is not a comprehensive review. It concentrates primarily on the ideas explored during the Skylab Active Region Workshop. To that extent, it represents the interests and prejudices of the Workshop participants, particularly those who participated in the Working Group on Mechanisms of Energy Supply. It is not intended as a comprehensive summary of all of the scientific results of the Working Group; these are to be found in the publications of the participants in the Working Group. Instead, an attempt is made to integrate the results of the Working Group into a broader context.

10.2 PERSPECTIVE

10.2.1 Generalities

It is probably foolhardy to discuss the energy supply of solar active regions by themselves. The solar convective motions evidently supply energy to the overlying layers of the solar atmosphere ubiquitously. Heating and acceleration occur in such diverse regions as: coronal holes; the quiet corona; coronal active region loops; the distant solar wind; polar and active region faculae; plages; the quiet chromosphere, network chromosphere, and the transition regions below both coronal holes and the quiet corona; spicules, "jets," and macrospicules; and the chromosphere, transition region, and corona above sunspots. The possibility certainly exists that the same underlying mechanism is responsible for energy propagation and dissipation in all of these regions, but that differences between regions are due to different physical parameters—different magnetic field strength and/or different geometries, such as open *vs* closed magnetic fields. In that case, isolating the active region problem from the question of how energy is supplied to all regions of the solar atmosphere would yield only a narrow and disjointed perspective on what is really an ubiquitous and unified phenomenon. On the other hand, it is possible that heating and acceleration occur via quite different physical processes in different regions of the solar atmosphere. If that is so, the study of active regions by themselves would be worthwhile. Only by viewing active regions in the context of the entire range of solar phenomena can the nature of the differences between the different types of regions be clarified.

The basic problem is that the energy balance of no single region of the solar atmosphere is understood. The corona, for example, has been known to be hot for over three decades; its heating mechanisms are still a complete mystery. It should not be expected, therefore, that even a concerted attack, such as the one waged during the workshop, would solve the problem of active region coronal heating. The best that can be hoped for is that our attempts to understand and model the heating of active regions will fill in one more piece of the puzzle.

Examination of the problem of the energy supply to active regions included certain preconceived notions (which were not carefully reexamined):

(1) The flare problem is held conceptually as separate from the active region energy supply problem. One reason is that the energy released in flares seems insufficient to account for the energy requirements of typical active regions (Orrall and Kopp, 1976). Another reason is that at least the "hot" coronal active region loops (see Webb, this volume, Chapter 7) appear to change only slowly on time scales of the order of their radiative cooling time, implying some form of quasi-continuous heating, which is presumably not associated with isolated catastrophic events such as flares. In addition, the basic morphology of individual loops can remain constant for hours, and the overall morphology of the active region corona can remain constant for days, implying the absence of gross rearrangements of the magnetic field structure, such as accompany flares.

(2) The "missing energy" of sunspots is not explicitly considered as part of the active region energy supply. One reason is that the nature of sunspots and their missing energy is not understood, and nothing can be gained without a full attack on the sunspot problem itself, which is not our intent. That new active regions can emerge before sunspots appear and that old active regions can persist after the sunspots have disappeared suggests that sunspots are not intimately connected to the active region energy supply. And the results of Nye and Hollweg (1979) suggest that the upward radiation of Alfven waves from sunspots is some three orders-of-magnitude too small to account for their missing energy; this means that the energy flux of upward-propagating Alfven waves from sunspots is too small to account for the energy requirements of any given active region as a whole, although it may be sufficient to account for the energy balance of those structures, such as some coronal active region loops, that are connected to sunspots.

(3) Heating and acceleration mechanisms that are associated with magnetic fields are given the strongest emphasis. One reason is that (nonmag-

netic) sound waves do not appear to be present with a sufficient energy flux to heat the atmosphere above the middle chromosphere (Athay and White, 1979a,b; White and Athay, 1979a,b). In addition, the transition region will be a strong reflector of upward-propagating sound waves, and the Brewster angle for sound waves incident on the transition region from below is only about six degrees (if the ratio of the coronal to the chromospheric sound speeds is ten), and it is hard to believe that sound waves can transport an appreciable energy flux into the corona from below. [It is true that sound waves are generated at the convection zone preferentially in the vertical direction (Kuperus, 1969), but this should be counteracted in part by refraction in the chromosphere.] Even more importantly, with the exception of sunspot photospheres, solar regions where the magnetic fields are the strongest are always observed to be the most strongly heated or accelerated. Network, plage, faculae, coronal active region loops, spicules, and the super-photospheric layers above sunspots are all regions of strong magnetic field, and all exhibit enhanced energy deposition (see, for example, Orrall and Kopp, 1976, for a review). The conclusion is almost inescapable: the magnetic field is at least indirectly, and probably directly, responsible for the heating or acceleration. In this context, it is useful to recall that an example exists in which theory and observation are both consistent with, and point strongly to, plasma heating and acceleration by magnetic fields: the solar wind (see Hollweg, 1978a, for a review). Finally, magnetic heating and acceleration mechanisms are given the strongest emphasis because the magnetic field seems to be the only quantity that is sufficiently variable to account for the variety of solar phenomena and the variability of rates of heating and acceleration. The granulation, supergranulation, five-minute oscillations, and nonthermal line broadening do not show strong variations across the solar disk, and variations in the velocity field do not appear to be a feasible means of answering questions such as why active regions are more energetic than quiet regions.

A thorough study of the magnetic field effects thus offers the greatest promise. And, since magnetic fields by themselves can do no work, it is necessary to consider the time-dependence explicitly, so that the induced electric field and quantities such as the Poynting vector can be calculated. The extent to which time-varying magnetic fields can heat and accelerate the solar atmosphere will be the principal subject of discussion in what follows.

10.2.2 Possible Clues

A. Waves. The dissipation of waves that propagate from the convection zone to the corona was historically the first mechanism proposed to heat the corona, and this basic idea has dominated ever since. Sound waves are not an attractive possibility for heating the corona and upper chromosphere, for the reasons given above. The gravity mode, too, is unattractive for heating these regions, because it is suppressed by the strong magnetic fields in those regions (Hollweg, 1979a), and by the high thermal conductivity of the coronal plasma, which tends to make fluctuations isothermal, thereby removing buoyancy forces. [These wave modes may be important in lower chromospheric heating, however; see for example, Leibacher (1977), B. Mihalas (1979), Stein and Leibacher (1974), Stein and Schwartz (1972), Ulmschneider (1971), Ulmschneider and Kalkofen (1977). But the solar five-minute oscillation, which is essentially a subphotospheric trapped acoustic wave (e.g., Leibacher, 1977), is probably not important in chromospheric or coronal heating (Hollweg, 1978b; Jordan, 1977; Ulmschneider, 1976), because there are extended regions of the solar atmosphere where it is evanescent.]

The fast and Alfven MHD modes, and the Alfvenic surface wave, are viable candidates for coronal and chromospheric heating, however, and both modes will be considered more carefully below. They have the appeal of being intrinsically magnetic in nature, and therefore, in principle, capable of explaining the enhanced energy deposition associated with strong magnetic fields on the sun.

The solar wind may provide a clue to the importance of the Alfven mode. Long-period ($\tau \approx$ hours) Alfven waves represent the dominant component of the microscale fluctuations in the solar wind at 1 AU. These waves have large amplitudes, $|\delta\mathbf{B}|/B_0 \approx 1$, but they nonetheless exhibit a special polarization state such that $|\mathbf{B}| = |\mathbf{B}_0 + \delta\mathbf{B}| \approx$ constant. The implication is that the waves have nonlinearly relaxed to that special polarization state, and that they have undergone nonlinear damping in the process, thereby heating the solar wind protons. This viewpoint is consistent with theoretical solar wind models (e.g., Hollweg, 1973, 1978a; Jacques, 1978). The damping mechanism is not currently understood. But the implication is: if Alfven waves in the solar wind can nonlinearly damp and heat the solar wind plasma, can they do so in the solar atmosphere as well?

In addition to the heating, there are reasons to believe that the solar wind, especially the high-speed streams, may be in part directly accelerated by Alfven waves, particularly between roughly 3 and 20 $R_\odot$ (see the above references). The acceleration occurs via the ponderomotive force. The implication is: if Alfven waves can accelerate the solar wind plasma, can they either contribute to supporting the solar plasma against gravity (i.e., alter hydrostatic equilibrium), or even drive flows such as spicules and macrospicules?

Finally, the waves observed at 1 AU almost always have their sense of propagation outward from the sun. An origin inside the Alfvenic critical point of the solar wind is implied, but it is tempting to speculate that the waves are of solar origin. Observations of fluctuations of the Faraday rotation of the 2.3 GHz signals from Helios A and B between 2 and 10 $R_\odot$ can be interpreted as being due to the Alfven waves, and their presence as close to the sun as 2 $R_\odot$ strengthens the possibility of a solar source. The success of solar wind theories that incorporate Alfven waves of solar origin also strengthens this viewpoint. The implied energy flux densities of the long-period waves at the coronal base are as large as 2×10^5 erg cm^{-2} s^{-1}, if the waves drive high-speed solar wind streams, and if the non-r^2 expansion factor of the coronal hole is $f_{max} = 4$; this value of f_{max} is smaller than the average value deduced by Munro and Jackson (1977) from observations of the edge of one polar coronal hole, but it may be appropriate near the center of a similar hole (G. Pneuman, private communication). The implication is: if the sun can radiate a significant flux of Alfven waves into a coronal hole at long-periods, to what extent can it radiate Alfven waves at other periods and in other regions of the solar atmosphere, and what will be the consequences of those waves? In this context, it is particularly worth noting that the flux of long-period Alfven waves into the solar wind may be roughly of the magnitude of the radiative and conductive losses from the quiet corona and coronal holes (see Withbroe, 1976, 1977). There is thus reason to believe that the sun's Alfvenic flux can be large enough to account for the heating of at least the quiet corona and coronal holes.

B. Loops and Structure. The active region corona is highly structured into coronal active region loops (Levine and Withbroe, 1977; Pye *et al.*, 1978; Reeves *et al.*, 1976; Rosner *et al.*, 1978b; Tousey *et al.*, 1973; Underwood *et al.*, 1976; Vaiana *et al.*, 1973a,b, 1976). In the steady state, the low beta of the coronal plasma implies that the coronal magnetic fields must be approximately force-free. Potential fields are a subset of force-free fields, but instances have been found where the observed magnetic field structures are apparently not potential (Krieger *et al.*, 1976). The suggestion has therefore been made (Rosner *et al.*, 1978a; Tucker, 1973) that the force-free parallel electric currents may be capable of heating active region loop structures. This idea is examined in more detail below.

Although strongly suggested, the presence everywhere of parallel currents on coronal loop structures is not proven. There are instances where currents are clearly present, but there are also examples where the agreement between calculated potential fields and the observed coronal loop structures is sufficiently good that there is no reason to require strong parallel currents (Poletto *et al.*, 1975; see also Levine and Altschuler, 1974). Parallel currents may be there, but they may be either too weak or too localized to be seen. On the other hand, the fact that heating occurs on some field lines and not on others may indicate that only some field lines are carrying current. Unfortunately, there appears to be no way to test whether parallel currents are a sine qua non of coronal active region loop heating because the energy requirements of a coronal active region loop can be satisfied even when the nonpotential component of the field is made arbitrarily small, if the current is concentrated into a sufficiently thin layer (see inequality Eq. 10.3.9 below); indeed, the currents must be concentrated to get heating.

The coronal active region loops are regions of enhanced emission, and therefore enhanced heating. It is tempting to conclude that all coronal heating is due to loop structure. However, this conclusion must be incorrect, since the long-lived open magnetic field structure of coronal holes and the solar wind also contains plasma with temperatures not greatly lower than the temperatures on coronal active region loops (about 10^6 K in the holes *vs* about 2.5×10^6 K in the loops). Loop structures are evidently not necessary for coronal and solar wind heating.

It is true that the volumetric heating in coronal active region loops is greatly in excess of that required in coronal holes, and one might be tempted to conclude that the closed magnetic structure per se is responsible for the enhanced heating. But a comparison between quiet (closed) corona and coronal holes shows that the net energy requirements (including the solar wind) may actually be larger in the holes than in the quiet corona (Withbroe, 1977). Closed *vs* open fields cannot be the whole story in coronal heating. Similarly, the chromosphere below coronal holes is indistinguishable from that below quiet corona (Withbroe, 1977), suggesting that closed *vs* open fields cannot be the whole story in chromospheric heating, either.

Of course, these arguments do not imply that parallel currents do no heating. They are merely intended to show that a complete understanding of the overall problem of the heating of the solar atmosphere must involve something else besides, or in addition to, parallel currents. And parallel currents can play no direct role in the acceleration of the solar atmosphere, because they are force-free.

An alternative explanation for the observed enhanced heating on coronal active region loops is that the propagation, dissipation, and generation of waves is different on closed magnetic structures, especially the active region loops, than on open ones. This possibility (Habbal *et al.,* 1979; Hollweg, 1979b) is explored in some detail below. In addition, regardless of whether the fields are open or closed, the strong local gradients may influence the very nature of the waves and introduce new modes, i.e., "surface waves," which can also carry energy and heat the plasma. This viewpoint (Ionson, 1978; Wentzel, 1978, 1979a,b,c) is also explored below.

Finally, the observed highly structured nature of the active region corona (and chromosphere for that matter) suggests the possibility that magnetic energy may be dissipated via small-scale reconnections. This is in essence the coronal heating theory proposed by Levine (1974a,b), and it is discussed in more detail below.

C. Large-Scale Reconnection. It is common in solar physics, particularly in the context of solar flares, to invoke reconnection of magnetic fields as an energy source. In the case of coronal active region loops, at least, there seems to be no evidence that large-scale reconnection is responsible for the loop heating. The reason is simply that the observations show that the basic coronal active region magnetic field configuration can be essentially unchanged for days or longer, and there is no reason to believe that the energy in the large-scale fields is being dissipated. The same reasoning applies to coronal holes, which may be stable for months; see Hundhausen (1977). Similarly, the photospheric layers of solar magnetic flux tubes may also require heating, but since the flux tubes are presumably monopolar, there appears to be no grounds for invoking large-scale reconnections as the energy source. Finally, comparing coronal X-ray structures with photospheric magnetograms and with potential field extrapolations of those magnetograms (Poletto *et al.,* 1975) does not suggest that reconnection of oppositely directed fields (on adjacent loops, for example) should play an important role in coronal heating (see also Rosner *et al.,* 1978a).

The above arguments do not rule out, however, the possibility that small-scale reconnections may serve as a mechanism for locally dissipating magnetic energy into heat, without disrupting the structure of the large-scale field. Small-scale reconnections may be one more consequence of the highly structured corona observed over active regions, as is discussed in Section 10.2.2.B, above.

D. Cool Cores. Some coronal active region loops may have cool cores, with core temperatures close to chromospheric values, although the hottest loops (about 2.5×10^6 K) probably do not have cool cores (Foukal, 1975, 1976; Krieger, this volume, Chapter 7; Levine and Withbroe, 1977; Withbroe, this volume, Chapter 8). It has been suggested that cool cores may serve as a test of the various mechanisms that have been proposed for heating the loops. For example, cool cores may be a natural consequence of anomalous dissipation of parallel currents (Rosner *et al.*, 1978a) or of surface-wave dissipation (Ionson, 1978), both of which are postulated to occur in a thin cylindrical "sheath" surrounding a loop.

On the other hand, cool cores may also be produced by dissipation of fast waves that impinge on the loop from outside if the dissipation is so strong that the waves are never able to penetrate into the interior of the loop (Habbal *et al.*, 1979). Cool cores could also result if the energy enters the loop in the form of twists that propagate as Alfven waves, since the Poynting flux may be largest away from the loop center because the twisting velocities may be larger and the reflections that occur between the convection zone and the corona may be weaker there (Hollweg, 1979b). In addition, Priest (1978) has shown from very basic principles that cool cores can be a consequence of the thermal balance being modified by the stretching or twisting of a loop, at least if the energy input is proportional to plasma density. The implication is that the presence or absence of cool cores may not be a very useful test of the heating mechanisms in coronal active region loops, since there seems to be an abundance of theoretical possibilities for their explanation. (See Priest, this volume, Chapter 9, for a discussion of the stability of loops with cool cores.)

E. Flows and Violation of Hydrostatic Equilibrium. Some coronal active region loops, particularly those containing cooler material, show a tendency for the plasma to extend to greater heights than is predicted if the scale height is the value corresponding to the peak temperature of formation of the particular spectral line in which the corona is being viewed (Foukal, 1976; Webb, this volume, Chapter 7). The material is evidently not in hydrostatic equilibrium, as a consequence either of flows (see Priest, this volume, Chapter 9) or some form of nonthermal support, such as ponderomotive force. It has been suggested that this violation of hydrostatic equilibrium may serve as a test of the proposed loop heating mechanisms. For example, any wave heating mechanism will exert a ponderomotive force on the plasma, which can provide support or even drive flows, much in the way Alfven waves can drive high-speed solar wind streams. And mechanisms in which the heating occurs in a thin sheath surrounding the loop can also drive a flow pattern in which the flow is upward in the sheath and downward in the loop interior, analogously to an ordinary convection cell (Ionson, 1978). However, flows can also be simply a response to temporal changes in the heating rate (Craig, 1979; Levine and Withbroe, 1977), or a consequence of material being transported across field lines from a region where it is heated to a region where the heating is weak and the material can cool and fall (Foukal, 1976; Priest, 1978). In addition, Priest (1978) has shown that cool material can reach to the observed heights in the corona if the pressure at the loop base is uniform across the loop cross section. The implication is that flows and violation of hydrostatic equilibrium may not be very useful tests of the loop heating mechanisms, because there is a breadth of theoretical possibilities for their explanation.

F. Steady-State Heating. Coronal active region loops do change with time, exhibiting both gradual evolutionary changes on time scales of hours or days, and more-rapid changes on time scales of the order of the radiative cooling time or shorter (Webb, this volume, Chapter 7; Levine and Withbroe, 1977). However, the observed lifetimes of coronal loop structures are generally much larger than their calculated radiative cooling times, implying the existence of a quasi-stationary heating mechanism (Gerassimenko *et al.*, 1978; Reidy *et al.*, 1968; Rosner *et al.*, 1978b). This is particularly true for the hot coronal active region loops, whereas cool loops or compact loops may have a significantly time-varying heating function superimposed on top of the quasi-stationary one (Krieger, this volume, Chapter 7). The implication is that it may not be unreasonable to attempt to construct models of coronal active region loops in which the energy supply and the loop structure are assumed to be temporally steady. Such models will be the focus of this chapter.

10.2.3 The Problem is Global

Regardless of the mechanism by which energy is dissipated into heating or acceleration of the photosphere, chromosphere, and corona, the primary question must be: how does the energy come up from the convection zone? Clearly, the invention of an efficient dissipation mechanism is meaningless unless it can be shown that the convection zone can actually generate adequate energy in the form for which the dissipation mechanism is appropriate. Alternatively, the problem can be turned around: once it is shown that the convection zone can supply energy in some form, then it makes sense to ask whether the energy can actually be dissipated.

Similarly, local stability analyses and other local calculations may be of limited use if the energy input is globally controlled. For example, stability analyses are usually carried out assuming that the volumetric heating rate is either unperturbed, or that it depends on the local values of density, temperature, etc., in some prescribed way. However, if the energy flux entering a loop from the convection zone depends on nonlocal factors, such as the loop length or the magnitude of the "discontinuities" in Alfven and sound speeds at the transition region, then the inclusion of such global effects is mandatory, and analyses from which they are omitted may be of only limited usefulness.

The amount of energy that reaches a given point in the solar atmosphere from the convection zone is almost certainly globally controlled, and so too is the form of the energy. For example:

(1) The energy flux carried by Alfven waves that are generated at the top of the convection zone is strongly affected by reflections off of the entire structure of a magnetic flux tube, and by whether the field configuration is open or closed. This means that the energy reaching a given point in the solar atmosphere depends on what happens everywhere else on the field line that passes through that point. It also means that even if one were given complete information about the wave properties at one point, it would be impossible to predict the wave properties at any other point without taking the entire atmospheric structure into account; contrast this with the usual WKB theory, where such a prediction could be made via relations such as $\varrho_0^{1/2} < \delta \mathbf{v}^2 >$ = constant and $< \delta \mathbf{B}^2 > \varrho_0^{-1/2}$ = constant along a field line. In this sense, the Alfven waves are globally controlled (Hollweg, 1979b).

(2) Similarly, in the full-wave non-WKB case the propagation of fast and slow MHD modes will be globally controlled much in the manner of Alfven waves. But even in the WKB limit, these modes will be globally controlled via refraction, and to some extent via strong reflections at "discontinuities" such as the transition region. Since ray paths can be quite complicated even in simple atmospheric configurations (Habbal *et al.*, 1979; Osterbrock, 1961), it is apparent that the energy that reaches a given point in the solar atmosphere can depend strongly on conditions elsewhere. In this sense, these waves too are globally controlled.

(3) Surface waves may provide a means for dissipating energy that is carried from the convection zone (Ionson, 1978). But, by the same arguments as in (1) and (2), the surface waves too are almost certainly globally controlled.

(4) There are proposed mechanisms for heating the solar atmosphere that are not formulated explicitly in terms of waves. Examples are the heating of coronal active region loops by parallel currents (Rosner *et al.*, 1978a; Tucker, 1973) or coronal heating by collapse of magnetic neutral sheets (Levine, 1974a,b). These theories for the most part merely postulate the existence of the parallel currents or neutral sheets and then calculate their dissipation, but important aspects such as the details of their production at the convection zone, the constraints on the energy they can carry, and the ability of the convection zone to replenish the energy once it is dissipated are generally not investigated. It is the contention here that these aspects too must be subject to global control. For example, parallel currents can be generated by twisting magnetic flux tubes (Parker, 1974; Rosner *et al.*, 1978a). If the parallel currents are dissipated into heat, perhaps via anomalous resistivity, they must be replenished via further twisting motions in the convection zone. But such twisting motions propagate as torsional Alfven waves (Hollweg, 1979b), and the extent to which the energy that has been dissipated into heat can be replenished is subject to the same global con-

straints as are encountered in the Alfven wave heating mechanisms. (Indeed, the anomalous resistivity can be thought of as another mechanism for damping Alfven waves.) Similarly, Levine (1974b) thinks of his postulated magnetic neutral sheets as being due to twisting and untwisting of the coronal magnetic field pattern. If these twists propagate as torsional Alfven waves, they will be subject to the same global constraints.

There may be some observational evidence that the energy input into coronal active region loops is globally controlled. Figure (9a) of Rosner *et al.* (1978b) indicates that temperature T and pressure p in an ensemble of loops are roughly related by

$$T = T_0 p^{\alpha} \ , \tag{10.2.1}$$

where $0 \lesssim \alpha \lesssim 0.2$ and $T_0 \approx 2.3 \times 10^6$ K, if p is in dyne cm^{-2}. If T is identified with the maximum temperature at the top of a loop, equations (4.3) and (4.4) of Rosner *et al.* show that the volumetric heating E_H scales as

$$E_H = 9.8 \times 10^4 L_{cor}^{-5/6} [2.74 \times 10^9 \, T_0^{-3} L_{cor}]^{7/(18\alpha - 6)} \text{erg cm}^{-3}\,\text{s}^{-1} \ , \tag{10.2.2}$$

where L_{cor} is half the length of the coronal loop, and E_H has been assumed uniform. If $0 \lesssim \alpha \lesssim 0.2$, $E_H \propto L_{cor}^{(-2)-(-3.75)}$. If the energy input flux propagates parallel to the loop, then the energy flux density entering the bottom of the loop is

$$E_H L_{cor} \propto L_{cor}^{(-1)-(-2.75)} \ . \tag{10.2.3}$$

This merely reproduces the known fact that longer coronal active region loops are less energetic than shorter loops. This has been interpreted previously in terms of active region evolution, the longer loops being older than the shorter ones (e.g., Rosner *et al.*, 1978b). However, if the heating is quasi-steady, relations (10.2.2) and (10.2.3) may indicate that the energy input is globally controlled. If the loop length is fixed by the magnetic field configuration, one must ask how the energy input "knows" how long the loop is. One possible explanation is that the energy input is globally controlled and somehow responds directly to L_{cor}. In that case, relations (10.2.2) and (10.2.3) may be offering some clues about the way energy gets into the corona from the convection zone, and how it dissipates. Alternatively, the energy input may merely be responding to some local parameter such as the magnetic field strength or filling factor, which is correlated with L_{cor}.

10.2.4 Goals of This Chapter

With the foregoing perspectives in mind, this chapter examines three questions of relevance to mechanisms of energy supply in active regions:

(1) How does energy come up from the convection zone?

(2) What are the processes by which the energy is dissipated into heat and momentum?

(3) For any postulated mechanism of energy transport and dissipation, is it possible to define what the observable consequences of that mechanism should be; i.e., is it possible to find "diagnostics" with which the presence or absence of some postulated mechanism can be observationally determined?

The following physical processes are examined here:

(1) the heating of coronal active region loops by anomalous dissipation of parallel currents in thin sheaths;

(2) the heating of coronal active region loops by minireconnections;

(3) the role of surface waves in active region loops;

(4) spatially dependent heating in the corona and the formation of coronal active region loops as influenced by refraction and Landau/transit-time damping of fast-mode MHD waves in the WKB limit; and

(5) the feasibility of propagating "twists" (Alfven waves) on magnetic flux tubes as a means of carrying energy from the convection zone into the chromosphere and corona on open and closed magnetic flux tubes, and the possible consequences of such twists vis-a-vis heating and acceleration.

These topics are covered in this review primarily because they were the principal interest of the Working Group, and that in turn was the case because they seemed the areas in which progress was most likely.

10.3 ANOMALOUS DISSIPATION OF PARALLEL CURRENTS IN THE CORONA

10.3.1 Perspective

The observations that suggest the possible importance of parallel currents on coronal active region loops are discussed above. These observations, and an earlier discussion by Tucker (1973), led Rosner *et al.* (1978a) to investigate the possibility that quasi-steady parallel currents, which are localized in thin sheaths surrounding cylindrical loops, can account for coronal active region loop heating. They envision the parallel currents in the corona to be produced by the upward convection of subphotospheric parallel currents during the process of flux tube emergence, or by the relaxation of torsional twists that are imposed on the flux tube by the convective motions. Whereas the former process may be particularly relevant to the initial heating phase of an emerging loop, the latter process is the one that is of interest in considerations of quasi-steady coronal active region loop heating.

Rosner *et al.* apparently were motivated to consider thin current sheaths, because Parker (1974) demonstrated the possibility that force-free field configurations can be generated in which the major part of the field variation occurs very close to the surface of the "flux rope" (e.g., Parker's Section IIb). However, Parker's sample solution for a force-free flux rope in which the magnetic field strength in the corona is weaker than that in the photosphere (see equations 44–49 of Parker), which corresponds to the situation envisioned by Rosner *et al.,* does not show any strong concentration of the current near the surface of the flux tube. Indeed, the opposite of field line concentration occurs, since field lines from the surface of the unexpanded part of the flux tube map into a finite volume of the expanded part.

If the parallel currents are generated by axisymmetrically twisting an axisymmetric vertical flux tube, it is in principle possible to calculate the process in detail, since the axisymmetric twists propagate essentially as Alfven waves (Hollweg, 1979b). If the twists are nonlinear, if the flux tube is not vertical, or if the axisymmetry is violated, coupling into other modes can occur, but this has been ignored in Hollweg's calculations. An Alfven wave can carry both perpendicular and parallel currents; the current parallel to the average field $\mathbf{B}_0$ is associated with the transverse gradient of the fluctuating wave quantities. Rather than calculating the parallel current, however, Hollweg investigated whether the Poynting flux associated with such axisymmetric twists could be large enough to supply the energy requirements of a coronal active region loop. The Poynting flux is parallel to $\mathbf{B}_0$, and does not depend on the transverse gradient. This in itself puts severe constraints on the ability of twists to supply the energy in the model of Rosner *et al.* Since the entire loop energy must enter the thin sheath along $\mathbf{B}_0$, and since $\mathbf{B}_0$ lies in the plane of the sheath, the energy flux *density* entering the sheath from the convection zone must be huge; for example, if the energy flux density averaged over the loop cross section is 10^7 erg cm^{-2} s^{-1}, then the energy flux density entering the sheath must be 10^{11} erg cm^{-2} s^{-1} if the loop radius is 1000 km and the sheath thickness is 5×10^3 cm. Hollweg's model indicates that Alfven waves that are driven by the convective motions cannot possibly supply this large a Poynting flux. Moreover, Hollweg looked at the Poynting flux that could be supplied to the entire loop by long-period twisting motions, with periods of the order of a few hours or so; if the Alfven waves in the solar wind can be used as a guide, these are the periods at which most long-period power can be expected to reside. Hollweg found that undamped long-period twists cannot supply a Poynting flux of 10^7 erg cm^{-2} s^{-1} or so. The conclusion is that long-period quasi-steady twists do not appear to be an effective means of supplying energy to a loop as a whole, and certainly not to a thin sheath.

The above considerations mean that the energy supply to loop sheaths is not understandable in terms of simple quasi-steady twists that originate in the convection zone, as envisioned by Rosner *et al.* This does not mean that other processes cannot get the energy into the sheath. One possibility, also mentioned by Rosner *et al.,* is that the coronal active region loop can act as the secondary circuit in a transformer, in analogy with tokamaks. This process has not been worked out in detail; in particular, the details of the energy transport from the convection zone to the loop sheath has not been considered, nor has the question of how the transformer results in a thin sheath. One possible problem is the following: the energy must enter the secondary of the transformer by propagating from the primary across the magnetic field into the secondary. Such propagation presumably must occur as a fast or slow MHD wave. But these modes do not appear to be a suitable means of getting energy into the corona (see Sections 10.2.1 and 10.6.1), and it is thus not clear if a transformer mechanism is energetically feasible.

The following section outlines the basic physical considerations associated with the sheath itself, independent of broader questions such as how the sheath forms and how the energy comes up from the convection zone. The focus is on the model considered by Rosner *et al.*, in which ion-acoustic turbulence in the sheath is responsible for the anomalous resistivity. This model is found to have self-consistency difficulties. An alternative model, in which the resistivity is due to electrostatic ion-cyclotron waves, is considered (Vlahos *et al.*, 1979).

10.3.2 Theory and Critique

The following paraphrases the arguments of Rosner *et al.* (1978a), and an extension and critique carried out by Vlahos *et al.* (1979) as part of the Working Group's efforts.

Equations (4.3) and (4.4) of Rosner *et al.* (1978b) give the total energy required by half of a (symmetrical) coronal active region loop of maximum temperature T, minor radius R, pressure p, and semilength L:

$$U_{req} = 2.3 \times 10^{13}\, p^2 R^2 T^{-5/2} L \tag{10.3.1}$$

(cgs units are used throughout). The energy dissipated in the sheath of thickness ΔR is

$$U_H = 3.6 \times 10^{19}\, \eta (R/\Delta R)(\Delta b)^2 L\ , \tag{10.3.2}$$

where η is the loop resistivity and Δb is the magnetic field jump across the sheath due to the sheath current. Setting $U_H = U_{req}$ gives

$$\Delta R = 1.6 \times 10^{6}\, \eta (\Delta b)^2 p^{-2} T^{5/2} R^{-1}\ . \tag{10.3.3}$$

The classical resistivity in the sheath is

$$\eta_c \approx 10^{-7}\, T_{sh}^{-3/2}\ , \tag{10.3.4}$$

where T_{sh} is the sheath electron temperature, which can differ from that in the rest of the loop, and the Coulomb logarithm has been taken to be 20. If the classical resistivity is to apply, the electron-proton drift in the sheath must not be large enough to excite instabilities; i.e., the drift velocity must be less than the critical drift V_c, and

$$\Delta R > \frac{c\Delta b}{4\pi e n_{sh} V_c}\ , \tag{10.3.5}$$

where n_{sh} is the electron concentration in the sheath. Combining Eqs. 10.3.3–10.3.5 yields

$$\Delta b > 4300 R p^2 p_{sh}^{-1} V_c^{-1} (T_{sh}/T)^{5/2} \ . \tag{10.3.6}$$

Inserting any reasonable values yields a huge lower limit on Δb. Classical resistivity does not work in the corona.

Rosner *et al.* assumed ion-acoustic turbulence in the sheath and took

$$\eta_{ia} \approx 2.2 \times 10^{-6} (V_{d,sh}/V_{e,sh}) n_{sh}^{-1/2} \ , \tag{10.3.7}$$

where $V_{d,sh}$ and $V_{e,sh}$ are the electron drift and thermal speeds (this value is smaller than Tucker's by a factor between 10^{-2} and 10^{-4}). If ion-acoustic turbulence is to be set up, then

$$\Delta R < c \Delta b / 4\pi e n_{sh} V_c \tag{10.3.8}$$

and

$$\Delta b < 1.4 \times 10^{18} p^2 n_{sh}^{-1/2} V_{e,sh} V_c^{-2} T^{-5/2} R \ . \tag{10.3.9}$$

If $n_{sh} = 10^9$ cm^{-3}, $p = 3$ dyn cm^{-2}, $T = 2.5 \times 10^6$ K, $R = 10^8$ cm, $V_{e,sh} = 2.7 \times 10^9$ cm s^{-1} (corresponding to $T_{sh} = 2.5 \times 10^7$ K $= 10T$), and $V_c = V_{sound}/2^{1/2} = 4.5 \times 10^7$ cm s^{-1}, then $\Delta b < 5.4$ G and $\Delta R < 600$ cm. These values are not unreasonable, Δb being smaller than the field strength and ΔR being larger than the electron gyroradius and Debye length. It should be noted that the upper limit on ΔR scales as $(\eta V_c^2)^{-1}$ The factor V_c^{-2} can be quite small for instabilities other than the ion-acoustic. If T_e is not much greater than T_p in the sheath, then the ion-acoustic mode cannot exist, and other modes must come into play. For example, Vlahos *et al.* have extended the work of Rosner *et al.* to include the electrostatic ion-cyclotron mode, for which $V_c = 12V_s$. If $T_e = T_p = 2.5 \times 10^6$ K in the sheath, then $V_c = 3.4 \times 10^8$ cm s^{-1}. The factor V_c^{-2} is smaller by a factor 60 in this case than in the example that follows Eq. 10.3.9, and the upper limit on ΔR is correspondingly smaller. However, η for the electrostatic ion-cyclotron mode is not the same as η_{ia}, and Vlahos *et al.* find that reasonable sheath thicknesses can result. Finally, note that the sheaths are well below any instrumental resolution; paradoxically, even though the anomalous dissipation theory is motivated by observations of coronal structures, it requires structures on a scale for which there is no evidence.

The assumption of ion-acoustic turbulence requires $T_e \gtrsim (8\text{–}10)T_p$ in the sheath (Jackson, 1960), in which case $V_c \approx V_{sound}/2^{1/2}$, where $V_{sound} = (2K(T_e + T_p)/m_p)^{1/2}$. The thermal equipartition time due to Coulomb collisions is

$$t_{eq} = 13 T_{sh}^{3/2} n_{sh}^{-1} \ , \tag{10.3.10}$$

if the Coulomb logarithm is 20. If $T_{sh} = 2.5 \times 10^7$ K and $n_{sh} = 10^9$ cm^{-3}, then $t_{eq} = 27$ min. Since this time is comparable to the radiative cooling time, it seems likely that any process that is able to keep the corona hot in the first place could also maintain the electrons hotter than the ions, if the electrons are preferentially heated.

If T_e is not much bigger than T_p in the loop [specifically, if $T_e \lesssim (8\text{–}10)T_p$], then other instabilities can be excited before the ion-acoustic mode. The electrostatic ion-cyclotron and Alfven modes are likely candidates (see Hollweg and Smith, 1976; Smith and Hollweg, 1977; and references therein). In the case of the electrostatic ion-cyclotron mode, $V_c \approx 12V_{sound}$ if $T_e = T_p$ in the sheath. The Alfven mode frequently has a lower value of V_c than the electrostatic ion-cyclotron mode. The Alfven instability could be the most relevant one in processes of this type, but it has not yet been investigated.

Vlahos *et al.* have pointed out that the continued existence of a thin sheath may be impossible once anomalous ion-acoustic resistivity is established in the sheath. The reason is that the ion-acoustic turbulence in the sheath results in enhanced diffusion in the sheath, so that the sheath broadens as the electrons drift along it. The time required for electrons to drift a distance L is L/V_d, during which time they spread a distance

$$\delta \approx (2DL/V_d)^{1/2} \; , \tag{10.3.11}$$

where D is the cross-field diffusion coefficient. The value of D is calculated by assuming that an electron jumps one gyroradius ϱ_{ce} in one collision time. Then

$$D = \varrho_{ce}^2 \nu_{eff} \; , \tag{10.3.12}$$

where ν_{eff} is the effective collision frequency for electrons. Writing

$$\eta = m_e \, \nu_{eff}/e^2 n_e \tag{10.3.13}$$

gives

$$\delta \approx 5.3 \times 10^8 (Lp_e\eta/\alpha)^{1/2} B^{-1} (T_e + T_p)^{-1/4} \; , \tag{10.3.14}$$

where $\alpha = V_d/V_{sound}$ and B is the magnetic field strength in the sheath. If $\eta = \eta_{ia}$, this becomes

$$\delta \approx 13 L^{1/2} p_e^{1/4}/B \; . \tag{10.3.15}$$

If $L = 10^9$ cm, $p_e = 3$ dyn cm^{-2}, and $B = 100$ gauss, then $\delta \approx 5400$ cm, which is an order-of-magnitude larger than the upper limit on ΔR found previously. Evidently, the turbulent diffusion will quench the anomalous ion-acoustic resistivity.

Alternatively, suppose the current sheath is associated with an MHD wave propagating at V_A. The thickening of the sheath then obeys resistive diffusion, and

$$\delta \approx (2DL/V_A)^{1/2} \ , \tag{10.3.16}$$

where

$$D = c^2\eta/4\pi \ . \tag{10.3.17}$$

If η_{ia} is used, then

$$\delta \approx 38(L/B)^{1/2}(V_d/V_e)^{1/2} \ . \tag{10.3.18}$$

If $L = 10^9$ cm, $B = 100$ gauss, and $V_d/V_e = 10^{-2}$, then δ is about 1.2×10^4 cm. This is again more than an order-of-magnitude thicker than the upper limit on ΔR found above, and it appears again that diffusion can quench the ion-acoustic instability.

Another difficulty pointed out by Vlahos *et al.* is that the thin current-carrying sheaths can be disrupted by the double tearing mode (Furth *et al.*, 1973) in a time on the order of a few seconds.

The problem may be even more severe than the above estimates imply. According to Papadopoulos (1977), the existence of a steady-state anomalous resistivity due to ion-acoustic turbulence requires that the ions not be heated more rapidly than the electrons; otherwise, T_e/T_p will decrease and quench the instability. If $T_e/T_p \gg 1$, the condition for steady-state η_{ia} is

$$\frac{T_e}{T_p}\frac{V_c}{V_d} < 1 \ . \tag{10.3.19}$$

In that case, the right-hand sides of Eqs. 10.3.8 and 10.3.9 are smaller by the factor $(T_p/T_e)_{sh}$. The upper limit on Δb is reduced by $(T_p/T_e)_{sh}$, and the upper limit on ΔR is reduced by $(T_p/T_e)^2_{sh} \approx 10^{-2}$. For the numerical values following Eqs. 10.3.8 and 10.3.9, one requires $\Delta b < .54$ G and $\Delta R < 6$ cm, if $(T_e/T_p)_{sh} = 10$. This value for ΔR is still larger than the electron gyroradius, but the diffusion arguments presented above indicate that it cannot be realized.

Thus, there seems to be a real theoretical problem in understanding how the ion-acoustic turbulence can be confined to sheaths of the required thinness. Perhaps the heating is pulsed, with thin sheaths being formed by some undetermined process and heating occurring before the sheath is disrupted. But this viewpoint does not seem attractive, because then not only does the loop energy have to be supplied in a very small fraction of the loop volume, but in a fraction of the available time as well.

Vlahos *et al.* have examined whether electrostatic ion-cyclotron turbulence can solve the problem. They find that on short coronal active region loops, the electrostatic ion-cyclotron waves can simultaneously supply the energy required by the loops and also satisfy the requirements $\Delta b < B$ and $\delta < \Delta R$. On long-loops, however, the saturation level of the waves is low;

this occurs because the saturation power is proportional to $\omega_{ce}^2/\omega_{pe}^2$, which may be 1 in the short loop sheaths but 0.01 in the long loop sheaths (ω_{ce} and ω_{pe} are the electron cyclotron and electron plasma frequencies, respectively). This means that η is low, and the upper limit on Δb, which is proportional to η^{-1}, becomes larger than B; this effect is aided by the fact that B is smaller in the longer loops, but it is counteracted by the possibility that p^2R (see Eq. 10.3.9) is smaller on the longer loops, even if $R \sim L$, since $p \sim L^{-1}$. (Actually, the argument vis-a-vis $\Delta b/B$ is not very strong. Equation 10.3.9 yields only an upper limit on Δb. The actual value of Δb could be forced to be smaller than the upper limit by driving the instability beyond threshold, and ΔR would similarly have to be reduced, in proportion to Δb^2.) It must be noted, however, that the precise nonlinear behavior of the electrostatic ion-cyclotron mode is subject to some question. Papadopoulos (1977) states that this mode does not lead to a steady-state anomalous resistivity, but rather to "ion hot spots." Vlahos *et al.* summarize some more recent work that may indicate that the electrostatic ion-cyclotron mode can lead to a suitable anomalous resistivity.

A quantity of interest is the ratio of the sheath electric field to the Dreicer field,

$$\left(\frac{E}{E_D}\right)_{sh} = 1.2 \times 10^{13} \frac{p^2 R T^{-5/2} T_{sh}}{\Delta b n_{sh}} \tag{10.3.20}$$

Inserting the lower limit on Δb for ion-acoustic turbulence, Eq. 10.3.9, yields

$$\left(\frac{E}{E_D}\right)_{sh} > \frac{8.6 \times 10^{-6}\, T_{sh} V_c^2}{n_{sh}^{1/2} V_{e,sh}} \tag{10.3.21}$$

If $T_{sh} = 2.5 \times 10^7$ K, $V_c = 4.5 \times 10^7$ cm s^{-1}, $n_{sh} = 10^9$ cm^{-3}, and $v_{e,sh} = 2.7 \times 10^9$ cm s^{-1}, then $(E/E_D)_{sh} > 5100$. This value and the right-hand side of Eq. 10.3.21 will be larger by the factor $(T_e/T_p)_{sh}$ if (10.3.19) is taken into account. A more relevant quantity is $(E/E_{D,anom})_{sh}$, where $E_{D,anom} = E_D \eta_{anom}/\eta_c$ is the effective Dreicer field when classical Coulomb collisions are not responsible for the resistivity. From Eqs. 10.3.3, 10.3.4, 10.3.8, and 10.3.20, there results

$$\left(\frac{E}{E_{D,anom}}\right)_{sh} > 3.9 \times 10^{-7}\, T_{sh}^{-1/2} V_c \ , \tag{10.3.22}$$

independent of the form of η_{anom}. If Eq. 10.3.19 is taken into account, the right-hand side of Eq. 10.3.22 is larger by the factor $(T_e/T_p)_{sh}$. If $T_{sh} = 2.5 \times 10^7$ K and $V_c = 4.5 \times 10^7$ cm s^{-1}, then $(E/E_{D,anom})_{sh} > 3.5 \times 10^{-2}$, where Eq. 10.3.19 has been taken into account with $(T_e/T_p)_{sh} = 10$. Even though $E > E_D$, the fact that $E < E_{D,anom}$ means that the entire electron population will probably not run away, and heating of the electrons can occur. [However, if $\omega_{ce} > \omega_{pe}$, the system will be one dimensional and ion-acoustic turbulence will not be established (Boris *et al.*, 1970; Papadopoulos, 1977). Instead, the Buneman instability turns on and off and there is no steady-state anomalous resistivity, but rather a pulsed one. The bulk electron drift velocity and the electron thermal speed both increase linearly with time (on the average) at half the free-streaming rate. Runaway, in effect, does occur, and the system in this case does not behave as assumed by Rosner *et al.* If $\omega_{ce} < \omega_{pe}$, however, ion-acoustic anomalous resistivity can result. There is no good way to estimate n_{sh} and therefore ω_{pe} in the sheath, but it appears from some typical numbers that ω_{pe} and ω_{ce} might be comparable in the sheaths of short loops, and it is not clear what happens (for example, if $B = 100$ G, $\omega_{pe} = \omega_{ce}$ when $n_e = 10^9$ cm^{-3}). On long loops, $\omega_{ce} < \omega_{pe}$ probably pertains, and ion-acoustic anomalous resistivity can be set up.]

The potential drop along the loop can be huge. From Eq. 10.3.21, for ion-acoustic turbulence,

$$E > \frac{1.1 \times 10^{-15} V_c^2 n_{sh}^{1/2}}{V_{e,sh}} \tag{10.3.23}$$

If Eq. 10.3.19 is taken into account, the right-hand side of Eq. 10.3.23 must be multiplied by $(T_e/T_p)_{sh}$. If $V_c = 4.5 \times 10^7$ cm s^{-1}, $n_{sh} = 10^9$ cm^{-3}, and $V_{e,sh} = 2.7 \times 10^9$ cm s^{-1}, $E > 2.6 \times 10^{-4}$ statvolt cm^{-1} = 7.8 volt m^{-1}, where Eq. 10.3.19 has been invoked with $(T_e/T_p)_{sh} = 10$. If the loop length is 10^9 cm, the potential drop is > 78 MeV. (The problem could become relativistic if $\omega_{ce} > \omega_{pe}$ and runaway occurs.)

Rosner *et al.* argue that potential differences of this size could in fact be produced as a consequence of the changes of magnetic flux during the growth and decay phases of active regions. Their arguments only involve the overall flux evolution, however. They have not shown that the evolution of any given system of coronal loops is consistent with such large parallel potential drops. The relative stability of the topology of the coronal loop systems suggests that such large potential drops may not be realizable. The suggestion here is that the evolution of specific loop systems should be examined to ascertain whether the required potential drops are in fact feasible, and that this be done in the context of the electrostatic ion-cyclotron turbulence model of Vlahos *et al.* as well as the ion-acoustic model of Rosner *et al.*

The final problem with the theory concerns how the thermal energy can be extracted from the sheath. Rosner *et al.* argue that even Bohm diffusion is not appropriate since it can be valid only in the presence of turbulence, i.e., in the sheath itself but not outside the sheath. Rosner *et al.* argue that **E** × **B** drifts could work, but it seems to us that **E** × **B** correspond to nothing more than a sheath that moves along with the magnetic field lines (**E** × **B** drifts satisfy the frozen-in theorem), and no extraction of heat from the sheath occurs. The waves that are generated in the sheath can also propagate out of the sheath into the loop interior, and take their energy along. Vlahos *et al.* have examined this possibility. They find that in the large coronal active region loops the damping length of the ion-acoustic waves is too small, by two to four orders of magnitude, for the waves to penetrate into the interior, and the loops cannot be heated. However, on the shorter loops they believe that electrostatic ion-cyclotron waves can be the relevant source of turbulence and anomalous resistivity, and these can in fact penetrate some 10^3 km into the loop, which is comparable to the radii of the short loops. The shorter loops could conceivably be heated by sheath-generated electrostatic ion-cyclotron waves.

The overall conclusion of this section is that the ion-acoustic model of Rosner *et al.* has self-consistency difficulties. The electrostatic ion-cyclotron model of Vlahos *et al.* may work on short coronal active region loops, but not on the long loops. The electrostatic ion-cyclotron model may itself have difficulties, insofar as the sheaths may in this case too be disrupted by the double tearing mode, and insofar as it has not been shown that the required potential drops can be consistent with the observed evolution of specific coronal active region loop systems.

10.3.3 Observables

The foregoing theory suggests several observable diagnostics. They are intrinsically limited by the extreme thinness of the sheaths. There is probably no way to test if $(T_e/T_p) > (8\text{–}10)$ in the sheaths. Even if $T_e = 2.5 \times 10^7$ K in the sheaths, the hard X-ray signature will be below detectable limits unless $n_{sh} > 10^{11-12}$ cm^{-3} (D. Spicer, private communication). Direct resolution of such thin sheaths is out of the question.

That $(E/E_{D,anom})$ is not very small suggests that electrons in the tail of the distribution function can run away. This, coupled with the large electric fields in the sheaths, suggests that nonthermal radio radiation can be produced. Vlahos *et al.* predict a flux in excess of 100 sfu (solar flux units) between 100 and 300 MHz. Thus, simultaneous radio and X-ray measurements should be undertaken to investigate the possibility that hot loops are intense radio emitters. [It should be borne in mind that nonthermal radio emission is not necessarily a signature of sheaths. The solar nonthermal radio emission is not energetic, and it could conceivably be a by-product of almost any heating process for which E/E_D or $E/E_{D,anom}$ is not very small. It is even conceivable that nonthermal radio emission can be produced by some energetically weak process that has nothing to do with the mechanism that is heating the loops. The basic point is that E_D is small in the corona, and runaways and nonthermal emission should be easily produced. Thus, if no nonthermal emission from coronal active region loops is found, this could serve as a "negative check" on the anomalous current dissipation models.]

Sheath heating is consistent with cool loop cores, but for the reasons given above, we do not regard cool cores as an effective diagnostic.

If the energy is carried out of the sheath by the waves themselves, then the presence of the waves in the sheath interior could alter the transport coefficients there. In particular, the parallel thermal conductivity could be modified, and this would show up as modifications of the scaling laws of Rosner *et al.* (1978b). However, the sensitivity of the data to this effect seems to be too weak for any conclusions to be drawn (J. Ionson, private communication).

Finally, heating in a thin sheath could lead to a convective layer of some 30-km thickness (see Section 10.5). If observed, such a convective layer could be an indirect indication of heating in a thin unresolved sheath.

10.4 MINIRECONNECTIONS IN LOOPS

10.4.1 Perspective

Rosner *et al.* (1978a) pointed out that a single current sheath, such as discussed in Section 10.4, could be subject to tearing instabilities and disrupted into a multitude of current sheets. Nonetheless, they computed the sheath heating on the basis of there being a single sheath surrounding the coronal active region loop, with the heating due to ion-acoustic anomalous resistivity. This was really a most favorable case, because if they considered N separate sheaths, each providing one N^{th} of the total heating, they would have found that the upper limits on Δb and ΔR are smaller by the factor N^{-1}; i.e., the requirement vis-a-vis $\delta/\Delta R$ would have been even more severely violated. (In the electrostatic ion-cyclotron model of Vlahos *et al.,* 1979, there is an upper limit on N, above which the requirement $\delta/\Delta R < 1$ fails to be satisfied.)

An alternative viewpoint has been suggested by Rosner *et al.* (1978a), but the details (Galeev *et al.,* 1979) are not yet available. It is proposed that the coronal active region loops are "magnetically heated," with the dissipation occurring via minireconnections of current sheets distributed throughout the loop. The thicknesses of the current sheets are in the range 0.1–1 km.

A major question again is: how does the energy come up from the convection zone; i.e., how are the current sheets generated and replenished? Since the current sheets are essentially sheared magnetic fields, it may be possible to think of them in terms of Alfven waves or fast-mode MHD waves. The feasibility of energy entering coronal active region loops in either of these forms is discussed below. If the current sheets postulated by Rosner *et al.* are really magnetic field shears associated with Alfven or fast waves, then the reconnections envisioned by Rosner *et al.* can in effect be thought of as a mechanism for Alfven and fast-mode damping.

A difficulty immediately arises. It is known that static current sheets are unstable to tearing (see the review by Drake and Lee, 1977). But, as far as we are aware, the stability of dynamic propagating current sheets, i.e., waves, to tearing has not been investigated. One question is whether the tearing can occur on a time shorter than ω^{-1} (ω is the angular wave frequency); if the tearing instability does not grow on this time scale or less, then it seems reasonable that the wave field can change its configuration and quench the instability. Another question is what effect the wave-associated velocity field has on tearing. In sum, do Alfven and/or fast waves tear, and under what circumstances?

If the waves are fast waves propagating across the average magnetic field, then the thickness of the current sheet is roughly V_A/ω. If this is to be in the range 0.1–1 km, and if $V_A = 10^3$ km s^{-1}, then the wave period must be in the range 6×10^{-4} to 6×10^{-3} s. It seems unlikely that much wave power could exist at such high frequencies, at least if the waves are to be generated in the convection zone. And waves of such high frequencies would be heavily damped by Joule-viscous-frictional damping in the photosphere and chromosphere before reaching the corona.

Alfven waves, on the other hand, obey the dispersion relation $\omega = k_{\parallel}V_A$ ($k_{\parallel}$ is the wavenumber parallel to $\mathbf{B}_0$). The transverse wavenumber does not play a role. This means that the current sheet thickness does not constrain ω, if the thickness of the current sheet is associated with the transverse scale $k_{\perp}^{-1}$ of the Alfven wave. The question is whether it is possible to get Alfven waves on coronal active region loops with $k_{\perp}^{-1} \approx 0.1$–1 km. If the waves come up from the convection zone with this short a transverse scale, they would be heavily damped by Joule-viscous damping in the photosphere and chromosphere (the transverse scales would be even smaller in the photosphere and chromosphere, because of the divergence of the magnetic field lines with height). Thus, direct generation of waves with such small values of $k_{\perp}^{-1}$ seems unlikely. But some sort of filamentation process that confines the waves to channels cannot be ruled out, although no specific calculations are available.

Hollweg (1979b) has argued that the loop energy requirements can possibly be supplied from the convection zone via high-frequency Alfven waves that are resonant with the loop (see Section 10.7.2). The waves are essentially linear, having $\delta B/B_0 \approx 10^{-2}$. (Because the waves are Alfvenic, these values do not depend on $k_{\perp}$.) The magnetic field shears are modest; this tends to reduce the growth rate of any tearing instability.

The basic point is that Alfven waves and the associated magnetic field shears may be able to supply the energies required by coronal active region loops, but it is not yet clear whether tearing can serve as a damping mechanism for the waves.

Of course, it is quite possible that the sheared magnetic fields and current sheets can exist in the corona in some form other than waves, but no specific suggestions for their nature and generation are available. Again, we must emphasize that any such suggestion cannot be complete unless the details of the energy flow from the convection zone to the corona are worked out.

10.4.2 Theory

Rosner (private communication) suggests that the tearing mechanism must be semicollisional (see Drake and Lee, 1977). Collisionless tearing nonlinearly saturates at too low a level and the heating is weak. Collisional tearing requires too small a growth rate. It can be shown, using the equations in Section II C of Drake and Lee, that the growth rate γ of the collisional tearing mode must satisfy

$$\gamma < \left(\frac{c}{\omega_{pe}}\right)^{8/3} \left(\frac{\omega_{ci}}{V_{th,i}}\right)^{2} \left(\frac{\xi}{V_{th,e}}\right)^{2/3} \nu_c^{5/3} , \tag{10.4.1}$$

where ω_{ci} is the ion gyrofrequency, ν_c is the electron-ion collision frequency given by equation (62) of Drake and Lee, and $\xi = B_0/\Delta b$ (in the notation of Drake and Lee, $\xi = l_s/a$). Some typical values for coronal active region loops are $\omega_{pe} = 3 \times 10^9$ s^{-1}, $\omega_{ci} = 3 \times 10^5$ s^{-1}, $V_{th,i} = (2KT_i/m_i)^{1/2} = 2 \times 10^7$ cm s^{-1}, $V_{th,e} = 9 \times 10^8$ cm s^{-1}, and $\nu_c = 50$ s^{-1} (the Coulomb logarithm is ≈ 20), and Eq. 10.4.1 requires $\gamma^{-1} > 10$ min if $B_0/\Delta b = 100$. It is possible that a stable current sheet can exist for this long a time in the corona without being dispersed by some other process such as wave propagation. The details of the collisional tearing mode depend heavily on the actual geometrical configuration of the current distribution (C. Chiuderi, personal communication, 1979). Perhaps it is premature to rule out the collisional tearing mode.

From Section II B of Drake and Lee, it can be shown that the growth rate of the semicollisional tearing mode is in the range

$$1 > \frac{\gamma}{\nu_c} > \frac{c^2 \omega_{ci}}{\omega_{pe}^2 V_{th,i} a} , \tag{10.4.2}$$

where $2a$ is the thickness of the current sheet. It can also be shown that

$$\frac{V_{th,i}}{\omega_{ci}} \left(\frac{\omega_{pe} l_e}{\xi c}\right)^{2/3} > a > \left(\frac{l_e}{\xi}\right)^{1/3} \left(\frac{c}{\omega_{pe}}\right)^{2/3} , \tag{10.4.3}$$

where $l_e = V_{th,e}/\nu_c$ is the electron mean-free-path (due to collisions with ions). Using the values following Eq. 10.4.1, we obtain $9.2 \times 10^4 > 2a > 530$ cm if $B_0/\Delta b = 100$. The corresponding range on the growth times is $0.02 \text{ s} < \gamma^{-1} <$ 10 min, the shorter growth times corresponding to the smaller values of a. The thickness $\triangle$ of the layer in which particle acceleration occurs is given by

$$\triangle = \left(\frac{\xi c}{l_e \omega_{pe}}\right)^{2/3} a . \tag{10.4.4}$$

For the values given above, we obtain $\triangle/a = 1.5 \times 10^{-3}$ and $67 > \triangle > 0.39$ cm; even for the slowest growing case, the heating layer is exceedingly thin. The magnetic energy that resides in the volume of width $2a$ is converted into internal energy of the plasma in the volume of width $\triangle$.

The semicollisional tearing mode can therefore have large growth rates. The real question is whether current sheets of the required thinness can form in association with whatever process carries energy from the convection zone into the corona.

10.4.3 Observables

Since the theory is not available at this time, the detailed predictions of the theory vis-a-vis observables cannot be commented upon. Presumably, the heating in layers of width $\triangle$ will involve large enough parallel electric fields to provide runaway electrons and nonthermal radio emission. And the tearing process itself will involve mass motions of the plasma, which may be recognizable as nonthermal line widths or even as non-Gaussian line profiles.

10.5 SURFACE WAVES ON LOOPS

10.5.1 Perspective

The previous two sections have been motivated by observations of a highly structured corona, but the theories require structures on scales much smaller than any that can be resolved. The situation is similar in the case of surface waves. It is presumed that gradients of V_A exist in the corona and that these gradients support Alfvenic surface waves. Somewhat in contradiction to their generic name, the surface waves can exist over essentially the entire cross section of a coronal active region loop (Wentzel, 1979a), and their existence does not require that the gradient in V_A be large (i.e., a "discontinuity" in V_A is not necessary). But the dissipation mechanism that Ionson (1978) has suggested for the surface waves does imply the existence of a very thin dissipation sheath, less than 1-km thick, inside the layer where V_A is changing (see also Wentzel, 1979b).

The importance of surface waves in heating coronal active region loops has been discussed by Ionson (1978) and Wentzel (1978, 1979a,b,c). In particular, they have considered a specific case, that of the Alfvenic surface wave, whose properties are looked at below. They envision a situation in which wave energy enters the loop from the convection zone in response to a large-scale shaking of a magnetic flux tube. The energy is assumed to come up as a surface wave, but it is important to note that the surface wave can actually exist over the entire cross section of the coronal loop if the index m is not too large (here m refers to a Fourier analysis in cylindrical coordinates, in which it is assumed that all quantities vary as $e^{im\phi}$). Thus, the entire cross section is available, and the energy flux *density* can be modest. For the Alfvenic surface wave, the energy flux is parallel to $\mathbf{B}_0$ when the waves are undamped. As long as m is not too large, the surface wave supplies energy to the loop in much the same way as an ordinary body Alfven wave. Hollweg (1979b) has shown that the loop energy requirements may very well be able to be supplied by body Alfven waves that originate in the convection zone, and it is therefore reasonable to suppose that the surface Alfven wave can likewise supply the required energy.

The Alfvenic body and surface waves are thus not too different vis-a-vis their ability to get energy into the loop, at least if m is not too large. The difference comes in the proposed mechanism of dissipation. The Alfvenic surface wave has been studied by Ionson, taking the finite thickness of the gradient of V_A into account. He proposes that a resonance between the Alfvenic surface and body modes can exist inside the layer where V_A is changing, and that the surface wave is thereby converted into a body Alfven wave. Ionson actually considers the body wave to be a kinetic Alfven wave, which can dissipate, so that in effect the surface wave dissipates and loop heating results. In the scenario proposed by Ionson, the energy flux of the surface wave acquires a component across $\mathbf{B}_0$ in order to provide the energy for heating.

It should be emphasized that the surface waves represent one mode by which a coronal loop can respond to shaking. In general, body and surface waves will be produced together. If the gradients of the medium are not too large, i.e., if there are no strong "discontinuities," the surface and body modes merge together. It is even possible to find body modes that do not directly respond to the structure of the medium, as in Section 10.7, below. What is ultimately required is a fully self-consistent analysis of the response of the structured solar atmosphere to motions in the convection zone.

10.5.2 Theory

The following brief summary of the surface wave heating theory is based on Ionson (1978) and Wentzel (1978, 1979a,b,c).

Consider first a true discontinuity in the y-z plane across which the magnetic field strength and density change. $\mathbf{B}_0$ is in the z-direction on both sides. Assume the thermal pressure to be zero on both sides. The background is uniform on both sides. The discontinuity supports a surface wave with dependence $\exp(\pm qx + ik_y y + ik_z z - i\omega t)$, where $q > 0$ and the $\pm$ is chosen so that the signal vanishes at $x = \pm\infty$. If $k_z^2 \ll k_y^2$, the dispersion relation is

$$\frac{\omega^2}{k_z^2} = \frac{B_{01}{}^2 + B_{02}{}^2}{4\pi(\varrho_{01} + \varrho_{02})} \tag{10.5.1}$$

Also,

$$q^2 = k_y^2 + k_z^2\left[1 - \frac{4\pi\varrho_0\omega^2}{B_0{}^2k_z{}^2}\right] \tag{10.5.2a}$$

$$\approx k_y^2 + k_z^2\left[1 - \frac{\varrho_0(B_{01}{}^2 + B_{02}{}^2)}{B_0{}^2(\varrho_{01} + \varrho_{02})}\right] \tag{10.5.2b}$$

$$\approx k_y{}^2 \tag{10.5.2c}$$

on each side. The time-averaged Poynting flux in the z-direction is

$$<P_z> = \frac{k_z}{\omega}<\delta V_x^2 + \delta V_y^2> \frac{B_0^2}{4\pi} \tag{10.5.3}$$

Since δV_x is continuous across the boundary, and since $\delta V_y \approx \pm i\delta V_x$, the Poynting flux is largest on the side with the largest B_0^2. If the discontinuity corresponds to the outer edge of a coronal active region loop, and if B_0 does not change across the discontinuity, then as much energy comes up outside the loop as inside (ignoring the cylindrical geometry). The average Poynting flux in the x-direction is

$$<P_x> = (B_0^2/4\pi) < \delta V_x(\pm q\delta V_x/i\omega) + \delta V_x(k_y\delta V_y/\omega)> \quad . \qquad (10.5.4)$$

Since δV_x and δV_y are 90 degrees out of phase, $<P_x> = 0$.

Ionson has argued that the body-wave/surface-wave resonance in a discontinuity of finite thickness leads to a complex ω, with

$$\frac{I_m\omega}{|R_e\omega|} = -\pi\left(\frac{\varrho_1}{\varrho_2} + \frac{\varrho_2}{\varrho_1} + 2\right)^{-1} \frac{|k_y|a\Delta V_A(4\pi\varrho_{av})^{1/2}}{B_0} \quad , \qquad (10.5.5)$$

where a is the thickness of the discontinuity and ΔV_A is the magnitude of the difference between the Alfven speeds on the two sides. $I_m\omega/|R_e\omega| \ll 1$ by assumption. Note that $I_m\omega$ does not depend in any way on there being a dissipative process, and that it exists even in a perfectly cold plasma. Ionson argues that $I_m\omega \neq 0$ results from there being an MHD resonance at the point where $\omega/k_z = V_A(x)$. Note too that $I_m\omega$ vanishes if there is a true discontinuity, $a = 0$. Because $I_m\omega \neq 0$, $<P_x>$ is no longer zero. (Quantities in Eq. 10.5.4 that were formerly 90 degrees out of phase are no longer so. Since $I_m\omega \neq 0$, q must be complex.) The surface wave energy flows into the plane where $\omega/k_z = V_A(x)$, where it vanishes in the MHD theory. However, Ionson suggests that the energy really goes into heating a thin layer, if non-MHD dissipative processes are included. The thickness of the layer is given roughly by

$$\varrho_{gi}\left(\frac{a}{\varrho_{gi}}\right)^{1/3} ,$$

where ϱ_{gi} is the ion gyroradius. The thickness turns out to be less than 1 km. This is really an upper limit, obtained by taking $a = R$ (R is the loop radius) for a long loop with $R \approx 10^9$ cm and $B = 10$ gauss. A short loop with $R = 10^8$ cm and $B = 100$ G would have a sheath thickness less than ≈ 0.1 km.

The frequency of the waves must be such that the damping length exceeds the loop length, and the wavelength $2\pi k_z^{-1}$ must be less than the loop length. For the long coronal active region loops, Ionson finds that waves with $\tau \approx 300$ s are suitable. For the short loops, periods of a few tens of seconds would be appropriate; the microturbulence could conceivably contain power at these periods and excite the waves. But it should be noted that no attempt has been made to consider the wave propagation problem globally; i.e., features such as possible loop resonances, reflections off of the transition regions at the ends of the loops, the effects of propagation through the lower solar atmosphere where the wave dispersion properties are strongly varying functions of height, etc., have not been considered. These features are of paramount importance for the Alfven wave (see below), and they can be expected to be similarly important for the Alfvenic surface wave.

Ionson goes on to suggest that the heating in the dissipation sheath will drive upward convection in a convection layer next to the sheath. He estimates the thickness of the convection layer to be about 30 km for the long coronal active region loops and some 3 km for the short loops; the layer thickness δ_c is given by

$$\delta_c \approx 11 T_6^{3/8} L_{10}^{3/8} B_1^{-1/2} \text{ km} , \tag{10.5.6}$$

where L is the loop length, T is the sheath temperature, and the subscripts are the powers of 10 to which the quantities (in cgs units) are normalized. The upward flow speed in the convection layer is roughly the free-fall speed based on the loop length, i.e., about 100 km s^{-1} for a long loop. The plasma rises in the convection layer, crosses the field lines via an electrostatic Rayleigh-Taylor instability near the top of the loop and thereby enters the loop interior, cools via radiation near the top of the loop, and subsequently falls in a cool core whose diameter is essentially the loop diameter. Ionson invokes the down flow as an explanation for "coronal rain" loops (Foukal, 1976, 1977). The interior temperature T_{int} is given in terms of the sheath temperature T_s as

$$\frac{T_{int}}{T_s} \approx \frac{2.4 \times 10^{-2}}{n_{e9} L_{10}^{1/2}} \ . \tag{10.5.7}$$

If $T_s \approx 3 \times 10^6$ K, then $T_{int} \approx 10^4$ K.

A possible problem with the theory is the predicted nonlinearity of the waves in the dissipation sheath. The amplitudes must become very large there, because the entire energy of the surface wave, which extends over a distance of the order of k_y^{-1}, must be thermalized in the exceedingly thin dissipating sheath. Ionson estimates the ratio of the magnetic field fluctuations inside and outside the sheath to be

$$\frac{|\delta B_{\text{sheath}}|}{|\delta B_{\text{outside}}|} \approx \left(\frac{\omega}{\nu_{ii}}\right)^{1/2} \left(\frac{a}{\varrho_{gi}}\right)^{2/3} \left[\frac{\pi \Delta V_A}{V_A}\left(\frac{\varrho_1}{\varrho_2} + \frac{\varrho_2}{\varrho_1} + 2\right)^{-1}\right]^{1/2} , \tag{10.5.8}$$

where ν_{ii} is the ion-ion collision frequency (this result follows from equating the energy loss rate of the entire surface wave to the viscous dissipation rate of the kinetic Alfven wave in the sheath). Since $(a/\varrho_{gi})^{2/3}$ is large, the velocity and magnetic field fluctuations in the sheath can become nonlinear, i.e., $\delta B/B_0 \approx \delta V/V_A > 1$; in addition, $\delta V/V_s \gg 1$, where V_s is the sound speed. This means that a nonlinear analysis is necessary. Reconnection, Kelvin-Helmholtz instabilities, and shock formation can conceivably occur. In addition, since the excursion of material near the sheath is much larger than the sheath thickness, an effective nonlinear broadening of the sheath may occur.

We must confess to having some doubts about the interpretation of the "surface wave dissipation rate," Eq. 10.5.5, discussed by Ionson (1978) and Wentzel (1979b). These authors begin with a pure MHD system that is strictly dissipationless, but they nonetheless obtain a nonzero value for ω_i (Eq. 10.5.5), which they interpret in terms of "dissipation of surface waves." The nonzero value of ω_i exists even when the plasma is perfectly cold. It is therefore difficult to see how their result can be interpreted in terms of dissipation. In fact, M. Lee (private communication, 1979) has called attention to a general energy principle for MHD eigenmodes. This energy principle (see, for example, Chandrasekhar, 1961, Chapter XIV) can be applied to surface eigenmodes, and leads to the result that ω^2 must be purely real for an eigenmode, i.e., ω must be purely real or purely imaginary; for the case considered by Ionson and Wentzel, ω^2 must in fact be purely real, in contrast to Eq. 10.5.5. Lee shows that Ionson's and Wentzel's result therefore does not refer to a surface eigenmode, or to dissipation thereof. Instead, Lee suggests that Eq. 10.5.5 must instead refer to redistribution of wave energy in space.

10.5.3 Observables

As is found below in connection with the Alfven and fast MHD wave modes, the Alfvenic surface waves can supply the required energies to the coronal active region loops with coronal velocities of the same order as are inferred from nonthermal line widths, i.e., tens of km s^{-1}. In this sense, the observations already confirm the possible importance of surface waves. More detailed confirmation would require simultaneous observations of δV_x and δV_y and the phase between them, and similarly for δB_x and δB_y, both interior and exterior to the coronal active region loops. It is unlikely that observations at the required level of detail are feasible.

The existence of "coronal rain" downflows and the possible presence of cool cores is certainly compatible with Ionson's basic theory [although any observed cool cores do not occupy most of the loop, as in Ionson's analysis (Levine and Withbroe, 1977)]. But, for the reasons given above, we do not regard either of these features as very useful diagnostics.

The most noteworthy feature of Ionson's analysis is the suggestion of a hot upflowing convection layer of some 3–30 km thickness. (The dissipating sheath itself is too thin to detect.) Future high-resolution observations may be able to confirm or refute the presence of at least the thickest of such layers. It is important to note, however, that a convection layer does not necessarily confirm the surface wave mechanism itself, since convective layers could result from any sheath heating mechanism, such as the anomalous resistivity mechanism of Section 10.3.

Finally, the kinetic Alfven wave, which is resonantly produced in the sheath, also has a parallel electric field that can, in principle, produce runaway electrons and nonthermal emission. The relevant calculations are not yet available, however.

10.6 CORONAL FAST WAVES

10.6.1 Perspective

Again motivated by the observed structure of the quiet and active region corona, Habbal *et al.* (1979) have noted that the structure should lead to refraction and selective regions of dissipation for body MHD wave modes. Since the slow mode is probably unable to supply the required energies to the corona, and since the Poynting flux of the Alfven mode is unaffected by refraction, Habbal *et al.* concentrate on the fast mode. They investigate, in particular, the possibility that coronal active region loops appear as distinct features because the fast wave energy is selectively deposited at certain preferred coronal sites, even though the energy enters the corona uniformly from below.

Unlike the Alfvenic surface and body waves, the fast waves can carry energy across the magnetic field. This gives them a certain advantage over the Alfvenic modes: the energy requirements of distinct coronal features can be supplied by waves of modest energy flux *density* entering the corona over a large surface area, if the energy can be focused into the right place. By contrast, the Alfvenic modes must carry all the energy only along the magnetic field lines that pass through the heated region, and large energy flux densities may be required.

The fast-mode waves have an advantage also over the slow (sound) mode: for a given observed rms velocity fluctuation, the energy flux density that can be carried by the fast mode is roughly (V_A/V_s) times the corresponding quantity for the sound wave. Since (V_A/V_s) is large in the upper chromosphere, the observed velocity fluctuation amplitudes do imply sizable fast wave fluxes, in contrast to the sound waves. (The Alfven mode also has this advantage.)

There are two major disadvantages to the fast mode, however:

First, consider the cold coronal plasma. The fast wave dispersion relation is $\omega^2 = k^2 V_A^2$. Splitting $\mathbf{k}$ into horizontal and vertical components gives

$$k_v^2 = (\omega^2/V_A^2) - k_h^2 \ . \tag{10.6.1}$$

If $k_h > \omega/V_A$, the waves are evanescent and carry no energy upwards. It turns out (Hollweg, 1978b) that the known solar motions have values of ω and k_h for which the fast waves in the corona should be evanescent. This argument is not airtight, because the existing ω-k_h diagrams refer to photospheric and chromospheric motions; what is really needed is an ω-k_h analysis of coronal (or upper transition region) motions. This argument can also break down if the corona is structured on scales comparable to or smaller than the fast-wave wavelength. It should be noted that active regions are characterized by larger values of V_A and presumably larger values of k_h than quiet corona or coronal holes. These trends work in the direction of enhancing the evanescent region of the ω-k_h diagram. If fast waves were important, one might then expect that active regions should be less energetic than other coronal regions, which is apparently not the case. Similarly, Hollweg (1978b) has argued that the centers of coronal holes might have larger values of V_A than the edges. This would tend to refract fast waves away from the centers. If coronal holes were heated or accelerated by fast waves, the centers should be less energetic than the edges, which is not the case.

An even more serious disadvantage is the possibility that fast waves that approach the transition region from below can be totally internally reflected and never reach the corona. (Total internal reflection is, of course, related to evanescence.) The Brewster angle is only 6 degrees if the ratio of the Alfven speeds above and below the transition region is 10. In that case, less than 0.5 percent of an isotropically distributed flux below the transition region can get into the corona. Further, ray-tracing (Osterbrock, 1961) suggests that the flux below the transition region will not be isotropic—in a sense that increases the fraction of rays that are totally reflected. We are again pessimistic about the role of fast waves in the corona.

It is not clear whether fast waves can explain why the longer coronal active region loops have smaller volumetric heating rates. The collisionless damping invoked by Habbal *et al.* increases with $\beta_p = 8\pi n_p K T_p / B^2$. Since longer loops have lower pressure, they could also have lower values of β_p and less damping and heating. On the other hand, longer loops have smaller magnetic field strengths, which works in the opposite direction. This needs to be investigated, along with the corresponding role of collisional damping, which will be important on the shorter active region loops (see below).

10.6.2 Theory

Habbal *et al.* assume waves with 3-s period. This period is chosen to keep the wavelengths small enough to allow a ray-tracing analysis to be used. It also allows reasonable damping lengths in the corona. It should be noted, however, that waves with a period of 3 s would probably be damped by Joule-viscous-frictional damping in the chromosphere (e.g., Osterbrock, 1961) and be unable to reach the corona from the convection zone.

They assume the (potential) magnetic field to be an offset dipole, and they assume hydrostatic equilibrium along field lines. They investigate an isothermal corona, but they allow a certain group of field lines to have higher or lower pressures than the surroundings. Their model is two-dimensional.

The waves are assumed to damp by Landau/transit-time damping (Barnes, 1966). This form of damping is collisionless and requires $\omega\tau_{pp} \gtrsim 1$, where ω is wave angular frequency and τ_{pp} is the proton-proton collision time (this assumes that the wave damping is via the Landau resonance of the ions, which is the case in the model of Habbal *et al.*). Using Braginskii (1965), this in turn requires

$$p_p \lesssim 3.9 \times 10^{-16}\, T_p^{5/2} \tau^{-1} \,, \qquad (10.6.2)$$

if the wave period is τ. This criterion is satisfied for the low-pressure loops considered by Habbal *et al.*, but it is not satisfied for the more compact higher-pressure loops. When $\omega\tau_{pp} \lesssim 1$, Hung and Barnes (1973a,b,c) have shown that the fast waves still damp, but by collisional processes such as Joule, viscous, and heat-conduction damping.

The interesting feature of Landau/transit-time damping is that it increases with β_p if β_p is small. This occurs because the number of resonant particles increases with β_p in that case. In the isothermal model of Habbal *et al.*, β_p is largest where V_A is smallest. However, the fast waves refract toward regions of lower V_A, and thus toward regions of larger damping. This means that the fast waves can lead to stronger heating in some regions of the corona than in others. The effect is enhanced by a "positive feedback" process in which local heating increases β_p, which in turn increases the heating rate, and so on. A "catastrophic" situation can occur in which once fast-mode waves begin heating the plasma, they dump all their energy in a small volume of space. Habbal *et al.* suggest that such processes can account for the highly structured nature of the active region corona in a natural way. They even suggest that cool cores are a natural consequence of coronal heating by fast waves: waves impinging on a cylindrical structure from the outside could dump all their energy on the outside of the cylinder and be unable to penetrate to the interior, resulting in a cool core.

Habbal *et al.* demonstrate via some simple examples that fast waves can lead to selective heating of coronal structures, in ways that can be compatible with observed structures; but detailed models of particular structures were not intended.

10.6.3 Observables

The model of Habbal *et al.* is primarily schematic, and detailed predictions of observables are not intended.

The general level of the observed coronal and transition region nonthermal line widths is consistent with a flux of fast-mode waves that is large enough to provide the coronal energy requirements. But it needs to be demonstrated that the nonthermal velocities are in fact fast-mode waves. One way to do this would be to observe simultaneously as many of the fluctuating wave quantities as possible, i.e., $\delta\mathbf{V}$, $\delta\mathbf{B}$, δn, and δT, as well as $\mathbf{B}_0$. In a low-β plasma, a fast wave has $\delta\mathbf{V}$ perpendicular to $\mathbf{B}_0$, δn and δT in-phase with $\delta|\mathbf{B}|$, and $\delta\mathbf{B}$ perpendicular to $\mathbf{k}$ but in the plane containing $\mathbf{k}$ and $\mathbf{B}_0$. Coronal observations of sufficient detail to verify these correlations seem unlikely in the near future. However, observations using Stokes polarimetry could conceivably search for the magnetic fluctuations of fast waves below the transition region.

An alternative approach would be to produce an ω-k_h diagram for the base of the corona (or even in the upper chromosphere). This approach cannot demonstrate the presence of fast waves. But, if it can be shown that the observed motions lie in the evanescent region, then fast waves can be ruled out.

Similarly, knowledge of the temporal power spectra will be needed in order to assess the damping lengths and the extent to which the wave refraction can be treated by ray-tracing techniques. If the wavelengths are not small compared to the spatial scales of the background, then full solutions of the wave equation will probably be required, at which point the analysis will be closely related to that employed for the surface waves.

Fast-mode MHD waves have parallel electric fields, and may therefore be expected to lead to some runaway electrons and nonthermal emission. The observational consequences have not been computed, however.

Finally, it should be mentioned that refinements to the theoretical calculations of Habbal *et al.* require more detailed observational knowledge of how $\mathbf{B}_0$, β_p, and V_A are structured in the corona, both inside and outside of the loops.

10.7 ALFVEN WAVES ABOVE THE CONVECTION ZONE

10.7.1 Perspective

It is fashionable to talk about coronal "magnetic heating" by "twists," "shears," and "parallel currents associated with twists or shears." The twists and shears are generally envisioned as being generated in the convection zone by twisting the footpoints of magnetic flux tubes. It seems not to be realized, however, that twisted and sheared magnetic fields are really nothing more than Alfven waves, obeying a specific set of wave equations. The idea of coronal heating by twists, etc., is not new, having been discussed by Alfven (1947) and numerous workers subsequently. It is our feeling that the twists, etc., should be studied using the wave equations that they obey as the starting point.

Alfven waves are noncompressive and therefore particularly amenable to detailed calculations. Questions such as how the energy comes up from the convection zone can therefore be addressed in detail; this stands in contrast to the other investigations summarized in this chapter. On the other hand, it will be seen that pure twists are particularly difficult to dissipate, especially in the corona. The situation with Alfven waves is therefore the converse of the other investigations summarized here: the energy flow can be investigated in detail, but not the damping.

In addition to their tractability, Alfven waves offer other advantages and disadvantages. There is never evanescence or total internal reflection of Alfven waves, in contrast to sound and fast waves. They can exist only in the presence of a magnetic field, and their energy always propagates along $\mathbf{B}_0$, in qualitative agreement with observations indicating that much heating of the solar atmosphere is magnetically associated. The energy automatically gets to where the magnetic fields are. On the other hand, the channeling of energy along $\mathbf{B}_0$ requires large Poynting fluxes, perhaps in excess of 10^7 erg cm^{-2} s^{-1} for the shorter coronal active region loops.

We believe that Alfven waves should not be ruled out as candidates for heating and acceleration of the solar atmosphere, because the solar wind probably stands as a precedent illustrating how they fill these two roles in nature. Moreover, the solar wind data probably shows that the Sun does radiate energetically interesting fluxes of Alfven waves. In addition, the model described below can probably account for the sun's generation of the waves seen in the solar wind, and we therefore have some confidence that it can be applied to other frequencies and other magnetic field configurations.

10.7.2 Theory and Results

Hollweg (1979b) has investigated the propagation of axisymmetric twists on an axisymmetric potential magnetic flux tube whose axis is vertical; the field expands in a fleur-de-lis pattern. The wave velocity is horizontal and no coupling to gravity occurs. Because of the sheared nature of the flow, each field line behaves independently of the others, and no propagation of information across field lines occurs, regardless of the extent to which the background density and magnetic field are structured in the nonazimuthal directions. Hollweg's equations are linearized, and damping is not included. Near the axis of the flux tube, the twists obey:

$$\left(V_A{}^2 \frac{\partial^2}{\partial_s{}^2} + \omega^2\right) B_0^{1/2} \delta V = 0 \qquad (10.7.1)$$

$$i\omega\, \delta B = B_0^{1/2} \frac{\partial}{\partial s} (B_0^{1/2} \delta V) \;, \qquad (10.7.2)$$

where s is distance along a field line, and harmonic motions at angular frequency ω have been assumed.

Because Eqs. 10.7.1 and 10.7.2 are one-dimensional, their numerical solution is straightforward. This means that the problem can be solved self-consistently in the entire region above the convection zone. Departures from the short-wavelength WKB limit can be studied in detail; non-WKB solutions are mandatory, because no frequency of interest gives waves whose wavelengths are short enough to satisfy the WKB approximation. Hollweg investigates the consequences of an imposed velocity fluctuation at the "top" of the convection zone (i.e., at zero height in the HSRA/VAL atmospheres) with emphasis on the energy flux that results. The density and magnetic field structures of the atmosphere are realistically modeled, but the calculations are confined to near the flux tube axis. Wave propagation on open field lines in and below a coronal hole, and on closed field lines in and below coronal active region loops, are investigated.

A. Open field. Long-period ($\tau > 10$ min or so) waves are predicted to carry an energy flux density $\langle P_s \rangle$ equal to about a few times 10^4 erg cm^{-2} s^{-1} (P is Poynting flux, the subscript s indicates the value at zero height, and the angle brackets indicate the average value obtained by taking the filling factor of the flux tubes into account). This value is at the low end of the "observed" range deduced from solar wind studies. Larger values of $<P_s>$, sufficient to drive high-speed solar wind streams, can probably be produced on field lines that are not near the axis of the flux tube, but the detailed structures of the flux tubes are not known with sufficient accuracy to warrant detailed computational support for this point. (Because of the greater length of nonaxial field lines, dV_A/ds is smaller on them, with the result that the wave reflections should be less severe and greater energy fluxes should result; but this effect could be counteracted by weaker field strengths away from the axis.) Larger values of $<P_s>$ could also result if the field strength at the base of the model is greater than the assumed value of 1500 G. But the value of $<P_s>$ is quite insensitive to the filling factor.

Alfven waves exert forces on the solar atmosphere, due to the $(\delta\mathbf{J} \times \delta\mathbf{B}/c) - (\varrho_0 \delta\mathbf{V} \cdot \nabla\delta\mathbf{V})$ ponderomotive force; this force may drive the high-speed solar wind streams. Hollweg suggests that the long-period waves might drive macrospicules, and they may account for the 5–10 arcsec "corrugation" of the transition region below coronal holes. Fully nonlinear dynamical models of the wave propagation are required to test this suggestion, however.

At high frequencies, 10 s $\lesssim \tau \lesssim$ 5 min, the model predicts $<P_s>$ about equal to a few $\times\ 10^6$ erg cm^{-2} s^{-1}. If the waves can be dissipated, fluxes of this magnitude can account for chromospheric and coronal heating. The energy is there; the problem is in the damping.

The high-frequency waves should exert upward forces on the chromosphere in excess of $\varrho_0 g$ (g is gravitational acceleration). Hollweg suggests that spicules may be a consequence, but fully nonlinear calculations are again needed.

How the waves damp is not understood. Hollweg has rescaled Osterbrock's (1961) estimates of the Joule, viscous, and frictional damping rates and found that Joule-frictional damping can be important in the middle chromosphere only for waves with $\tau \lesssim 10$ s; the damping lengths scale as τ^2. Unless most of the wave power is at such high frequencies, which seems unlikely, Joule-frictional damping will not be important on open field lines.-tively, Hollweg's estimates of the damping coefficients are crude. Estimates based on detailed flux tube models are needed.)

Hollweg suggests in addition that the waves could damp nonlinearly. If $<P_s> \approx 10^6$–10^7 erg cm^{-2} s^{-1}, moderate nonlinearities are required below the corona, in the sense that $<\delta V^2>/V_A^2 \approx 1$. Large relative shears, of the order of V_A, could develop between "adjacent" field lines, and these could be unstable to a Kelvin-Helmholtz instability. Such an instability could result in energy "cascading" to higher frequencies, which can be dissipated via Joule-frictional damping and/or shock formation. Similarly, Hollweg finds $<\delta B^2>/B_0^2 \approx 1$ in the photosphere and chromosphere. "Adjacent" field lines can be sheared with respect to one another, suggesting the possibility that the waves can be unstable to collisional tearing. Hollweg estimates that wave energy can be dissipated in this fashion if the transverse scale of the shears is less than some 14 km.

The ponderomotive force can also lead to nonlinear damping of the Alfven wave by driving compressional modes (i.e., fast or slow modes), which are more amenable to damping than the Alfven mode. The compressional modes can damp via Joule, viscous, frictional, or heat-conduction damping, or via shock formation if their amplitudes are large enough. The ponderomotive force can probably deposit a significant fraction of the Alfven wave energy into compressive modes in the chromosphere on open field lines. The ponderomotive force is weak in the photosphere, however. And it is difficult to get a significant fraction of the Alfven wave energy into the compressive modes in the corona, because the first-order Alfvenic velocity and magnetic field fluctuations are linear (in the sense that $<\delta V^2>/V_A^2$ and $<\delta B^2>/B_0^2$ are small) in the corona. On the other hand, if Alfven waves can nonlinearly drive shocks or compressive modes in the chromosphere, it is possible that these shocks, or driven compressive modes, can subsequently enter the corona, where they can heat the corona via viscous and heat-conduction damping, or shock dissipation. Hollweg, Jackson, and Galloway (1980, work in progress) have recently been carrying out a numerical investigation of the nonlinear propagation of Alfvenic "twists" on open magnetic flux tubes. They find that waves with periods of less than a few minutes, and velocity amplitudes greater than about 0.5 km s^{-1} in the photosphere, can nonlinearly steepen to form fast and slow shocks in the chromosphere. Chromospheric heating is one consequence. Moreover, some of the energy in shocks and driven compressive modes penetrates into the corona and heats that region. Finally, the waves exert a large enough ponderomotive force on the upper chromosphere and transition region to drive spicules. The model holds much promise for heating and accelerating the solar atmosphere.

Another damping mechanism involves "geometrical MHD wave coupling" (Hollweg and Lilliequist, 1978). This effect is linear, but it exists only in virtue of the finite wavelengths; i.e., it vanishes in the WKB limit. What geometrical MHD wave coupling does is convert energy from the Alfven to fast and slow modes as the Alfven waves undergo refraction and/or try to propagate around curved magnetic field lines. In the low-β coronal plasma, coupling into the fast mode is preferred. If the fast mode subsequently damps via viscosity or Landau/transit-time damping, heating of the corona can result. However, the viscous damping of fast waves is strongest for fast waves propagating across $\mathbf{B}_0$, but the geometrical coupling from Alfven waves is weakest for cross-field propagation.

A final interesting possibility is the following: Hollweg has shown that Alfven waves can supply the energies required for coronal heating. The difficulty arises in the damping, which is weak in the corona because the waves are compressive only in second order in the wave amplitudes. The Alfvenic surface mode, however, obeys a dispersion relation that is very similar to that of the pure Alfven mode. In particular, only $k_\parallel$ appears in the dispersion relation (e.g., Eq. 10.5.1), and this mode too is therefore not subject to evanescence or total internal reflection. The conclusion is that if the Alfven wave can supply the required energies to the corona, then so too can the Alfvenic surface wave. The Alfvenic surface wave has the advantage of being weakly compressive even in first order, and therefore subject to viscous and heat-conduction damping in the corona. This damping occurs throughout the volume occupied by the wave, and is distinct from the sheath effect proposed by Ionson (1978). We have even made a preliminary estimate of this damping. Consider a uniform magnetic field $\mathbf{B}_0$. Let the plasma density and pressure be ϱ_0 and p_0, except in a cylinder of radius R, where they are $\varrho_i(>\varrho_0)$ and p_i. If $B_0^2/8\pi \gg (p_0,p_i)$, then the wave dispersion relation is given by Eq. 10.5.1, as long as $Rk_z \ll 1$. Let $\vec{\xi}$ be the displacement of an element of plasma due to the wave. Then from Wentzel (1979a), it is easy to show that

$$\frac{\nabla \cdot \vec{\xi}}{\xi_\theta} = \pm \frac{ir}{m} k_z^2 \left(\frac{\varrho_i - \varrho_0}{\varrho_i + \varrho_0}\right) , \tag{10.7.3}$$

where cylindrical coordinates (r,θ,z) are employed, and it is assumed that all wave quantities vary as $e^{im\theta}$; the upper (lower) sign refers to $r < (>) R$. The quantity $\nabla \cdot \vec{\xi}$ is a measure of the compressibility of the plasma, and therefore of the viscous and heat-conduction damping. The viscous damping comes mainly from the protons in the coronal plasma. If $\omega_{gp}\tau_p \gg 1$ and $\omega\tau_p \lesssim 1$ (ω_{gp} is the proton gyrofrequency and τ_p is the proton collision time), then the volumetric heating rate due to viscosity is $\eta_0(\nabla\cdot\delta\mathbf{V})^2/3$, where $\eta_0 = 0.11\ T_{p6}^{5/2}$ (cgs) if the Coulomb logarithm is 20, and where it has been assumed that $\delta V_z \approx 0$ for the Alfvenic surface wave (Braginskii, 1965). Also from Braginskii, the volumetric heating rate due to electron heat conduction is $\varkappa_{\|e}(\nabla_\|\delta T_e)^2/T_e$, where $\varkappa_{\|e}$ is the electron heat conductivity parallel to $\mathbf{B}_0$ (it has been assumed that $\varkappa_{\perp e}$ is negligible). If the Coulomb logarithm is 20, then $\varkappa_{\|e} = 9.11 \times 10^8\ T_{e6}^{5/2}$ (cgs). If the heat conduction is weak, then δT_e is given by the usual adiabatic relation; this requires that the wave period be shorter than the thermal conduction time, which is roughly $3n_e\varkappa/(2\varkappa_{\|e}k_z^2)$, where $\varkappa$ is Boltzmann's constant. If the damping is weak, the damping length L_z is simply given as the Poynting flux in the z-direction divided by the volumetric heating rate. We obtain

$$L_{z5} = \frac{\tau^4 B_1^5 T_6^{-1} p_{av}^{-3/2} m^2(\varrho_i + \varrho_0)^2}{[0.215 + 0.0358\, p_{av} B_2^{-2}] r_8^2 (\varrho_i - \varrho_0)^2} \tag{10.7.4}$$

The subscripts are again the powers of ten to which the quantities (in cgs units) are normalized. The quantity p_{av} is the average of the pressures inside and outside the cylinder. In deriving Eq. 10.7.4, $T_e = T_p$ has been assumed, and the ratio of specific heats has been taken to be 5/3. The damping length is very sensitive to wave period τ and magnetic field strength, because the compressive component of the wave depends on these quantities (Eq. 10.7.3). This sensitivity means that the damping lengths are huge in coronal active region loops, and this damping mechanism is unsuitable there. In coronal holes, however, we might have $B_0 = 5$ gauss, $T_6 = 1$, $p_{av} = 0.02$ dyne cm^{-2}, $r = 5000$ km (based on a reasonable mean separation between photospheric flux tubes), and $(\varrho_i - \varrho_0)/(\varrho_i + \varrho_0) = 1$ if the cylinders are surrounded by vacuum. If $\tau = 20$ s, and if $m = 1$, then $L_z = 0.2$ solar radii. If the wave periods are short enough, this mechanism may be effective in heating coronal holes. Thus, further analysis, relaxing the above assumptions, seems to be warranted.

B. Closed loops. Hollweg has modeled Alfven wave propagation on a symmetric coronal active region loop of 33,400 km total length (i.e., from convection zone to convection zone). The field strength is 1500 G at the convection zones, and 100 G in the corona. The coronal pressure is 6 dyne cm^{-2}. The loop has been artificially straightened out, to avoid coupling into gravity and other modes.

Long-period waves, $\tau \gtrsim 10$ min, are probably unable to supply the large energies required by coronal active region loops. This means that quasi-static twists, and the associated parallel currents, are not attractive candidates for coronal active region loop heating. This conclusion is confirmed by Querfeld and Hollweg (1979), who used coronal emission line polarimetry to search for long-period fluctuations in the magnetic field direction in a quiet coronal region near 1.3 $R_\odot$. They found that for periods longer than 0.5 h, an upper limit of 10^3 erg cm^{-2} s^{-1} was implied by the results.

The most interesting feature exhibited by Alfven wave propagation on closed loops is the appearance of "loop resonances." These result from wave energy being stored in the corona, as a result of reflections off the steep gradients in V_A at each end of the coronal loop. The resonant wave periods are

$$\tau_{res} = \frac{2L_{cor}}{mV_{A,cor}} , \qquad (10.7.5)$$

where $m = 1,2,3$, etc., and L_{cor} is the length of the coronal part of the flux tube. A short loop might have $L_{cor} = 2 \times 10^4$ km, $B_0 = 100$ G, and $n_p = 10^{10}$, giving $\tau_{res} = 20$ s, 10 s, 7 s, etc. Longer loops and/or loops with smaller Alfven speeds would have longer resonant periods. Duebner's (1976) results indicate that the microturbulence contains power at these resonant periods, and thus the loop resonances could be excited. The remarkable feature about the loop resonances is that they act like a "window" whereby a large energy flux can pass almost unimpeded from the convection zone into the corona. At resonance, the energy flux density is essentially the WKB value, $\varrho_0 < \delta V^2 > V_A$; if $B_0 = 1500$ G, $\varrho_0 = 3 \times 10^{-7}$ gm cm^{-3}, and $\delta V_{rms} = 1$ km s^{-1} near the top of the convection zone, then the WKB flux density is 2.3×10^9 erg cm^{-2} s^{-1}. If the magnetic field is 100 G in the corona, then $< P_s > = 1.5 \times 10^8$ erg cm^{-2} s^{-1}. This value must be diluted by the fraction of the frequency domain that is available to excite the resonances. Hollweg's results indicate that 0.1 might be a typical value of this fraction, leading to an energy flux 1.5×10^7 erg cm^{-2} s^{-1} into a short loop. This is roughly the required energy. The loop resonances act as a window for several reasons. Normally, upward-propagating wave energy is reflected off the steep Alfven speed gradient, and not much energy reaches the corona. However, when the loop resonances are excited, the reflections at the top of the transition region cancel the reflections occurring below, and a large energy flux can enter the corona. (The effect is similar to the behavior of antireflection coatings on camera lenses.) In addition, because the reflections below the transition region are cancelled, not much energy is stored in the photosphere and chromosphere, and this means that the wave dissipation is weak below the transition region, again allowing energy to penetrate to the corona from the convection zone.

There are several ways in which the loop resonances could account for the observed fact that longer loops require smaller values of $<P_s>$. The longer loops generally have greater diameters and smaller field strengths. This means that the filling factor decreases with increasing loop size. Since the WKB flux density at the base of the loop is (presumably) independent of filling factor, the value of $<P_s>$ entering the corona decreases with decreasing filling factor and thus decreases with increasing length. If the length/diameter ratio of a loop is constant, then the filling factor could vary as L^{-2} if all loops have the same total magnetic flux. In that case, $<P_s> \propto L^{-2}$. This result is consistent with Eq. 10.2.3 above. Other factors can affect $<P_s>$, however. Longer loops presumably have larger values of τ_{res}; if the power in the convective motions varies with period, a corresponding variation in $<P_s>$ could result. There may also be some tendency for the loop resonances to become narrower on longer loops, and this by itself would result in a negative $<P_s>$-L correlation. Finally, what really matters for coronal heating is the energy that is actually dissipated in the corona. The dissipation could vary with wave period, and therefore with L_{cor} in virtue of Eq. 10.7.5.

Outside of the loop resonances, high-frequency ($\tau \lesssim 10$ min) undamped waves are not able to carry a significant energy flux, because the reflections are severe. However, this means that a lot of energy is stored below the transition region, in an (approximately) standing wave. The energy storage leads to significant damping in the photosphere and chromosphere. But damping will tend to reduce the reflections and allow more energy to propagate. It is possible that a balance can exist in which an energetically significant energy flux can propagate upward from the convection zone and be dissipated in the photosphere and chromosphere, but a model that includes damping self-consistently is necessary to test this possibility.

In contrast to the open field lines, Alfven waves on closed flux tubes can lead to net forces that can be either upwards or downwards. It is difficult to draw any general conclusions about the consequences of these forces, except to say that they may account for the complexity of active region motions.

The basic considerations vis-a-vis damping are much the same on closed flux tubes as on open. Coronal damping is particularly difficult. The waves in the corona are generally linear, and it does not appear to be feasible to nonlinearly couple enough energy into the compressive modes. Perhaps geometrical MHD wave coupling into the fast mode, with subsequent viscous dissipation, is the most feasible process. The loop resonances always have wavelengths comparable to or shorter than the loop radii of curvature, and the geometrical coupling should be operative. Clearly, full solutions to the equations of motion on curved-structured magnetic fields are required, at which point the theory should merge into that employed for surface waves.

10.7.3 Observables

The general level of nonthermal line widths in the corona is compatible with what theory requires if the coronal active region loop energies are to be supplied by loop-resonant Alfven waves. What is needed are detailed observations of transverse (to $\mathbf{B}_0$) motions on the coronal loops themselves, and on the magnetic flux tubes in the photosphere and chromosphere.

The fact that the velocities are transverse to $\mathbf{B}_0$ suggests an indirect method for their detection: observations of transverse displacements of material. If the corona is structured, it should be possible to observe distinct "features" moving in response to the waves. For example, a 25 km s^{-1} velocity amplitude moving at a 100-s period would have a net excursion of about 1 arcsec; the excursion scales as velocity times period. Since Alfven wave velocities tend to increase with decreasing density, the observations should be made as far as possible above the solar surface. (It should be noted that transverse motion of distinct structures probably requires an analysis that includes the spatial gradients in detail. Such an analysis will resemble that employed for surface waves.)

Alfven waves also contain transverse magnetic field fluctuations. Except for the loop resonances, the transverse fluctuations may be comparable to $|\mathbf{B}_0|$ in the photosphere and chromosphere. Stokes polarimetry may be able to detect such fluctuations. In the corona, the transverse fluctuations will be small, but it may be possible to detect fluctuations in the direction of the total field via coronal emission line polarimetry (Querfeld and Hollweg, 1979); it may even be feasible to detect the loop resonances, if $\tau_{res} > 10$ min or so.

If the Alfven waves are more easily generated near the edges of the flux tubes, because the reflections are weaker there or because the twisting velocities are greater, cool cores could result in the coronal loops. Again, cool cores do not seem to be a very useful diagnostic.

If Alfven waves in the corona damp via coupling into compressive modes, parallel electric fields would be produced, along with some runaway electrons and nonthermal emission. Because the damping is not understood, no calculations can be carried out at present.

The most interesting prediction of the Alfven wave model is the possibility that loop resonances can be excited by the convective motion, and that the resonances may be able to supply the required loop energies. The prediction of distinct resonant periods (Eq. 10.7.5) suggests that power spectral analyses of coronal velocity and magnetic field fluctuations should be undertaken. And if the loop-resonant Alfven waves couple into compressive modes, it may be possible to detect distinct frequencies in power spectra of intensity fluctuations.

Finally, the Alfven mode always carries energy only along $\mathbf{B}_0$. This is in contrast to the fast mode in a low-β plasma, which can carry energy across the field. One possible observational way to detect this difference is to search for the correspondence between heated magnetic structures in the chromosphere, and the coronal active region loops. If the chromospheric-coronal correspondence is strong, in the sense that thin coronal active region loops map directly onto similarly thin chromospheric structures, i.e., field line by field line, then the most likely explanation would seem to be that both the coronal and chromospheric structures are being heated by energy that is being channeled directly along $\mathbf{B}_0$. On the other hand, if coronal loops do not match well with chromospheric structures, i.e., if there is no evidence for the same field lines being heated in the chromosphere and corona, then one possible interpretation would be that the heating occurs via energy transport across field lines. The former case would tend to argue in favor of Alfven waves and rule out fast waves, and conversely for the latter case. (The former case could also indicate the efficacy of field-aligned motions, such as spicules or the slow mode.)

10.8 DYNAMICS OF FILAMENT FORMATION

10.8.1 Perspective

The importance of coronal structure has been repeatedly emphasized above. For the most part, it has been assumed that the observed structure is passively determined by the magnetic field configuration, and perhaps too by the globally-determined heat input.

Chiuderi and Van Hoven (1979a,b) have examined a *dynamic* model that describes the temporal evolution of one type of structure, viz. coronal filaments. Their basic viewpoint is that a sheared magnetic field structure can control the heat balance in a spatially varying fashion, via the magnetic inhibition of cross-field heat conduction, and that starting from a uniform equilibrium, a localized region of thermal instability (e.g., Field, 1965) can result where $\mathbf{k} \cdot \mathbf{B}_0 = 0$. The localization of the instability results in a "knife-blade" filament structure.

They believe that the filament formation may be able to trigger solar flares. The filament itself cools in the course of the thermal instability, and the electrical conductivity of the plasma is reduced as a consequence. This in turn increases the growth rate of the collisional tearing instability of the sheared magnetic field configuration. Chiuderi and Van Hoven suggest that the nonlinear development of the thermal instability could be accompanied by a tearing instability in which the magnetic energy of the sheared field is deposited into thermal energy and mass motion of the plasma as a flare.

Acknowledgements

This review would not have been possible without the benefit of participation in the Skylab Active Region Workshop. The author is grateful to Prof. F. Q. Orrall for the invitation to participate, and for his subsequent guidance and encouragement. The author must also acknowledge with gratitude the patience with which the members of the Working Group explained to him much physics with which he was initially unfamiliar. He is especially grateful to Drs. Chiuderi, Golub, Habbal, Ionson, Papadopoulos, Rosner, Spicer, Van Hoven, and Vlahos. He is grateful also to Dr. Chiuderi for a very helpful critical reading of this manuscript. The National Center for Atmospheric Research is sponsored by the National Science Foundation.

REFERENCES

Alfven, H.: 1947, *Monthly Notices Roy. Astron. Soc.* **107,** 211.
Athay, R. G., and White, O. R.: 1979a, *Astrophys. J. Suppl.* **39,** 333.
Athay, R. G., and White, O. R.: 1979b, *Astrophys. J.* **229,** 1147.
Barnes, A.: 1966, *Phys. Fluids* **9,** 1483.
Boris, J. P., Dawson, J. M., Orens, J. H., and Roberts, K. V.: 1970, *Phys. Rev. Lett.* **25,** 706.
Braginskii, S. I.: 1965, *Rev. Plasma Phys.* **1,** 205.
Chandrasekhar, S.: 1961, *Hydrodynamic and Hydromagnetic Stability,* Clarendon Press, Oxford.
Chiuderi, C., and Van Hoven, G.: 1979a, University of California, Irvine, UCI Physics Dept. Report No. 79-1.
Chiuderi, C., and Van Hoven, G.: 1979b, *Astrophys. J. Lett.,* 15 August.
Craig, I.: 1979, Paper presented at Skylab Active Region Workshop.
Deubner, F.: 1976, *Astron. Astrophys.* **51,** 189.
Drake, J., and Lee, Y. C.: 1977, *Phys. Fl.* **20,** 1341.
Field, G. B.: 1965, *Astrophys. J.* **142,** 531.
Foukal, P.: 1975, *Solar Phys.* **43,** 327.
Foukal, P.: 1976, *Astrophys. J.* **210,** 575.
Furth, H. P., Rutherford, P. H., and Selberg, H.: 1973, *Phys. Fluids* **16,** 1054.
Galeev, A. A., Rosner, R., Serio, S., and Vaiana, G. S.: 1979 (in preparation).
Gerassimenko, M., Solodyna, C. V., and Nolte, J. E.: 1978, *Solar Phys.* **57,** 103.
Habbal, S., Leer, E., and Holzer, T.: 1979 (preprint).
Hollweg, J. V.: 1973, *Astrophys. J.* **181,** 547.
Hollweg, J. V.: 1978a, *Rev. Geophys. Space Phys.* **16,** 689.
Hollweg, J. V.: 1978b, *Geophys. Res. Lett.* **5,** 731.
Hollweg, J. V.: 1979a, *Solar Phys.* (in press).
Hollweg, J. V.: 1979b (in preparation).
Hollweg, J. V., and Lilliequist, C. G.: 1978, *J. Geophys. Res.* **83,** 2030.
Hollweg, J. V., and Smith, D. F.: 1976, *J. Plasma Phys.* **15,** 245.
Hundhausen, A. J.: 1977, in J. Zirker (ed.), *Coronal Holes and High-Speed Wind Streams,* Colorado Associated University Press, Boulder, Colorado, Ch. 7, p. 225.
Hung, R. J., and Barnes, A.: 1973a, *Astrophys. J.* **180,** 253.
Hung, R. J., and Barnes, A.: 1973b, *Astrophys. J.* **180,** 271.
Hung, R. J., and Barnes, A.: 1973c, *Astrophys. J.* **181,** 183.

Ionson, J. A.: 1978, *Astrophys. J.* **226,** 650.
Jackson, E. A.: 1960, *Phys. Fluids* **3,** 786.
Jacques, S. A.: 1978, *Astrophys. J.* **226,** 632.
Jordan, S.: 1977, in E. Hansen and S. Schaffner (eds.), *Proceedings of the November 7–10, 1977 OSO-8 Workshop,* University of Colorado, Boulder, Colorado, p. 340.
Krieger, A. S., de Feiter, L. D., and Vaiana, G. S.: 1976, *Solar Phys.* **47,** 117.
Kuperus, M.: 1969, *Space Sci. Rev.* **9,** 713.
Leibacher, J. W.: 1977, in E. Hansen and S. Schaffner (eds.), *Proceedings of the November 7–10, 1977 OSO-8 Workshop,* University of Colorado, Boulder, Colorado, p. 311.
Levine, R.: 1974a, *Astrophys. J.* **190,** 447.
Levine, R.: 1974b, *Astrophys. J.* **190,** 457.
Levine, R., and Altschuler, M.: 1974, *Solar Phys.* **36,** 345.
Levine, R., and Withbroe, G.: 1977, *Solar Phys.* **51,** 83.
Mihalas, B.: 1979, 'Internal Gravity Waves in the Solar Atmosphere', University of Colorado, Boulder, Colorado, Ph.D. Thesis.
Munro, R. H., and Jackson, B. V.: 1977, *Astrophys. J.* **213,** 874.
Nye, A., and Hollweg, J. V.: 1979 (in preparation).
Orrall, F., and Kopp, R.: 1976, in R. M. Bonnet and Ph. Delache (eds.), *The Energy Balance and Hydrodynamics of the Solar Chromosphere and Corona, IAU Colloquium No.* **36,** p. 457.
Osterbrock, D. E.: 1961, *Astrophys. J.* **134,** 347.
Papadopoulos, K.: 1977, *Rev. Geophys. Space Phys.* **15,** 113.
Parker, E. N.: 1974, *Astrophys. J.* **191,** 245.
Poletto, G., Vaiana, G. S., Zombeck, M. V., Krieger, A. S., and Timothy, A. F.: 1975, *Solar Phys.* **44,** 83.
Priest, E. R.: 1978, *Solar Phys.* **58,** 57.
Pye, J. P., Evans, K. D., Hutcheon, R. J., Gerassimenko, M., Davis, J. M., Krieger, A. S., and Vesecky, J. F.: 1978, *Astron. Astrophys.* **65,** 123.
Querfeld, C. W., and Hollweg, J. V.: 1979 (in preparation).
Reeves, E. M., Timothy, J. G., Foukal, P. V., Huber, M. C. E., Noyes, R. W., Schmahl, E. J., Vernazza, J. E., and Withbroe, G. L.: 1976, *Prog. Astro. Aero.* **48,** 73.
Reidy, W. P., Vaiana, G. S., Zehnpfennig, T., and Giacconi, R.: 1968, *Astrophys. J.* **151,** 333.
Rosner, R., Golub, L., Coppi, B., and Vaiana, G. S.: 1978a, *Astrophys. J.* **222,** 317.
Rosner, R., Tucker, W. H., and Vaiana, G. S.: 1978b, *Astrophys. J.* **220,** 643.
Smith, D. F., and Hollweg, J. V.: 1977, *J. Plasma Phys.* **17,** 105.
Stein, R. F., and Schwartz, R. A.: 1972, *Astrophys. J.* **177,** 807.
Stein, R. F., and Leibacher, J. W.: 1974, *Ann. Rev. Astron. Astrophys.* **12,** 407.
Tousey, R., Bartoe, J. D. F., Bohlin, J. D., Brueckner, G. E., Purcell, J. D., Scherrer, V. E., Sheeley, N. R., Jr., Schumacher, R. J., and Vanhoosier, M. E.: 1973, *Solar Phys.* **33,** 265.
Tucker, W. H.: 1973, *Astrophys. J.* **186,** 285.
Ulmschneider, P.: 1971, *Astron. Astrophys.* **14,** 275.
Ulmschneider, P.: 1976, *Solar Phys.* **49,** 249.
Ulmschneider, P., and Kalkofen, W.: 1977, *Astron. Astrophys.* **57,** 199.
Underwood, J. H., Chapman, G. A., Janssens, T. J., Landecker, P. B., Mayfield, E. B., McKenzie, D. L., Vorpahl, J. A., Walker, A. B. C., Jr., Milligan, J. E., de Loach, A. C., Hoover, R. B., McGuire, J. G., and Wilson, R. M.: 1976, *Prog. Astro. Aero.* **48,** 179.
Vaiana, G. S., Davis, J. M., Giacconi, R., Krieger, A. S., Silk, J. K., Timothy, A. F., and Zombeck, M.: 1973a, *Astrophys. J.* **185,** L47.
Vaiana, G. S., Krieger, A. S., and Timothy, A. F.: 1973b, *Solar Phys.* **32,** 81.
Vaiana, G. S., Krieger, A. S., Timothy, A. F., and Zombeck, M.: 1976, *Astrophys. Space Sci.* **39,** 75.
Vlahos, L., Ionson, J. A., and Spicer, D. S.: 1979 (preprint).
Wentzel, D. G.: 1978, *Rev. Geophys. Space Phys.* **16,** 757.
Wentzel, D. G.: 1979a, *Astron. Astrophys.* (in press).
Wentzel, D. G.: 1979b, preprint.

Wentzel, D. G.: 1979c, *Astrophys. J.* **227,** 319.
White, O. R., and Athay, R. G.: 1979a, *Astrophys. J. Suppl.* **39,** 317.
White, O. R., and Athay, R. G.: 1979b, *Astrophys. J. Suppl.* **39,** 347.
Withbroe, G.: 1976, in R. M. Bonnet and Ph. Delache (eds.), *The Energy Balance and Hydrodynamics of the Solar Chromosphere and Corona, IAU Colloquium No.* **36.**
Withbroe, G.: 1977, in J. B. Zirker (ed.), *Coronal Holes and High-Speed Wind Streams,* Colorado Associated University Press, Boulder, Colorado, Ch. 5, p. 145.

11. FUTURE OBSERVATIONS OF ACTIVE REGIONS

Randolph H. Levine

11.1 INTRODUCTION

11.1.1 With New Eyes

Because of the acquisition of unprecedented solar data from the Apollo Telescope Mount instruments on Skylab, we have seen the sun, literally and figuratively, in a new light. As often happens, this has produced advances in our understanding of the natural world that have redefined the framework of our thinking. Although new insights necessarily lead to new questions, a paradoxical consequence of progress is that we may find ourselves asking some of the same questions that we asked before. It is important to realize that these old questions are not really old; the new perceptual and conceptual framework that underlies the study of the solar atmosphere has redefined them and provided a new challenge to their solution.

For example, the most classic question in the physics of the solar atmosphere, that of finding the source of nonthermal energy input to the corona, is still asked, but from an entirely different point of view than that of ten years ago. The basic schematic shift that has redefined this and other problems of solar physics (as well as creating new puzzles) is the realization of the overriding influence of the magnetic field on the solar atmosphere. The entire topology of the atmosphere is ordered by the magnetic field, and its importance to the nontransient physical processes in the plasma is only beginning to be understood. The realization of the centrality of the magnetic field has played a role in the recent advances in the study of the solar cycle, the emergence of activity on a wide variety of spatial scales, and the structure of the interplanetary medium. It is also a unifying thematic concept* in much of the recent work on the physics of solar active regions.

The recent progress in the study of solar active regions has reflected this new point of view and has been highlighted by the Skylab Solar Workshop on Active Regions. A significant amount of the Skylab data on active regions can be successfully interpreted by detailed theoretical models. However, rather different structural models often can explain the same data. The initial stage of success in interpreting and modeling the observations has enabled us to define the importance of some appropriate new observations, and what we might expect to gain from them.

---

*The role of thematic constructions in the work and progress of science has been emphasized by Holton (1973).

The following analysis of what we might realistically hope to learn about active regions is based on the perspective of the Workshop and reflects the interests of its participants and the limitations of its scope. The most central emphasis is that the coronal portion of an active region is composed of a collection of magnetically confined loops of plasma. Thus, for example, the question of the heating of the corona is now the question of the heating of the individual magnetic structures that make up the corona.

11.1.2 The Problems of the Future

Because of the progress made in understanding active regions in greater detail, we are in a position to be more demanding both of the instruments we use and of the tools for extracting information from the data. Delineating the specific need for these greater demands is one of the purposes of this chapter. The intent of having the participants in the Skylab Workshop Series on Active Regions speak on this subject is to provide specific examples of progress that can be expected, and to suggest which of the many types of conceivable observations might best be used in the study of solar active regions.

In our process of seeking to understand more of the physics of active regions, we have found two classes of impediments.

(1) Newly discovered or better understood limitations to the interpretability of observations (e.g., spectral diagnostics) can be minimized only through an understanding of the difficulties and careful planning of optimal observational characteristics.

(2) Many questions are unsolved merely because we have not yet looked (or been able to look) for the answers. These are problems for which specific characteristics of spatial, spectral, and temporal resolution can usually be defined.

The following sections of this chapter discuss these issues from within the framework of the Skylab Active Region Workshop. Section 11.2 examines the limits to our theoretical understanding of even the best observational data. Overcoming these limitations, either by developing new techniques or by optimizing experimental characteristics with respect to theoretical limits, promises dividends that are independent of instrumental advances. Sections 11.3 through 11.7 are an overview of possible areas of new inquiry. These are divided along the lines of key areas identified by the separate Working Groups of the Workshop. The final section, 11.8, is a brief summary of the entire chapter intended to refresh the memory of those reading this volume in detail and to serve as an independent summary, which can be read separately by those wishing a less detailed account.

11.2 LIMITS TO INTERPRETATION

11.2.1 Temperature Diagnostics and Temperature Sensitivity

Perhaps the most basic way in which spectral observations of the transition region and corona are interpreted is in terms of temperatures. The optically thin radiation consists of a series of continua and emission lines from excited states of the most abundant solar species. Because the lines are formed by collisional excitation, the (electron) temperature of the plasma will tend to favor certain transitions over others requiring significantly more or less than the mean thermal energy. These lines are each said to be formed near a characteristic temperature and spectroheliograms or filtergrams made in a given line (or set of lines) are commonly used to indicate plasma temperatures, at least to first order.

The difficulty, of course, is that the radiation is not only formed at the characteristic temperature. The actual range of temperatures over which significant emission in a given line can occur is often large (see Chapter 6 of this volume). Thus, for example, while Mg X (625 Å) is most favorably formed at around $\log T = 6.15$, a plasma of the same density would emit within a factor of one-half of the maximum intensity at any temperature between $\log T = 6.0$ and $\log T = 6.3$. In general, the range of temperatures including significant emission from a given line increases exponentially with the most favorable formation temperature of the line. Thus, the very high temperature X-ray lines are formed over temperature ranges orders of magnitude larger than those over which chromospheric EUV lines are formed.

It is clear that a single spectroheliogram or filtergram cannot provide reliable temperature information in this way, because the temperature at which the ions are actually radiating can be very different from the most favorable formation temperature of the transition to which the observation is sensitive. It is, however, reasonable to expect that multispectral observations using lines formed over a variety of temperature ranges can overcome this difficulty to some extent. Observations in a series of lines whose formation temperatures span the expected range of plasma temperatures can significantly remove the ambiguities that are necessarily present. Unfortunately, it is not true that the more lines one uses the better one is able to discern temperatures. This will be discussed in detail in Section 11.2.3, which deals with the analytical form of this type of analysis.

Related problems occur in the use of broadband filters in X-ray telescopes. Although more complicated in practice, the analysis is conceptually similar to that for spectrally resolved line emissions. Instead of observing the emission from a single line, the filter technique provides the summed emission over several lines within the passband of the filter.

In either case, use of observations in a single line (or filter) whose temperature sensitivity covers the expected range of temperatures is used to give information on the total emission measure of the plasma. Pairs of lines or filters can be used to establish what is essentially a color temperature, as an approximation to the actual (electron) plasma temperature. The difficulty in the first of these cases is that the temperature range in a typical solar active region is large enough so that even the most favorably sensitive single spectral line can give an approximation to the total emission measure that is off by more than a factor of two. In the second instance, the color temperatures determined by ratios of line or filter observations tend to be relatively insensitive to variations in actual plasma temperatures. In addition, analysis of Skylab data has shown that the plasma along the line of sight of a single spatial resolution element is often distributed over a wide temperature range. Thus, the emission observed in separate lines or filters may be formed in macroscopically distinct elements of plasma.

11.2.2 Density Diagnostics and Density Sensitivity

Determinations of the electron density in various parts of the solar atmosphere have been based primarily on the intensity ratio between density-sensitive triplets and singlets. Although this type of technique is in principle a powerful tool (the ratio technique used on pairs of lines from the same ion eliminates many factors whose value is uncertain and greatly reduces the dependence on instrumental calibration), its use has been limited by uncertainties in the atomic parameters, the magnitude of observational errors compared with the sensitivity of a given ratio, and the fact that such ratios are also unavoidably temperature sensitive to some extent. (See Chapter 6.)

It is also possible to use an emission line or filter whose temperature sensitivity is maximum near the most likely temperature of a particular active region. By assuming that the line or filter is sensitive to all of the active region plasma, a line-of-sight density can be determined (e.g., Levine and Pye, 1980). Alternatively, Withbroe *et al.* (1971) showed that low-resolution Mg X (625 Å) spectroheliograms correlated with white-light observations of the corona, and they interpreted this as indicating that variations in the Mg X emission were primarily due to electron density variations.

Emission lines whose temperature sensitivity makes them attractive candidates for such a procedure include Mg X (625 Å) and O VII (18.63 Å). Unfortunately, determinations of the temperature distribution of active region plasma that use a variety of lines (see Section 11.2.3 below) show that the range of temperatures at which there is a significant amount of plasma is much larger than the sensitivity range of any single emission line or any filter used to date. For this reason, such determinations, which use a single line or filter, are often in disagreement with more elaborate calculations by a factor of two or more.

Theoretical calculations of the density sensitivity of various line ratios can be extremely complex. It is necessary to include all possible radiative and collisional processes in a calculation that accounts for all of the energy levels of a particular ionization series. Examples of this type of work are the papers by Gabriel and Jordan (1972), Nussbaumer (1972), and Dupree *et al.* (1976) for Beryllium-like ions and by Flower and Nussbaumer (1975) and Vernazza and Mason (1978) for Boron-like ions. For the Beryllium sequence, line ratios of C III and O V have proved to be most useful, whereas in the Boron sequence the most promising candidates are lines of Mg VIII and of

Si X, although several other useful pairs are pointed out by Vernazza and Mason (1978). Several additional pairs of lines are unsuitable for use only because one or both lie near other strong emission lines. Many of these lines could be reevaluated for use in density measurements if suitably improved spectral resolution capabilities were available.

In addition to techniques that seek to determine the electron density directly by taking advantage of line ratios which are not sensitive to temperature, recent calculations of ionization equilibrium for commonly used chromospheric, transition region, and coronal lines have revealed a significant density sensitivity in their ionization balance. Vernazza and Raymond (1979) have shown that the relative abundances of Lithium-, Beryllium-, and Boron-like ions are density dependent at electron densities as low as 10^6 cm^{-3}. The differences in these new results from previously used ionization equilibrium calculations (e.g., Allen and Dupree, 1969; Jordan, 1969; Summers, 1974) reflect the inclusion of metastable levels in the atomic calculations and the density sensitivity of the dielectronic recombination process.

Although these new ionization balance calculations complicate the use of many popular emission lines as temperature diagnostics, they open the possibility of using new or existing observations in a variety of lines as density indicators. One example of a procedure to accomplish this is given by Vernazza and Raymond, but many other algorithms could be devised. There appears to be no reason why such procedures cannot be tailored to the scientific and design constraints of a particular instrument.

11.2.3 Differential Emission Measure

The temperature sensitivity of optically thin solar emission lines, which was discussed above in Section 11.2.1, can be exploited analytically to determine the distribution of temperatures within a volume of observed plasma. For each line, the total intensity, which is the integral of the ionization fraction (a function of temperature explicitly and position implicitly) times the square of the electron density over the observed volume, can be expressed alternatively as an integral over temperature of the ionization times a function $Q(T)$, called the differential emission measure, which is an indicator of the emission measure (electron density squared) within an infinitesimal temperature range. The exact analytical statement of this definition is somewhat more complicated, but for most nonpathological cases the above interpretation is sufficient.

A schematic mathematical statement of the concept of differential emission measure is

$$\int n_e^2 dV = \int Q(T)\, dT \ , \qquad (11.2.1)$$

where T is the electron temperature, n_e is the electron density, and V is the volume containing the emitting plasma. Physically, the quantity $Q(T)$ is an indication of the amount of emitting material near temperature T. In practice, $Q(T)$ can be obtained directly from theoretical models of individual plasma structures or empirically from an appropriate set of observed spectral intensities whose emission characteristics as a function of temperature are known. One view of the complementary nature of these two approaches is given by Vaiana and Rosner (1978). The implications of theoretical derivations of differential emission measure are discussed in Section 11.6. The remainder of this section will be devoted to a discussion of the empirical derivation of $Q(T)$.

In general, the differential emission measure can be derived if we may write the observed emission intensities in the form

$$I\,(\mathrm{erg\ s^{-1}\ str^{-1}}) = C \int n_e^2\, \Gamma(T)\, dV \ , \tag{11.2.2}$$

where $\Gamma(T)$ represents the temperature-dependent contributions to the spectral intensities. This is true for individual spectral lines that are collisionally excited, in which case $C = 1.73 \times 10^{-16}\, Af$, where A is the abundance (relative to hydrogen) of the element producing the emission, f is the absorption oscillator strength, and $\Gamma(T) = gG(T)$, where g is the Gaunt factor and $G(T)$ represents the ionization properties of the line (usually ionization equilibrium; see, e.g., Pottasch, 1965). A similar relationship is also true for those broadband spectral filters whose major transmitted energy is proportional to n_e^2, such as the S-054 Skylab X-ray telescope (Vaiana *et al.*, 1977) and SOLRAD-9 (Dere *et al.*, 1974).

Derivation of the differential emission measure $Q(T)$ requires the inversion of a set of integral equations similar to Eq. (11.2.2), using an appropriate set of spectral characteristics and line intensities. The complexity and reliability of this inversion process depends on the characteristics of the spectral lines (or filters) being used (Craig, 1977). For example, if the contribution functions $gG(T)$ are widely separated (i.e., lines formed over very different temperature ranges are used) the inversion is stable but has little temperature resolution. On the other hand, using lines formed over temperature ranges that are significantly overlapped allows the possibility of better temperature discrimination, but can result in an ill-conditioned numerical problem whose solution is unstable with respect to small perturbations of the input data (Craig and Brown, 1976). This is especially true of X-ray lines, whose contribution functions tend to be very broad and closely spaced with respect to temperature.

Whatever method is chosen to perform the inversion of the integral equation for differential emission measure (e.g., Withbroe, 1975; Parkinson, 1975; Pye *et al.*, 1978; Levine and Pye, 1980), there are two ways of increasing the reliability of the solution. First, the spectral lines used must be chosen from those available to optimize the temperature resolution of the solution without significantly decreasing its stability. For lines whose contribution functions overlap greatly, this is difficult to do and the more important criterion of a stable solution often dictates a rather coarse temperature resolution. The simultaneous inclusion of ultraviolet lines (formed in the high chromosphere and transition region) with coronal X-ray lines increases the stability of the overall solution, because the contribution functions of the ultraviolet lines are relatively narrow and more widely separated than those of the X-ray lines. A second procedure crucial to demonstrating the reliability of a solution for $Q(T)$ is to test the stability by randomly perturbing the input data within the limits of its experimental or statistical uncertainty in a series of solutions. This allows error bars to be placed on the solution that directly reflects the stability of the procedure being used.

In the design of instrumentation for the study of solar active regions, consideration of these factors can greatly enhance the quality of the eventual data interpretation. Inclusion of a variety of spectral lines or filters, and the ability to observe the same feature simultaneously in several lines or filters that are carefully chosen, can be shown in advance to lead to a higher degree of confidence in the results of a particular experiment. Recognition of the limitations of spectral diagnostics and attempts to overcome those limitations by features of experiment design or data analysis are factors that should be taken into consideration in the evaluation of proposed instrumentation programs.

11.3 BETTER MORPHOLOGICAL SYSTEMATIZATION

11.3.1 Gross Properties

Active regions as a whole have been studied extensively using the data from the Skylab/ATM experiments. In fact, our concept of active region may have been altered by the work on X-ray bright points (e.g., Golub *et al.*, 1976), which has shown that these small X-ray features must be considered part of a continuum of magnetic structures in the solar atmosphere that extends all the way to the largest solar active regions. This continuum is known to exist as a function of spatial scale, thermal energy content, and total magnetic flux (Golub *et al.*, 1979b). However, studies of average active region temperatures have not shown a continuous spectrum, but rather a tendency for average temperatures to lie near a few values, with the different classes corresponding to different properties (such as lifetime) of the active region (Little and Krieger, 1977). This morphological catalog needs to be extended, especially with the aid of improved diagnostics, in order to provide a reliable picture of the spectrum of active regions as a function of several important physical variables. In particular, the properties of active regions of different lifetimes (i.e., days to months) are in need of closer examination.

Of special importance in this area is further investigation of the relationship between X-ray bright points and XUV bright points (which have not been studied extensively) and the relationship between X-ray bright points and ephemeral active regions. These latter are small bipolar features seen in solar magnetograms that correlate well, but not completely, with X-ray bright points (Martin and Harvey, 1979; Golub *et al.*, 1977; Harvey and Martin, 1973). Because X-ray bright points have been used as an indicator of small-scale magnetic flux emergence on the sun, it is crucial to establish their relation to actual measurements of the flux at different times in the solar cycle, to eliminate the possibility of visibility or lifetime effects in the very intriguing result (Golub *et al.*, 1979a) that the number of bright points, and by inference the amount of magnetic flux at small scales, rises and falls 180° out of phase with the sunspot cycle. This result has given rise to the conjecture that the solar cycle is not an oscillation in the total amount of magnetic flux emerging on the sun, but that the amount of flux emerging through the photosphere in a given period of time remains fixed, and the cycle is a modulation of the spatial scale at which most of the flux emerges.

The morphology of emerging magnetic flux itself needs clarification. Little and Krieger (1977) suggest that the extremes of initial temperature of an isolated region of emerging flux correlate with the extremes of observed lifetimes of the feature and, in fact, emerging structures that are too cool never grow into active regions. The correlation of temperature with lifetime was not found to extend to active regions in general in a study by Mosher (1979) using OSO-8 data from a broadband, low-resolution X-ray instrument. Because of the important role played by emerging flux in replenishing the structural framework of the corona, furter detailed study of this process is imperative.

11.3.2 Structural Details

There is a great deal still to be learned about the structures that make up active regions and their emission characteristics over a wide range of temperatures. Active regions extend into the solar atmosphere in the form of collections of magnetically confined loops of plasma. It is these loops of plasma that form the active region corona, but not all of these loops contain plasma at coronal temperatures. There is apparently significant substructure in the temperature distribution of active region loops.

The observational distinction relevant to this statement is that some loops (the so-called "cool loops") are prominent in emission from lines formed in the transition region and low corona (e.g., Mg IX, 368 Å; Ne VII, 465 Å) while others ("hot loops") are primarily visible in what must be considered purely coronal lines (e.g., Fe XV, 284 Å; X-ray filters). There is also some evidence that the Mg X (625 Å) line, thought to be an essentially coronal emission, is not spatially coincident with structures seen in X-ray filter observations (Webb and Levine, 1979). Although there is some evidence that the cool loops are somewhat more variable and shorter-lived than the hot loops (Sheeley, 1980; Dere, 1979), the initial morphological distinction of two classes ("hot" and "cool") is not firmly established, and there is no understanding of the physical mechanism responsible for this apparent observational distinction.

Some active region loops, especially but not exclusively those over sunspots, have been observed to have significant emission in lines formed over a wide range of temperatures. Such loops are often visible in coronal and transition region lines at the same time. The observed spatial relationship of these emissions has suggested the interpretation of "cool cores" in some active region loops. In this picture, such loops may be thought of as being axially symmetric with a radial temperature gradient extending from low transition region temperatures along the loop axis to coronal temperatures on concentric shells some distance from the axis (Foukal, 1976, 1978; Levine and Withbroe, 1977).

Several basic problems exist in present attempts to understand these loops with cool cores. The most elementary morphological problem is the extent of the cool cores. Foukal (1976) has suggested that the core diameter is typically 3000–4000 km, which is the resolution limit of the S-055 EUV spectroheliometer whose data he used. Levine and Withbroe (1977) constructed empirical models of radial loop structure that suggested that the loop core was an order of magnitude smaller. There is a clear need for cospatial observations of loops with cool cores in various spectral lines and with a spatial resolution significantly better than the few arcseconds available on Skylab. Another problem in the understanding of cool cores is their very existence. Static MHD models (e.g., Van Hoven *et al.*, 1977) are not entirely satisfactory, and even extensions to magnetic geometries that are more realistic (Chiuderi and Van Hoven, 1979; Chiuderi *et al.*, 1980) do not explain the formation of stable cool cores in the first place. One possible explanation has been given by Priest (1978). There are, of course, suggestions that pressure gradients are associated with cool cores (Foukal, 1975), and this raises the possibility of mass interchanges or flows in an essentially dynamic system. As a morphological question, the observation of such flows and their time history is a singularly important goal. To be of use, such observations would have to observe the core and mass motions in a variety of spectral lines and over most of the length of the loop.

11.3.3 Structure of the Chromosphere

The connection between the active region corona and the underlying chromosphere is mainly through the footpoints of magnetic loops. This greatly complicates the study of the chromosphere in active regions, because a complete physical diagnosis of the basic structure of an active region requires simultaneous identification of entire loops in chromospheric, transition region, and coronal emission.

Except for flares, the actual response of the chromosphere to events higher in the atmosphere has not generally been studied. Thus, we do not know observationally the ways in which the chromosphere acts as a source or sink for energy and mass in the overall active region system. As previous chapters have discussed in more detail, the chromosphere is thought to play a significant modulating and stabilizing role in the physics of active region loops. Comparisons of observed decay times of brightness variations in coronal loops with theoretical conductive and radiative cooling times (Krieger *et al.*, 1975; Gerassimenko *et al.*, 1978; Peres *et al.*, 1978) suggest that loop energetics are dominated by the balance between energy deposition and conduction (ultimately to the chromosphere) and by the inhibition of conduction, as when chromospheric evaporation occurs (see Section 11.5). Observations of mass and/or energy interchange between an active region loop and its chromospheric footpoint will provide major evidence of processes and conditions in the coronal loops.

At this time, there is no generally accepted means of identifying specific loop footpoints in the chromosphere. A great deal of study is needed to determine whether there is in fact a chromospheric signature of loop footpoints. In addition, there are many loops with one or more footpoints in weaker plage regions, or even in quiet sun areas. For these cases, the relation of the loop footpoints to the network (and to spicules, downflows, and other phenomena associated with the network) is not known. In addition, the distinction in contrast between chromospheric faculae and pores may be related to the total flux in the magnetic bundles thought to occur in association with each of these features (Frazier, 1978). It is true that above a certain magnetic flux the energy deposition mechanism changes (causing a radiative deficit we see as the distinction between dark pores and bright faculae); this will have important consequences for the coronal and transition region structures above.

And, although it is not the direct subject of our present interest, we must note that the magnetic nature of the network itself has not been observationally delineated. It is also important to recall here that our definition of active regions must extend to the so-called ephemeral active regions, which are associated with X-ray bright points. While some of these features are known to be cospatial with parts of the chromospheric network, their overall relation to the network has not been studied. It is entirely possible that ephemeral active regions/X-ray bright points, which have lifetimes not far from those of supergranule cells, will be shown to play an important role in the evolution of the network pattern.

11.3.4 Characterizing Flows

The largest single void in our observations of active region structures is that we have not been able to study flows in a systematic and exhaustive manner. Recent rocket-borne instruments (Brueckner *et al.*, 1977; Brueckner and Bartoe, 1978), as well as OSO-8 (White and Athay, 1979), have shown a wide variety of flow phenomena in chromospheric and transition region emissions. A broader approach is necessary, however, in order to study the entire problem of coronal dynamics above active regions. It will be necessary to characterize flows both temporally (i.e., steady or transient) and spatially (direction and position in loop). Another capability of great value in future observations will be the ability to observe flows simultaneously in lines formed at a variety of temperatures. In this way, all of the mass, energy, and momentum flow can be accounted for. Significant constraints on viable theories can be expected to result from such observations. Flows are discussed more generally in Section 11.5.

11.4 TEMPORAL VARIATIONS

The question of temporal variations in active regions can now be reassessed in terms of our latest theoretical and observational understanding of active region structures. Because it is not possible to interpret unresolved or averaged observations properly, it is of fundamental importance to resolve and observe individual structures within active regions. Once the individual structures are resolved, an analysis of their temporal behavior will contribute to better understanding in several areas.

The basic morphological questions about temporal evolution, as they apply to individual active region structures, are still not completely answered. How do active region structures vary in time as a function of peak temperature, size, or density? Are some classes of active region structures more variable than others? Answers to these and other questions will pinpoint areas for further physical investigation. For example, it has been suggested from the analysis of Skylab data that loops more easily visible in lines formed at cooler temperatures (e.g., Ne VII as opposed to Fe XV) are also more variable in time (Sheeley, 1980). If this proves true in general then there are likely to be basic differences in the physical processes in loops distinguished by disparate temporal variation. Some of these differences might be in the energy supply or in the response of the chromosphere to the energy being conducted downward from the upper portions of the loop.

A very fundamental question that can be addressed by systematic observations of time variations in active region structures is the character of the physical state of the system. Are variations in active region structures consistent with a transition from one steady state to another, or are we dealing with a continually dynamic system? Some time-dependent models (Craig, 1979) suggest that perturbations of quasistatic models of coronal loops result in mass and energy flows that eventually cease, but that also leave the loop in a steady state different from the initial state. This suggests that it is entirely possible to observe active region loops undergo transitions from one steady state to another, with the amount of energy and mass change during the transition period related to the initial and final states in a predictable way.

On the other hand, stability analyses of quasistatic loop models (Habbal and Rosner, 1979; Antiochos, 1979) suggest that these models of coronal loops, which have generally been successful in their overall explanations (Rosner *et al.*, 1978; Pye *et al.*, 1978; Withbroe, 1978; Emslie and Machado, 1979; Hood and Priest, 1979; Wragg and Priest, 1980; Antiochos and Sturrock, 1978; Vesecky *et al.*, 1978; Craig *et al.*, 1978; Landini and Monsignori-Fossi, 1975; Shmeleva and Syrovatskii, 1973), are stable with respect to density perturbations that exclude flow.* All of the significant variations in

*The apparently disparate results of Habbal and Rosner (1979), who claim stability for the simple model loops, and Antiochos (1979), who finds the opposite, can be reconciled by recognizing that the chromospheric boundary conditions used in the two analyses are different. Habbal and Rosner take the lower boundary to be the point in the chromosphere at which $\nabla T = 0$. They impose this condition as a function of T at all times, so the spatial position of

response to this type of perturbation occur in the chromosphere. Thus, observations of the temporal state of the chromosphere associated with an evolving loop (preferably coincident with observations of the coronal and transition region portions of the loop) can give a direct observational key to theoretical discrimination between physical processes.

It is also possible that because of the heating mechanism (discussed in more detail in Section 11.6), the presence of cool material in the loop core, or because of some other factor, most active region loops are in a continually dynamic state that is physically distinct from the quasistatic or time-dependent models that have been and are being used. Observations of sufficient resolution and that can detect mass motions simultaneously in several temperature regimes (including the chromosphere, if possible) should be able to distinguish this possibility and provide the data necessary for its empirical and theoretical modeling.

11.5 FLOWS

The recent history of the study of mass motions on the sun has followed the general pattern that the closer one looks the greater variety of mass motions one sees. (See Chapters 4 and 5.) This has been true in the case of the very largest scale motions, the global oscillations of the sun, as well as of the smallest observed scales, on the order of one arcsecond. Using high-resolution observations, which are of more interest to the study of active regions, Brueckner *et al.* (1979) have recently shown that there are significant flows (as indicated by Doppler shifts) in the quiet sun that can only be detected with a spatial resolution better than one arcsecond. Whether the same requirement is true of active regions is not known, but the existence of magnetic flux concentrations in the photosphere, as well as the inferred scales of emission features within active region loops, argues strongly that high spatial resolution be sought in instruments that propose to study flows in active regions.

There are many types of flow possible in an active region. In fact, as Chapter 9 indicates, the theory of flows in active regions can be expected to develop much more in the future, leading to many new plausible mechanisms for driving mass motions. Among the presently named mechanisms suggested as causing flows in active regions are chromospheric evaporation, the pressure siphon, wave pressure, precipitous draining, thermal instability, and MHD instability.

---

the lower boundary can fluctuate. Antiochos sets the lower boundary at the spatial point where $\nabla T = 0$ in the unperturbed model, and keeps it at that spatial position at all times. His result is that loop stability requires significant conductive flux through this lower boundary. Habbal and Rosner merely let the boundary "float" to achieve stability. If they were to consider a single spatial position at the loop base they, too, would find that a significant conductive flux develops in response to perturbation. The point of both analyses is that the chromosphere can be expected to be in continual dynamic response to perturbations of the coronal loop.

Evaporation of material from the chromosphere can occur whenever it is burdened with the deposition of more energy than it can radiate or conduct away. Assuming that the excess is deposited from above, the upper chromosphere is heated and the material can begin to flow up into the loop from which the excess energy was deposited. The equations for this process are similar to solar wind equations, with the physical solutions being subsonic in the cases that have been studied. Although evaporation has been studied mostly in the context of post-flare loops (Antiochos and Sturrock, 1978; Peres *et al.*, 1978), any increase in heating in a steady-state loop can cause such an upflow. This may be one of the major mechanisms causing loops to fill with material.

Conversely, a decrease in heating will cause material to drain from a loop due to lack of a suitable equilibrium at the newly established temperature. Such events may have been observed by Levine and Withbroe (1977) and by Schmahl (1979). Foukal (1975, 1978) has claimed that the inferred scale height of transition region and chromospheric emission in active region loops rooted in sunspots is inconsistent with hydrostatic equilibrium. If true, this would result in a downflow of material, which Foukal postulates is occurring continually in the cores of these loops. The major difficulty with this explanation of the longitudinal extent of the cool cores is that the observed lifetimes of loop cores require that material be continually supplied to the core to maintain the observed scale height in the presence of such a downflow.

Siphon flows have been postulated to occur when there is a pressure difference between the two ends of a loop. Such a situation might occur if one loop footpoint were in a sunspot and the other in a plage or network region, for example. These flows have been modeled by Cargill and Priest (1980) and by Noci (1979). Some possible consequences of siphon flow are stationary shocks and brightness differences between the footpoints of a loop as seen in emission from lines formed at hotter temperatures.

Various thermal instabilities and MHD instabilities, such as the kink instability, have been shown to occur under certain conditions in coronal loops. The conditions for these instabilities, which are highly specific to each mechanism, are discussed in Chapter 9.

There are three important aspects to the proper study of these and other flow mechanisms:

(1) The temporal coverage must be sufficient to follow the complete history of a transient flow, or to establish that a flow is more likely steady. Most of the mechanisms mentioned above can be studied within a nominal spacecraft daylight orbit of an hour or so. However, some mechanisms, such as evaporation, may require more extended, though not necessarily continuous, observations.

(2) The plasma must be observed simultaneously in emission from different temperature regimes and over a wide field of view in order to identify properly which portion of the plasma is flowing, where it initiates, and where it ceases. For several likely flows it is important that the temperature coverage include observations of the material in the chromosphere because of the central role it can play as a source and a sink of plasma.

(3) Observations of flow should be coupled with observations capable of providing plasma parameters. There are many theories, such as siphon flow, evaporation, or wave pressure that could produce similar flows. Distinguishing which mechanism is operating, and therefore finding the important physics of active region flows, requires knowledge not only of the flowing material itself, but of the physical parameters within the structure containing the flow.

11.6 CAN WE DISTINGUISH AMONG POSSIBLE HEATING MECHANISMS?

The heating of the solar corona is one of the classic unsolved problems of astrophysics. It remains both classic and unsolved but, as mentioned in the introduction to this chapter, the formulation of the problem is now very different from what it was ten years ago. Heating the corona means heating magnetic flux tubes. More specifically, it means supplying nonthermal energy throughout individual magnetic structures in order to maintain the plasma at the observed high temperatures. There are three stages to this process: generation of nonthermal energy, propagation of the energy to appropriate parts of the corona, and dissipation of the energy. As Chapter 10 points out, none of these processes can be considered to be understood anywhere on the sun, despite the fact that many mechanisms have been proposed and several are presently under examination.

Although much of the progress in understanding coronal heating will have to come from theoretical studies, there are several areas in which observers should be aware of potential contributions to solving the heating problem. Some of these areas are discussed below.

11.6.1 Scaling Laws

The simplest quasistatic models of coronal loops (Rosner *et al.*, 1978; Emslie and Machado, 1978; Withbroe, 1978; Vesecky *et al.*, 1978; Craig *et al.*, 1978; Hood and Priest, 1979) are all formulated independently of an assumed heating mechanism. This is not accidental, because the heating mechanism operative on the sun is not known. By invoking local energy balance, or

$$E_H + E_R = -\nabla \cdot \mathbf{F}_c \qquad (11.6.1)$$

at each point, where $\mathbf{F}_c$ is the thermal conductive flux, E_R is the total radiative loss, and E_H is the heating function, i.e., the local input of nonthermal energy, the local heating function can be eliminated from the equations if some assumption about its spatial or parametric dependence is made (see, for example, Chapter 8). The integral of the local heating function, which is simply the total energy being supplied to the loop, can then be determined by noting that the energy input must be equal to the radiative losses over the whole loop plus the conductive flux at the lower boundary.*

An important consequence of analytic models resulting from these conditions is the existence of scaling laws that relate the overall plasma parameters of such loops in a systematic way. One of these scaling laws has the form

$$pL = C\,T_{max}^3 , \tag{11.6.2}$$

where L is the length of the loop (or the half-length in some formulations), p is the (constant) pressure, and T_{max} is the temperature at the top of the loop. The constant C is determined by various assumptions about the heating function. In particular, it is dependent on some suitable spatial or parametric average of the heating function over the loop. Other factors that can influence the value of C are the boundary condition for $\mathbf{F}_c$ at the base (Craig *et al.*, 1978), and magnetic flux tube divergence (Levine and Pye, 1980). (Both of these can also have an effect on the exponent of T_{max} in Eq. 11.6.2, with flux tube divergence causing a greater range of variation.)

These scaling laws suggest that statistical studies of coronal loops might be able to distinguish (a) whether these structures follow such a scaling law, and hence whether they might reflect some of the physical assumptions in the models, and (b) whether all loops, or classes of loops, follow a scaling law with sufficient accuracy to allow a determination of the constant in Eq. 11.6.2 and thus possibly rule out some types of heating mechanisms. The first of these possibilities has been examined most extensively by Rosner *et al.* (1978), who show a general agreement with a $pL \sim T_{max}^3$ dependence. This has proven to be an important demonstration, because the data used in their study included large-scale loop structures, active region loops, X-ray bright points, and large prominence-related flare loops. There are two problems with interpreting this result as supporting energy balance models of coronal loops. First, the predicted loop lengths are only significant to within about a factor of two and tend to be slightly larger than the observed lengths. Second, the observed temperatures of each type of structure (e.g., active region loops, X-ray bright points) lie within a fairly narrow range. Thus, there is a trend for post-flare loops to be hotter and denser than active region loops and the average of each group falls along an appropriate scaling law. But there is significant variation from the scaling law within each group. Some possible reasons for this are discussed by Hood and Priest (1979). There is clearly a need to understand much more about the differences that can occur among morphologically similar structures.

The second example mentioned above for statistical studies of scaling laws, attempting to narrow down possible heating mechanisms, has not been possible yet. Due mostly to uncertainties in the determinations of p, L, and T_{max}, a sufficiently narrow range of empirical scaling has not been demonstrated.

*For stability reasons (Rosner *et al.*, 1978) the upper boundary in these models is taken to be the loop apex and the temperature is assumed to reach a maximum there. Thus the conductive flux is zero at this boundary and the full loop is symmetric about its center.

11.6.2 Unobservables

Although this chapter deals with observations of active regions, it is necessary to point out that features of some proposed coronal heating mechanisms cannot possibly be observed in the near future. Several theories, such as surface MHD waves, anomalous current dissipation, tearing mode instability, and other MHD and plasma instabilities, require energy dissipation to occur in sheaths whose thickness is of the order of a few kilometers (i.e., a few thousandths of an arcsecond) for solar conditions. While such features cannot be observed directly, their indirect effects may include phenomena within the range of instrumental sensitivity. For example, anomalous current dissipation, as well as most instability mechanisms, either require or produce a high degree of turbulence. To detect this, consideration should be given to observations of δv_{rms} at high spatial resolution. Another effect that is an indirect consequence of sheath dissipation is possible radio emission from X-ray emitting loops due to runaway electrons produced by the strong electric fields associated with many plasma mode heating mechanisms (Vlahos *et al.*, 1979). It should also be remembered that energy generated in thin sheaths must be propagated and dissipated throughout a significant fraction of a coronal loop. Mechanisms for these processes are beginning to be studied and may have observable consequences. For example, Vlahos *et al.* (1979) suggest that localized energy deposition cannot maintain the entire volume of larger loops at coronal temperatures, but may be adequate for structures within X-ray bright points.

11.6.3 Wave Properties

Many variations now exist on the original suggestion that the corona is heated by (acoustic) waves dissipating as they travel upward in the solar atmosphere. Some of the wave modes that have been studied under coronal assumptions are Alfven waves, fast mode MHD waves, and various plasma-coupled modes such as ion-acoustic and ion-cyclotron waves. Variations such as Alfvenic surface wave modes have also been proposed.

Because the propagation and dissipation of these modes is strongly dependent on the excited frequencies and wavelengths, the most important observations needed by theorists are of the actual motions associated with wave modes in the solar atmosphere. In the language of the theorist, what is needed is a k-ω diagram at the base of an active region. Observationally, this means two-dimensional Doppler information as a function of time. This type of observation is conceptually similar to the observations of solar p-mode oscillations in the photosphere (Rhodes *et al.*, 1979) which require two-dimensional velocity information to probe solar structure below the surface. There are several complications to using similar types of studies to investigate the outer atmosphere. Perhaps the most important are the need to observe in appropriate spectral regimes, to relate observed motions to a sufficiently specific height range or structural region, and the possibility of quasi-periodic flows within magnetic loops contributing to the signal.

11.6.4 Transient Energetic Processes

Flares and other transient energetic processes are an important topic of solar study. The suggestion that the low energy end of the spectrum of such events might contribute to the heating of the corona was made by Levine (1974), although in the context of an unstructured corona. Foukal (1976, 1978), as well as Levine and Withbroe (1977), noted that transient events seem to occur somewhere in an active region continually. Aside from searching for individual energetic events (which might prove difficult if the energy output from each is too small), they might be investigated by looking for hard X-rays or by evaluation of energetic particle data.

It is well known that solar flares populate the corona and interplanetary medium with energetic protons, electrons, and heavy nuclei. Some flare events, however, appear to have decay times associated with their energetic particles suggesting that the total energy required to accelerate the particles is not likely to be released at the time of the flare. This implies a continual acceleration of particles, and is often accompanied by a shift in the apparent source of the particles away from the flare site (Gold *et al.*, 1975). There are also recurrent (Roelof and Levine, 1978) and impulsive (Dulk and Suzuki, 1980) phenomena associated with energetic particles that suggest possible continual acceleration. Alternatively, coronal loops have been suggested as a storage medium for energetic particles (Newkirk and Wentzel, 1978), with the storable particle flux dependent on the excitation of plasma-coupled and MHD wave modes.

Because transient events are known to increase the temperature and density of active region structures (e.g., post-flare loops, X-ray bright points), their effect, as well as that of less easily noticed events, on the overall energetics of an active region deserves further study. This would require instruments capable of high sensitivity and good time resolution, as well as coordination with instruments located elsewhere in the heliosphere.

11.7 THE ROLE OF THE MAGNETIC FIELD

11.7.1 Field Divergence

The divergence of the magnetic field in the solar atmosphere can be considered to occur in two stages. A bundle of field lines in the high chromosphere with cross-sectional area A will in general have an area $A' > A$ higher in the atmosphere. We will refer to this as the coronal divergence of field lines. Lower down, however, we know that the field in the photosphere is concentrated into intense knots of flux that are at most a few hundred kilometers in diameter (Stenflo, 1978). Further, the field is not everywhere dense, but at the photospheric level exists essentially only in the knots. Thus, there must be a significant divergence of the field above the photosphere, where the external pressure is insufficient to maintain the concentration of the field lines. This will be called the chromospheric divergence of the field.

The role of the coronal divergence of the field has been investigated by Withbroe (1978), Vesecky *et al.* (1978), Levine and Pye (1980), and Wragg and Priest (1980). Within the context of quasistatic models of loops, these authors all find that field divergence increases the temperature gradient along the loop, leading to increased differential emission measure relative to a nondivergent loop with the same maximum temperature. A larger temperature gradient also suggests that the emission from lines formed near the maximum loop temperature will be less uniform than in a nondivergent loop. Additionally, the enhanced gradient will cause the temperature to reach chromospheric levels in a shorter distance; so increased field divergence means a shorter loop for a given T_{max}. This is reflected in a modification to the $pL \sim T_{max}^3$ scaling law for quasistatic loops (Levine and Pye, 1980). Observational evidence for this effect may have been found by Withbroe (1978) in post-flare loops, as pointed out by Levine and Pye (1980).

The chromospheric divergence of the field is not as well understood as the coronal divergence, and it may have a more profound effect on the physics of the loops above. The role of chromospheric divergence is discussed in Chapter 3 of this volume and has been pointed out by Kopp and Kuperus (1968) and Gabriel (1976) for the chromospheric network and by Kopp and Orrall (1976) and Rosner and Vaiana (1978) for the lowest regions of coronal holes. These studies are in the context of open magnetic structures (implicitly or explicitly). For closed structures, the chromospheric divergence of the field suggests two obvious areas of inquiry: the structure of the chromosphere itself and the variations in illumination among loops.

The efficient manner in which the magnetic field confines plasma should modify our picture of the chromosphere. The chromospheric divergence of the field produces an extremely inhomogeneous mass distribution, especially in the lower chromosphere. This is often expressed as a filling factor, modifying 1- or 2-dimensional radiative transfer models of the chromosphere. According to Spruit (1979), the filling factor determined from the ratio of magnetic field to magnetic flux can vary from 0.03 to 0.14 through the low chromosphere. Ayres (1979), using weighted averages of facular and quiet sun models for the Calcium lines, finds values near 0.25 in the middle chromosphere, as do Heasley and Allen (1980), using the wings of Magnesium II. Because these are horizontally averaged results, however, it is not possible to state results reflecting the variation in the filling factor. Thus, it is not clear whether variations in the filling factor are due to a spatial dependence of the field divergence or to inherently different structures, although the former is usually assumed. For example, Lindsey and Heasley (1980) have found that a filling factor of 0.1 fits their infrared data, but this cannot be attributed unambiguously to geometry or to the variability of different features. As is the case for the coronal divergence, the chromospheric divergence of the magnetic field will modify the heat conduction. Because the structure of the chromosphere is very sensitive to the location of the temperature minimum, relatively small variations in the field divergence (which could be due to purely geometric aspects of the field distribution) might have a significant effect on the chromospheric portions of individual flux tubes.

There is no demonstration that the coronal organization into discrete flux tubes based on plasma properties (mainly temperature and density) is in any way related to the photospheric and chromospheric organization into discrete flux tubes based on magnetic flux. Whereas it is often assumed that each coronal loop is connected to one (or a few) photospheric field bundles (and vice versa), this has not been shown observationally. An effort to do this would probably require extremely high resolution simultaneously in the transition region and throughout the chromosphere. Multiple viewing angles might also prove necessary to eliminate redundant interpretations.

In fact, Chapman suggests in Chapter 3 that such a relation does not exist; that is, he suggests that there is a merging height above which all bundled photospheric fields have joined to form a layer of uniformly dense magnetic field. In either case, there already are explicit models of flux tube divergence that can be compared with careful observations of the brightness distribution with height. As shown by Owocki and Auer (1980) and others, interpretation of the models and observations may require fully two-dimensional radiative transfer in some cases. Two issues that can be explored in such studies are the extent of the emission relative to the extent of the diverging flux tube and what currents are necessary to account for the divergence found.

Better understanding of the topology of the chromospheric field is central to the goal, expressed well by Delache (1973), of including "another kind of physics, not just radiative transfer" in a description of the chromosphere. The most obvious other kinds of physics to include in theories and to attempt to observe are mass flow and specific local heating mechanisms. Both of these are strongly modulated by the magnetic field structure.

The reason that we tend to speak of magnetic loops as discrete entities is that they appear to exhibit variations in their illumination, which produce the observed contrast. In the corona, however, it is difficult to imagine a mechanism whereby the field lines could recognize these structural distinctions: with field lines everywhere in the corona there is no a priori way for them to aggregate into discrete structures. The existence of thin sheaths of energy dissipation, as required by some heating theories (Section 11.6), could provide such an organization. However, this organization is known to exist in the photosphere and chromosphere as the coronal field converges into tight bundles in the photosphere. The ways in which the structures in the chromosphere are reflected in the organization of the corona are in need of exploration. Because the chromosphere is largely optically thick, and because it varies from a largely plasma-dominated regime near the bottom to a mostly field-dominated regime at the top, this will be both theoretically and observationally difficult.

11.7.2 Field Topology and the Role of Currents

The topology of the magnetic field above an active region is determined primarily by the distribution of magnetic flux in the photosphere. The strength, distribution, and separation of flux of each polarity determines the basic scale and orientation of the structures in the corona. Because measurements of the magnetic field throughout the corona overlying an active region are not possible, verifying the relation of photospheric and coronal field topologies must be done indirectly. Two assumptions must be made. The first is that the structural organization of coronal emissions represents a division along magnetic field lines: this is merely a restatement of the belief that coronal loops are magnetic flux tubes. The second assumption concerns the field itself and specifies the current distribution.

Given a photospheric magnetogram, an extrapolated magnetic field distribution can be calculated above the photosphere using any of several methods (Levine, 1975). While some studies have found that the observed coronal structures are consistent with a potential (or current-free) extrapolated field (Poletto *et al.*, 1975; Foukal, 1975), several instances have been found where it is necessary to assume that significant currents are present in order to reproduce the observed topology (Levine, 1976; Rust *et al.*, 1975).

Because it is not possible for $\mathbf{J} \times \mathbf{B}$ forces to be maintained in the corona, any steady currents must be force-free, or along the magnetic field direction at each point. The force-free field is expressed as $\nabla \times \mathbf{B} = \alpha\mathbf{B}$, where α is, in general, a function of space and time. The studies of force-free coronal fields based on observed magnetic fields have all been done with the simplifying assumption that α is a constant over the domain of study (Nakagawa and Raadu, 1972; Chiu and Hilton, 1977). This may be an unacceptable restriction in several cases. Levine (1976) showed several examples of active regions where the topology of the field could only be explained by values of α that changed sign, indicating currents flowing in opposite directions, over the active region. In order to study the role of currents in shaping the corona it will be necessary to develop methods of extrapolating the field structure that can represent force-free fields with varying α. Such a procedure has been developed by Sakurai (1979), but it has only been applied to simple test cases so far.

A useful approach to the problem of determining the role of currents in the gross structure of the solar atmosphere is to attempt to parallel the physical constraints in the theoretical modeling. Thus, for example, the general fact that the two ends of a field line remain connected needs to be reflected in a model. As the ends of the field line are moved about, currents are generated and the topology responds. Thus, the structure of the field reflects not only the boundary condition of the photospheric field distribution but also the history of footpoint motions. As in the case of models of the plasma structure of coronal loops, perturbing the system and then restoring it to its original positions may result in a configuration different from the initial one. In this case, moving footpoints and then moving them back may produce a change in the current distribution and thus in the magnetic topology. Until we can reliably determine large-scale currents in the corona, we will not be able to investigate the general role they play in active regions. To pursue this type of study, it will be necessary to maintain facilities for high-resolution magnetograph observations, and to develop new expertise in observing photospheric velocity fields.

11.8 SUMMARY

This chapter has reviewed some of the areas in which future observations of solar active regions are likely to prove fruitful. The viewpoint has been that of the participants in the Skylab Solar Workshop on Active Regions, which emphasized plasma physics and the organizing role of solar magnetic field structures. The areas in which we expect to make progress in future observations are summarized briefly below, organized under topics reflecting the main instrumental characteristics associated with each area. In practice, of course, each of these areas requires an appropriate combination of instrumental capabilities for proper study.

Spatial Capability:

- Improved morphological study of active region structures.
- Structure of X-ray bright points.
- Study of emerging structures in the photosphere, chromosphere, and corona.
- Study of the cool cores of some active region loops.
- Study of the internal structure of active region loops in different temperature regimes.
- Analysis of the response of the chromosphere to variations in coronal loops.
- Study of magnetic flux concentrations in the photosphere and their extension and spreading through the chromosphere and transition region.
- Search for small-scale phenomena associated with transient heating or with instabilities.
- High-resolution observations of the photospheric magnetic field and its evolution.

Spectral Capability:

- Isolation of suitable lines for use in improved spectral diagnostic techniques.
- Simultaneous identification of active region structures in the chromosphere, transition region, and corona.
- Identification of loop footpoints in the chromosphere.
- Morphology and evolution of "hot" and "cool" loops.
- Direct observation of flows in loops, preferably simultaneously in the chromosphere, transition region, and corona.
- Investigation of scaling laws for loop heating mechanisms.
- Study of oscillations relevant to wave propagation in the transition region and corona.
- Study of the photospheric velocity field and its relation to the motion of loop footpoints.

Temporal Capability:

- Study of active regions with a variety of lifetimes.
- Relate X-ray bright points to the evolution of the chromospheric network.
- Identify steady states in loops (if present) and the process of transition from one steady state to another.
- Follow the complete history of a flow in all relevant temperature regimes.
- Search for transient phenomena relevant to loop energetics.
- Study of footpoint motions to determine the possible current distribution and the role of currents in the active region corona.

REFERENCES

Allen, J. W. and Dupree, A. K.: 1969, *Astrophys. J.* **155,** 27.
Antiochos, S. K.: 1979, *Astrophys. J. Lett.* **232,** L125.
Antiochos, S. K. and Sturrock, P. A.: 1978, *Astrophys. J.* **220,** 1137.
Ayres, T.: 1979, presentation to Skylab Workshop on Active Regions.
Brueckner, G. E. and Bartoe, J.-D. F.: 1978, *Bull. Am. Astron. Soc.* **10,** 416.
Brueckner, G. E., Bartoe, J.-D. F., and VanHoosier, M. E.: 1977, in E. Hansen and S. Schaffner (eds.), *Proceedings of the November, 1978, OSO-8 Workshop*, Univ. of Colorado, Boulder, p. 380.
Brueckner, G. E., VanHoosier, M. E., and Bartoe, J.-D. F.: 1979, *Bull. Am. Astron. Soc.* **11,** 397.
Cargill, P. J. and Priest, E. R.: 1980, *Solar Phys.* **65,** 251.
Chiu, H. Y. and Hilton, H. H.: 1977, *Astrophys. J.* **212,** 873.
Chiuderi, C., Elhaudi, G., and Torricelli-Ciamponi, G.: 1980, *Astron. Astrophys.* (submitted).
Chiuderi, C. and Van Hoven, G.: 1979, presentation to Skylab Workshop on Active Regions.
Craig, I. J. D.: 1977, *Astron. Astrophys.* **61,** 575.
Craig, I. J. D.: 1979, presentation to Skylab Workshop on Active Regions.
Craig, I. J. D. and Brown, J. C.: 1976, *Astron. Astrophys.* **49,** 239.
Craig, I. J. D., McClymon, N., and Underwood, J. H.: 1978, *Astron. Astrophys.* **70,** 1.
Delache, P.: 1973, in C. Jordan and E. A. Avrett (eds.), *Stellar Atmospheres*, NASA Sp-317, p. 207.
Dere, K. P.: 1979, presentation to Skylab Workshop on Active Regions.
Dere, K. P., Horan, D. M., and Kreplin, R. W.: 1974, *J. Atmos. Terr. Phys.* **36,** 989.
Dulk, G. A. and Suzuki, S.: 1980, *Astron. Astrophys.* (in press).
Dupree, A. K., Foukal, P. F., and Jordan, C.: 1976, *Astrophys. J.* **209,** 621.
Emslie, A. G. and Machado, M. E.: 1979, Center for Astrophys. Preprint no. 1034.
Flower, D. and Nussbaumer, H.: 1975, *Astron. Astrophys.* **45,** 349.
Foukal, P. V.: 1975, *Solar Phys.* **43,** 327.
Foukal, P. V.: 1976, *Astrophys. J.* **210,** 575.
Foukal, P. V.: 1978, *Astrophys. J.* **223,** 1046.
Frazier, E. N.: 1978, *Astron. Astrophys.* **64,** 351.
Gabriel, A. H.: 1976, *Phil. Trans. Roy. Soc. London Ser. A* **281,** 339.
Gabriel, A. H. and Jordan, C.: 1972, in E. W. McDaniel and M. R. C. McDowell (eds.), *Case Studies in Atomic and Collision Physics*, Vol. 2, North Holland, Amsterdam, Ch. 4.
Gerassimenko, M., Solodyna, C. V., and Nolte, J. T.: 1978, *Solar Phys.* **57,** 103.
Gold, R. E., Roelof, E. C., Nolte, J. C., and Krieger, A. S.: 1975, *Proc. 14th Intl. Cosmic Ray Conf. (Munich)*, 1975.
Golub, L., Davis, J. M., and Krieger, A. S.: 1979a, *Astrophys. J. Lett.* **229,** L145.
Golub, L., Krieger, A. S., Harvey, J. W., and Vaiana, G. S.: 1977, *Solar Phys.* **53,** 111.

Golub, L., Krieger, A. S., and Vaiana, G. S.: 1976, *Solar Phys.* **50,** 311.

Golub, L., Maxson, C., Rosner, R., Serio, S., and Vaiana, G. S.: 1979b, *Bull. Am. Astron. Soc.* **11,** 408.

Habbal, S. and Rosner, R.: 1979, *Astrophys. J.* **234,** 1113.

Harvey, K. L. and Martin, S. F.: 1973, *Solar Phys.* **32,** 389.

Heasley, J. and Allen, M. S.: 1980, *Astrophys. J.* (submitted).

Holton, G.: 1973, *Thematic Origins of Scientific Thought: Kepler to Einstein*, Harvard University Press, Cambridge.

Hood, A. W. and Priest, E. R.: 1979, *Solar Phys.* **64,** 303.

Hood, A. W. and Priest, E. R.: 1979, *Astron. Astrophys.* **77,** 233.

Jordan, C.: 1969, *Monthly Notices Roy. Astron. Soc.* **142,** 501.

Kopp, R. and Kuperus, M.: 1968, *Solar Phys.* **4,** 712.

Kopp, R. and Orrall, F.: 1976, *Astron. Astrophys.* **53,** 363.

Krieger, A. S., Chase, R. C., Gerassimenko, M., Kahler, S. W., Timothy, A. F., and Vaiana, G. S.: 1975, *IAU Symp.* **68,** 103.

Landini, M. and Monsignori-Fossi, B. C.: 1975, *Astron. Astrophys.* **42,** 213.

Levine, R. H.: 1974, *Astrophys. J.* **190,** 457.

Levine, R. H.: 1975, *Solar Phys.* **44,** 365.

Levine, R. H.: 1976, *Solar Phys.* **46,** 159.

Levine, R. H. and Pye, J. P.: 1980, *Solar Phys.* (in press).

Levine, R. H. and Withbroe, G. L.: 1977, *Solar Phys.* **51,** 83.

Lindsey, C. and Heasley, J.: 1980, *Astrophys. J.* (submitted).

Little, S. J. and Krieger, A. S.: 1977, *Bull. Am. Astron. Soc.* **9,** 341.

Martin, S. F. and Harvey, K. L.: 1979, *Solar Phys.* **64,** 93.

Mosher, K.: 1979, presentation to Skylab Workshop on Active Regions.

Nakagawa, Y. and Raadu, M. A.: 1972, *Solar Phys.* **25,** 127.

Newkirk, G. A., Jr., and Wentzel, D. G.: 1978, *J. Geophys. Res.* **83,** 2009.

Noci, G.: 1979, presentation to Skylab Workshop on Active Regions.

Nussbaumer, H.: 1972, *Astron. Astrophys.* **16,** 77.

Owocki, S. and Auer, L. N.: 1980 (in preparation).

Parkinson, J. H.: 1975, *IAU Symp.* **68,** 45.

Peres, G., Serio, S., Vaiana, G. S., and Rosner, R.: 1978, in M. J. Rycroft and D. J. Stickland (eds.), *COSPAR: Space Research*, Vol. XVIII, New York, Pergamon Press, p. 341.

Poletto, G., Vaiana, G. S., Zombeck, M. V., Krieger, A. S., and Timothy, A. F.: 1975, *Solar Phys.* **44,** 83.

Pottasch, S. R.: 1965, *Bull. Astron. Inst. Neth.* **18,** 8.

Priest, E. R.: 1978, *Solar Phys.* **58,** 51.

Pye, J. P., Evans, K. D., Hutchson, R. J., Gerassimenko, M., Davis, J. M., Krieger, A. S., and Vesecky, J. F.: 1978, *Astron. Astrophys.* **65,** 123.

Rhodes, E. J., Deubner, F.-L., and Ulrich, R. K.: 1979, *Astrophys. J.* **227,** 629.

Roelof, E. C. and Levine, R. H.: 1978, *EOS* **58,** 1204.

Rosner, R., Tucker, W. H., and Vaiana, G. S.: 1978, *Astrophys. J.* **220,** 643.

Rosner, R. and Vaiana, G. S.: 1977, *Astrophys. J.* **216,** 141.

Rust, D. M., Nakagawa, Y., and Neupert, W. M.: 1975, *Solar Phys.* **41,** 397.

Sakurai, T.: 1979, *Publ. Astron. Soc. Japan* **31,** 209.

Schmahl, E. J.: 1979, *Bull. Am. Astron. Soc.* **11,** 434.

Sheeley, N. R., Jr.: 1980, *Solar Phys.* (submitted).

Shmeleva, O. P. and Syrovatskii, S. I.: 1973, *Solar Phys.* **33,** 341.

Spruit, H.: 1979, presentation to Skylab Workshop on Active Regions (quoted in more detail in Chapter 3 of this volume).

Stenflo, J. O.: 1978, *Rep. Prog. Phys.* **41,** 865.

Summers, H. P.: 1974, Culham Laboratory, Internal Memo No. 367.

Vaiana, G. S. and Rosner, R.: 1978, *Ann. Rev. Astron. Astrophys.* **16,** 393.

Vaiana, G. S., Van Speybroeck, L., Zombeck, M. V., Krieger, A. S., Silk, J. K., and Timothy, A. F.: 1977, *Space Sci. Instr.* **3,** 19.

Van Hoven, G., Chiuderi, C., and Giachetti, R.: 1977, *Astrophys. J.* **213,** 869.

Vernazza, J. E. and Mason, H. E.: 1978, *Astrophys. J.* **226,** 720.

Vernazza, J. E. and Raymond, J. E.: 1979, *Astrophys. J. Lett.* **228,** L89.

Vesecky, J. F., Antiochos, S. K., and Underwood, J. H.: 1978, *Astrophys. J.* **233,** 987.

Vlahos, L., Ionson, J. A., and Spicer, D. S.: 1979 (in preparation).

Webb, D. and Levine, R. H.: 1979, presentation to Skylab Workshop on Active Regions.

White, O. R. and Athay, R. G.: 1979, *Astrophys. J. Suppl.* **39,** 317 (also p. 333, p. 347).

Withbroe, G. L.: 1975, *Solar Phys.* **45,** 144.

Withbroe, G. L.: 1978, *Astrophys. J.* **225,** 641.

Withbroe, G. L., Dupree, A. K., Goldberg, L., Huber, M. C. E., Noyes, R. W., Parkinson, W. H., and Reeves, E. M.: 1971, *Solar Phys.* **21,** 272.

Wragg, M. A. and Priest, E. R.: 1980, *Solar Phys.* (submitted).

APPENDIX

PARTICIPANTS
SKYLAB WORKSHOP ON SOLAR ACTIVE REGIONS

Antiochos, Spiro
Institute for Plasma Research
Stanford University
Via Crespi
Stanford, California 94305

Athay, R. Grant
High Altitude Observatory
P.O. Box 3000
Boulder, Colorado 80307

Auer, Lawrence
National Center for Atmospheric Research
High Altitude Observatory
P.O. Box 3000
Boulder, Colorado 80307

Ayres, Thomas R.
University of Colorado
Joint Institute for Laboratory Astrophysics
Boulder, Colorado 80302

Brueckner, Guenter E.
Code 7142
U.S. Naval Research Laboratory
Washington, DC 20375

Chapman, Gary A.
Department of Physics & Astronomy
California State University
Northridge, California 91330

Cheng, Chung-Chieh
Code 7149 CCC
Naval Research Laboratory
Washington, DC 20375

Chiuderi, Claudio
Osservatorio Astrofisico di Arcetri
Largo Enrico Fermi, 5
50125 Firenze, ITALY

Cook, John
Code 7144
U.S. Naval Research Laboratory
Washington, DC 20375

Craig, Ian J. D.
Dept. of Astronomy
The University of Glasgow
G12 800 SCOTLAND

Dere, Kenneth P.
E. O. Hulbert Center for Research
U.S. Naval Research Laboratory
Washington, DC 20375

Emslie, A. Gordon
Harvard College Observatory
60 Garden Street
Cambridge, Massachusetts 02138

Engvold, O.
Institute of Theoretical Astrophysics
University of Oslo
P.O. Box 1029
Blindern, Oslo 3
NORWAY

Monsignori-Fossi, Brunella
Osservatorio Astrofisico di Arcetri
Largo Enrico Fermi, 5
50125 Firenze, ITALY

Golub, Leon
Harvard College Observatory
60 Garden Street
Cambridge, Massachusetts 02138

Habbal, Shadia
Center for Astrophysics
60 Garden Street
Cambridge, Massachusetts 02138

Heasley, James
Institute for Astronomy
University of Hawaii
2680 Woodlawn Drive
Honolulu, Hawaii 96822

Hollweg, Joseph
National Center for Atmospheric Research
High Altitude Observatory
P.O. Box 3000
Boulder, Colorado 80307

Hood, Alan
University of St. Andrews
Dept. of Applied Mathematics
North Haugh, St. Andrews
KY16 9SS SCOTLAND

Ionson, James
Laboratory for Astronomy & Solar Physics
NASA/Goddard Space
Flight Center
Greenbelt, Maryland 20771

Jackson, Bernard
CSIRO, Division of Radiophysics
P.O. Box 76
Epping, NSW
2121 AUSTRALIA

Kopp, Roger
Los Alamos Scientific Laboratory
P.O. Box 1663, T-6
Los Alamos, New Mexico 87545

Krall, Kenneth
Space Science Laboratory
NASA/Marshall Space
Flight Center
Huntsville, Alabama 35812

Krieger, Allen
American Science &
Engineering, Inc.
955 Massachusetts Avenue
Cambridge, Massachusetts 02139

Kuperus, Max
The Astronomical Institute at Utrech
Sonneborgh Observatory
Zonneburg 2, Utrecht
THE NETHERLANDS

Landini, Massimo
Osservatorio Astrofisico di Arcetri
Largo Enrico Fermi, 5
50125 Firenze, ITALY

Lemaire, Philippe
Laboratoire de Physique
Stellaire et Planetaire
Boite Postale N 10
91370 Verries-Le Buisson
FRANCE

Levine, Randolph
Harvard College Observatory
60 Garden Street
Cambridge, Massachusetts 02138

Linsky, Jeffrey
Joint Institute for Laboratory Astrophysics
University of Colorado
Boulder, Colorado 80309

Lites, Bruce
Laboratory for Atmospheric & Space Physics
University of Colorado
Boulder, Colorado 80309

Machado, Marcos
Observatorio Nacional de Fisica Cosmica
Av. Mitre 3100
1663 San Miquel-Pcia de
Buenos Aires, ARGENTINA

Mason, Helen
Dept. of Applied Mathematics and Theoretical Physics
Silver Street
Cambridge GB39 EW
UNITED KINGDOM

Moe, Olav Kjeldsuth
Institute of Theoretical Astronomy
University of Oslo
P.O. Box 1029
Blindern, Oslo 3
NORWAY

Mosher, James
Lockheed Palo Alto Research Lab
B/202 D/52-12
3251 Hanover Street
Palo Alto, California 94304

Newkirk, Gordon
National Center for Atmospheric Research
High Altitude Observatory
P.O. Box 3000
Boulder, Colorado 80307

Noci, Giancarlo
Osservatorio Astrofisico di Arcetri
Largo Enrico Fermi, 5
50125 Firenze, ITALY

Noyes, Robert W.
Center for Astrophysics
Harvard College Observatory
60 Garden Street
Cambridge, Massachusetts 02138

Orrall, Frank
Institute for Astronomy
University of Hawaii
2680 Woodlawn Drive
Honolulu, Hawaii 96822

Owocki, Stan
National Center for Atmospheric Research
High Altitude Observatory
P.O. Box 3000
Boulder, Colorado 80307

Papdopoulos, Dennis
U.S. Naval Research Laboratory
Washington, DC 20375

Parkinson, John H.
University College London
Mullard Space Science Laboratory
Holmbury St. Mary, Dorking Surrey
UNITED KINGDOM

Phillips, Kenneth
Culham Laboratory
Abingdon, Axon
OX14 30B
UNITED KINGDOM

Priest, Eric R.
University of St. Andrews
Dept. of Applied Mathematics
North Haugh, St. Andrews
KY 16 9SS SCOTLAND

Roberts, Bernard
University of St. Andrews
Department of Applied Mathematics
North Haugh, St. Andrews
KY16 9SS SCOTLAND

Rosner, Robert
Center for Astrophysics
60 Garden Street
Cambridge, Massachusetts 02138

Sheeley, Neil R., Jr.
Space Research Division
Naval Research Laboratory
Code 714 ONS
Washington, DC 20375

Smerd, Stefan
CSIRO Division of Radiophysics
Epping NSW 2121
AUSTRALIA

Smith, Dean
Department of Astro-Geophysics
University of Colorado
Boulder, Colorado 80309

Spicer, Daniel S.
E. O. Hulbert Center for Space Research
U.S. Naval Research Laboratory
Washington, DC 20375

Spruit, Henk
National Center for Atmospheric Research
High Altitude Observatory
P.O. Box 3000
Boulder, Colorado 80307

Strong, Keith T.
Lockheed Palo Alto Research Lab
Space Science Laboratory
Department 52-12
Building 202
3251 Hanover Street
Palo Alto, California 94304

Tandberg-Hanssen, Einar
ES11, Space Sciences Laboratory
NASA Marshall Space Flight Center
Huntsville, Alabama 35812

Underwood, James H.
Institute for Plasma Research
Stanford University
Via Crespi
Stanford, California 94305

Van Hoven, Gerard
Department of Physics
University of California
Irvine, California 92717

Vernazza, Jorge
Center for Astrophysics
60 Garden Street
Cambridge, Massachusetts 02138

Vesecky, John
Stanford University
233 Durand Building
Stanford, California 94305

Vlahos, Loukas
University of Maryland
Physics Department
College Park, Maryland 20742

Webb, David F.
American Science & Engineering
955 Massachusetts Avenue
Cambridge, Massachusetts 02139

Withbroe, George L.
Harvard College Observatory
60 Garden Street
Cambridge, Massachusetts 02138

Zirin, Harold
California Institute of Technology
Hale Observatory, Bldg. 354-33
Pasadena, California 91125

Abundances, elemental, 129–130, 155
Acoustic waves, 202, 279–280
Active regions (AR)
  definition of, 17
  emergence of, 185–194
  emission measure analyses, 13, 177–181
  evolution, 13, 17–40, 185–187
  overall structure, 17–40
Alfven waves, 10–11, 202, 285, 280–281, 307–315
  in solar wind, 280
Alfvenic twists, 10, 285–286, 288, 307–315
Anomalous current dissipation, 9, 202, 287–296, 334
Arcade of loops, 166, 239
Arch filament systems, 30, 32
Atomic structure, 148–149
Autoionization, 151–152

Balmer continuum, 8, 64
Bipolar magnetic regions (BMR), 13, 17–40
  evolution of, 17–40
  statistical properties, 23–27
Bohm diffusion, 294
Bright points, 24–27, 186–187, 325–326, 335
  temporal behavior, 191–192
  variation over solar cycle, 190–194

Ca II infrared triplet, 8, 80
Calroc, 8, 74
Cayrel's nebular mechanism, 66
Chromosphere and photosphere
  active region (plage), 43–73, 79–81
    horizontally averaged models, 43–48
    inhomogeneous structure, 7, 48–60, 70–73
  quiet sun models, 45–48
  radiative losses from, 60–73
  within flux tubes, 46, 48
Chromospheric energy balance, 60–73, 80
Close coupling approximation (CC), 149–150
Collision cross sections
  electron-ion, 149–150
  proton-ion, 150–151
Collisional ionization, 151–154
Continua
  Balmer, 8, 64
  H minus, 8, 61, 64–72, 80
  infrared, 48
  Lyman, 64, 79–80, 136
  microwave, 45
  Paschen, 8, 64, 67
  Si I, 54–55
Cool cores, 14, 175, 218, 221, 283–284, 321, 326–327
Cool loops (see under Coronal plasma loops)
Coronal plasma loops, 5–14, 32–36
  dynamic formation, 234–237
  evaporation and draining, 205, 227–234
  heating (see Heating processes, coronal)
  phenomenology
  cool cores, 89, 217, 283, 314, 326–327, 331
  cool loops, 14, 166, 327
  connecting different active regions, 169
  emission measure, 177–181
  evolution with active regions, 34, 184–189
  flux loops vs. plasma loops, 32, 165
  footpoints 5–6, 13–14, 32–34, 166–170, 217, 327–328
  hot loops vs. cool loops, 32–33, 166, 326–327
  lifetimes, 11, 14
  radio emission from, 170–176
  spikes, 14, 34, 167–176
  temporal changes, 6, 13, 329–330
  Thermal structure, 182–183
  quasi-static models, 5–6, 14, 199–211
  basal heat flux, 12, 81, 202–204, 208–211
  comparison with observations, 206–211
  cross-sectional area, 200, 288–211
  differential emission measure, 6, 13, 208–211
  energy balance, 199–205, 213
  MHD stability, 11, 263–273
  scaling laws, 5–6, 176–177, 202–207, 213–214, 229, 332–333
  thermal stability, 6, 12–13, 213, 237–251, 273
  siphon flow, 11–13, 126, 217–227, 272, 332
Coronal holes, 277, 280–282
Coronal rain, 89, 216
Coulomb-Born (CB) approximation, 136, 149–150, 152
Culgoora, 14, 172–174

Diagnostics
  for electron density, 74, 141–148, 322–323
  for electron temperature, 134–141, 321–322
  specific for active regions, 138–139

Dielectronic recombination, 152–154
Differential emission measure (DEM), 6–7, 75–76, 177–181, 208–210, 323–325
  active region, 177–183
  definition, 132
  in transition region, 75–77
  methods of analysis, 132–134
  X-ray, 136–138
Distorted wave approximation (DW), 149–150
Draining of plasma loops, 6, 11–12, 89–90, 227–234, 272–273, 331
Dreicer field, 293

Emerging flux region (EFR), 187–190
Emission measure (see Differential emission measure)
Ephemeral region (ER), 13, 23–27, 186, 194, 326
EUV Bright points (see Bright points)
EUV Burst, 91–99, 216
Evaporation, chromospheric, 6, 11–12, 89–90, 205, 218, 227–234, 272–273, 331, 332
Evershed flow, 85–86, 215
Excitation, 129–131, 134–136, 140–151
Explosive events, 114–121, 126

Faculae, 7, 8, 43, 48, 49, 277
  contrast, 8
Facular, points 29
Fibrils 30–32, 39
  flows associated with, 87–88
Field transition arches, 30
Filament formation, 30, 315–316
Filigree, 7, 48–49
Filling factor, 7, 51–57, 70–72
  empirical determination, 52–57
  height dependence, 55-57, 81, 336–337
Flares, 338
  effect on AR energy balance, 278
Flows, 83–99, 113–126, 215–237, 328–332
Flux tubes (see Magnetic elements)
Footpoints of coronal loops (see Coronal plasma loops)
Force-free magnetic field, 50

Granular buffeting, 13, 217, 261
Granulation, 18

H alpha, 8, 30, 45
H beta, 30
H gamma, 30
H minus, 8, 61, 64–66, 70–72
H and K lines (Ca, II) 8, 30, 46–48, 57–60, 63–64, 68–69, 77
h and k lines (Mg, II) 8, 30, 45, 46–47, 63–64, 68–69, 77
Hale's law, 18
Heating processes, chromospheric, 60–73, 278–280
Heating processes, coronal, 8–10, 320, 334
  Alfven waves, 10, 202, 307–315
    in closed loops, 312–314
    in open fields, 308–312
    in solar wind, 280–281
    propagation, 307–308
  Alfvenic surface waves, 299–303
  global nature of heating problem, 284–286
  importance of magnetin field, 8, 278–279
  large-scale reconnection, 282–283
  MHD fast waves, 280, 285, 304–307, 313, 315
  observational signatures, 6, 8–10, 101–111, 206–210, 295–296, 299, 303, 306, 314–315, 334
  Parallel electric currents, 281–282, 287–296, 337–338
    anamalous dissipation of, 202, 289–296
  origin of, 287–289
  small-scale (mini-) reconnections, 296–299
  unimportance of gravity and nonmagnetic acoustic waves, 279–280
Horizontal radiative transfer, 8, 57–60
Hot loops (see Coronal plasma loops)
Hot wall effect, 59, 80
HRSA model, 56–57, 59
HRTS 8, 10–11, 52, 54–55, 74, 76, 113–127, 156–157

Infrared radiation of plages, 48
Ion-acoustic turbulence, 9, 290–292, 334
Ion-cyclotron turbulence, 9, 289–295, 334
Ionization equilibrium, 151–156, 323
  effects of diffusion and flows, 123–124, 155–156

K—ϖ diagram, 334
Kink instability, 263–269, 331
Kruskal-Shafranov condition, 268–269

Landau/transit time damping, 10, 305–306
Line blanketing, 66
Line cooling, 61–64

Line emission processes, 129–136
Line profiles, 102–111, 156–157
in chromosphere, 105–110
in corona, 110
nonthermal broadening, 9, 102–111, 156
saturation effects, 109, 158–159
Stark and Zeeman effect, 157–160
in transition region, 104–107, 110, 156–157
broadening by waves, 104
Line tying, 12, 263–265
Loop resonances, 10
Lyman alpha line emission, 9, 30, 45–47, 203–204
Lyman alpha/Lyman beta ratio, 78–80
Lyman continuum
absorption in norona and transition region, 136, 168–169
emission, 8, 79

Macrospicules, 277
Macroturbulence, 102–111, 156–160
Magnetic elements (intense flux tubes), 7, 11–13, 22, 45, 48, 73, 80, 335–337
force free vs. flux tube regime, 50
generation of spicules in, 261–262
instability in, 218, 252–257, 273
models, 7, 48–57
spicule generation in, 261–262
steady flows in, 257–260
Magnetic field
chromospheric tracers of, 30–32, 43
field line evolution, 34–38
inner network fields, 22–23
potential, 18–19, 39–40, 337–338
Magnetic flux, 17–40
emergence of, 192-196
Maunder's butterfly diagram, 18
Merging height, 7, 50–51, 335–339
Metastable levels, 140–141
MHD fast-mode waves, 10, 285, 289, 304–307
MHD stability, 11–12, 218, 263–273, 331
Microsurge, 215–216
Microturbulence, 75, 102–111, 156–160
Microwave continuum, 45, 176
Minireconnections, 9, 296–299
Moat, 29, 86, 215
Multidimensional radiative transfer, 57–60, 80
definitions of 1D, 1.5D and 2D, 57

Network, 13, 18, 215
Nonequilibrium, 213, 217–218
Nonpotential magnetic fields, observational evidence for, 281

Oscillations, 99–102
coronal, 9, 102
in plages, 100
in sunspot umbrae, 99–101
in transition region, 101
photospheric, 99
OSO-8, 8, 10–11, 46, 68, 74, 83–111, 114, 159, 326

Parallel currents, 9, 281–282, 287–296, 337–338
Paschen continuum 8, 64, 67
Photon channeling, 60
Plage (see also Chromosphere and Photosphere, active regions), 113–127, 277
motions, 121–123
Plumes, sunspot (see Sunspots)
Pores, 43
Pressure gradients within coronal loops, 264–265, 273

Quasi-static loop models (see Coronal plasma loops)

Radial pressure gradients, 12
Radiative losses, chromospheric, 8, 60–73
Radiative recombination, 153–154
Radio emission from active regions, 170–176
Rayleigh-Taylor instability, 9, 205
Reconnection, 9, 34–38, 282
Runaway electron, 9

Scaling laws, 5–6, 176–177, 332–333
Siphon flow, 11–13, 126, 217–227, 272, 332
Skylab, 8, 10–11, 45
Solar wind, 277, 280–281
SOLRAD, 137, 324
Spicules, 13, 88–89, 216, 277
generation in intense fields, 261–262
Spoerer's law, 18
Sprays, 90–91
Sunspot cycle, 18
Sunspot groups
classification, 29–30
evolution, 27–30
Sunspots, 23, 27–30, 43
corona above, 34, 175–176
flows over, 11, 87, 113–127, 175, 216
microwave emission, 176

"missing energy," 276
plumes, 14, 34
transition region above, 175
umbral oscillations, 99–101
Supergranular flow, 18, 85, 215
Surface waves, Alfvenic, 299–303
Surges, 90–91, 215

Tearing mode instability, 9, 270–272, 334
Temperature minimum, 27, 45, 64–72
Thermal conductivity, 201
Thermalization length, 8, 59
Thermal stability, 6, 12–13, 237–251, 331
effect of sheared magnetic field, 9, 11, 315
Transition region, 6, 8, 13
active region model, 77–79
density determination, 74
emission measure, 75–76
energy balance, 73–75
inhomogeneous model, 159–160
loop structure, 32–35, 75–77
motions, 87, 89–99
oscillations, 101—102

Upper chromosphere, 6, 61–64, 67–68, 73–79

VAL model, 45—46, 61

Wave and turbulent pressure, 201, 286, 332
Wilson depression, 50, 58–59, 80

X-ray bright point (XBP) (see Bright point)
X-ray diagnostics, 136–138